World Water Resources

Volume 6

Series Editor

V.P. Singh, Department of Biological and Agricultural Engineering & Zachry Department of Civil Engineering, Texas A&M University, College Station, TX, USA

This series aims to publish books, monographs and contributed volumes on water resources in the world, with particular focus per volume on water resources of a particular country or region. With the freshwater supplies becoming an increasingly important and scarce commodity, it is important to have under one cover up to date literature published on water resources and their management, e.g. lessons learnt or details from one river basin may be quite useful for other basins. Also, it is important that national and international river basins are managed, keeping each country's interest and environment in mind. The need for dialog is being heightened by climate change and global warming. It is hoped that the Series will make a contribution to this dialog. The volumes in the series ideally would follow a "Three Part" approach as outlined below:

In the chapters in the first Part *Sources of Freshwater* would be covered, like water resources of river basins; water resources of lake basins, including surface water and under river flow; groundwater; desalination and snow cover/ice caps. In the second Part the chapters would include topics like:

Water Use and Consumption, e.g. irrigation, industrial, domestic, recreational etc. In the third Part in different chapters more miscellaneous items can be covered like impacts of anthropogenic effects on water resources; impact of global warning and climate change on water resources; river basin management; river compacts and treaties; lake basin management; national development and water resources management; peace and water resources; economics of water resources development; water resources and civilization; politics and water resources; water-energy-food nexus; water security and sustainability; large water resources projects; ancient water works; and challenges for the future. Authored and edited volumes are welcomed to the series. Editor or co-editors would solicit colleagues to write chapters that make up the edited book. For an edited book, it is anticipated that there would be about 12–15 chapters in a book of about 300 pages. Books in the Series could also be authored by one person or several co-authors without inviting others to prepare separate chapters. The volumes in the Series would tend to follow the "Three Part" approach as outlined above. Topics that are of current interest can be added as well.

Readership

Readers would be university researchers, governmental agencies, NGOs, research institutes, and industry. It is also envisaged that conservation groups and those interested in water resources management would find some of the books of great interest. Comments or suggestions for future volumes are welcomed.

Series Editor:

V.P. Singh, Department of Biological and Agricultural Engineering & Zachry Department of Civil Engineering, Texas A&M University, TX, USA. vsingh@tamu.edu

More information about this series at http://www.springer.com/series/15410

Jose A. Raynal-Villasenor

Editor

Water Resources of Mexico

 Springer

Editor
Jose A. Raynal-Villasenor
Department of Civil and Environmental
Engineering
Universidad de las Americas Puebla
Cholula, Puebla, Mexico

ISSN 2509-7385 ISSN 2509-7393 (electronic)
World Water Resources
ISBN 978-3-030-40685-1 ISBN 978-3-030-40686-8 (eBook)
https://doi.org/10.1007/978-3-030-40686-8

This Springer imprint is published by the registered company Springer Nature Switzerland AG.
The registered company address is: Gewerbestrasse 11, 6330 Cham, Switzerland

Preface

A book like *Water Resources of Mexico*, which contains many related water information, is needed to understand the complex relationships between water and food production, energy generation, and environmental protection. Such book will be a required reference to obtain Mexico's water resources information for general and specific audiences.

The book contains contributions from a wide series of themes, from historical aspects of water use to perspectives of water for the future; so, any layman could enjoy reading the book *Water Resources of Mexico* because all the chapters in it have been written in simple terms that almost anybody could understand the contents of any of the chapters of such book.

That was the motivation that guided the production of the book *Water Resources of Mexico*, in which several Mexican distinguished water scientists have gathered to share their knowledge and experience on analyzing, researching, and managing the water resources of Mexico.

The very nature of occurrence of water in planet Earth, in the form of precipitation, infiltration, evaporation, groundwater, and surface water, is composed by the occurrence of very complex natural phenomena. Some of these phenomena are related with meteorological variables, hydrological variables, soil and moisture variables, and many other variables related with many fields of study.

Mexico, as a country with three oceans surrounding it, the Pacific Ocean, the Gulf of Mexico, and the Caribbean Sea, has a continued exposition to all types of cyclones, which are meteorological phenomena that produce every year a very important rainy season, which is not always a fortunate event but a devastating experience must of the time.

Given that 65% of the area of Mexico is above of 1000 m above mean sea level, it is no wonder to know that 74% of the population of Mexico is living above such elevation. Four out of the five biggest cities of Mexico, population-wise, are above 1000 m above mean sea level.

These conditions mentioned before make Mexico a unique country, water-wise, which had to deal since ancient times with the two extremes of the occurrence of water: floods and droughts. Such struggle between abundance and scarcity of water continues until today, now with the additional components of an important population growth that reached 130.6 million of people in year 2017 and a global climate change in development nowadays.

Cholula, Puebla, Mexico   Jose A. Raynal-Villaseñor

Introduction

Mexico is the 14th largest country of the world, with an extension of 1,964,575 km², and is a federated country with a political division composed by 32 states and has several characteristics that made it a very unique country. It has more than 10,000 km of coastline formed by the Pacific Ocean, the Gulf of Mexico, and the Caribbean Sea.

The precipitation processes in Mexico are mainly conformed by rainfall in the rainy season during the period from June to November; snowfall is mainly observed in the Northern states during Winter and at the top of the mountains and in the volcanos in Central Mexico. Hailstorms are produced during the rainy season. It is unusual to have tornados in Mexico, even there are historical records of the occurrence of these events since several hundreds of years ago. The incidence of cyclones, with all types such as tropical disturbances, tropical depressions, tropical storms, and hurricanes (Windows to Universe (2017)), is a meteorological event that occurs every year along the coastline of Mexico, formed by the Pacific Ocean, the Gulf of Mexico, and the Caribbean Sea.

Most of the Northern states of Mexico receive 50–250 mm of rain per year, whereas in the Southern states, such amount is 1000–4500 mm of rain per year. These extremes in the amounts of rain are reflected by the existence of large deserts in the Northern states of Coahuila, Chihuahua, and Sonora and tropical rainforests in the Southern states of Chiapas and Tabasco.

Because 65% of the area of Mexico is above 1000 m above mean sea level, it is no wonder to know that 74% of the population of Mexico is living above such elevation. Four out of the five biggest cities of Mexico, population-wise, are located above 1000 m above mean sea level: Mexico City, the largest city of Mexico (2250 m above mean sea level); Guadalajara, the second largest city in Mexico (1566 m above mean sea level); Puebla, the fourth largest city in Mexico (2135 m above mean sea level); and Leon, the fifth largest city in Mexico (1815 above mean sea level). Only Monterrey, the third largest city in Mexico, population-wise, is below 1000 m above mean sea level (544 m above mean sea level).

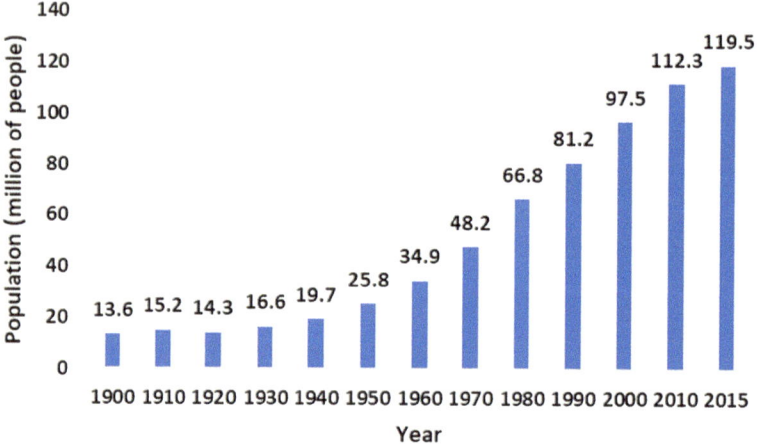

Fig. 1 Population growth (1900–2015), INEGI (2017)

Mexico grew in an important manner in the period 1920–1970, duplicating its population almost every 20 years (see Fig. 1). Since 1970, the population growth index has declined up to 2.2% nowadays, and it will continue declining to reach 2% in year 2030.

Mexico City, the largest city in Mexico, population-wise, has about 9 million people, and in a small area, roughly 1494 km², almost 14% of the population of Mexico is concentrated there, but its metropolitan area is about 9000 km² with a population of almost 26 million people, making the population concentration worse with about 20% of the population of Mexico living in the metropolitan area of Mexico City, causing huge problems now and in the future with regard to water supply, water treatment, transportation, power supply, food supply, housing, generation of pollution by cars and industries, trash generation and trash disposal sites, etc.

Brief Description of the Chapters

A brief description of the chapters is as follows:

In Chap. 1, the processes and mechanisms of rainfall are described from the particularities that Mexico has for being a country which is surrounded by water, in the west by the Pacific Ocean and in the East by the Gulf of Mexico and the Caribbean Sea.

In Chap. 2, the analysis of the groundwater of Mexico is reviewed in detail, given that it is one of the water components of the hydrological cycle that is severely endangered in Mexico, due to the exaggerated withdrawals to provide water for urban water supply and to sustain agriculture through ancient irrigation and modern irrigation techniques.

In Chap. 3, the hydrogeology of Mexico is carefully reviewed to understand the geological processes that created the rock formations that today foster the groundwater reservoirs.

In Chap. 4, the water-energy-food nexus is explored in Mexico. The paramount importance of this cycle is obvious in the light of food production and industrial growth.

In Chap. 5, the data models for river basin management in Mexico are analyzed by the application of modern techniques to the Bravo/Grande river basin, one of the most important watersheds in Mexico because it comprises 2019 km out of the 3185 km of the US-Mexico border, El Colegio de San Luis, AC (2017).

In Chap. 6, the domestic and industrial water use and consumption in Mexico is reviewed, given their importance in the conservation of the water resources of Mexico, now and for the future.

In Chap. 7, a proposal for the future water resources management in Mexico is being reviewed, due to the support it provided to the population regarding water supply and flood protection.

In Chap. 8, the wastewater treatment in Mexico is analyzed with a critical point of view, because it is a water sector that needs urgent attention given that from all the water used in Mexico for water supply and for industrial use, only 20% is being treated and reused.

In Chap. 9, the importance of climate change and water resources in Mexico is analyzed in the light that this phenomenon is putting a lot of pressure in the water resources of Mexico.

In Chap. 10, the water security and sustainability in Mexico is reviewed closely, due to their importance for the viability of Mexico as a country in the present and in the future.

In Chap. 11, the expected impacts on agriculture due to climate change in Northern Mexico is analyzed, given that this region of Mexico is where less water resources exist and in the Bravo/Grande river basin, all the existing water resources are already being allocated.

In Chap. 12, the dam operation policy in Mexico using regional flows canonical correlation analysis is explored by modern techniques that allow a more efficient operation of the hydraulic infrastructure during the hurricane season.

In Chap. 13, the most important hydrologic and hydraulic works of the Aztec Civilization are reviewed, given the importance that they played along the period of existence of the Aztec Empire. A brief review of the hydromythology of the Aztec Civilization is provided, too.

In Chap. 14, the spatial dependence of hydrological variables in a geographical and physical environment is introduced using directional variograms. It is explained how this mathematical function can represent spatial variability and form rainfall patters within a region through a kriging procedure.

In Chap. 15, the possible scenarios of global warming effects on the evaporation in Mexico are presented for several very important watersheds. For each one, five scenarios were constructed for increments of 1, 2, 3, 4, and 5 °C above current air temperatures.

References

INEGI (2017) Cuentame. http://cuentame.inegi.org.mx/poblacion/densidad.aspx? tema=P

INEGI (2017) Cuentame poblacion. http://cuentame.inegi.org.mx/poblacion/ habitantes.aspx?tema=P

El Colegio de San Luis, A. C. (2017) La Cuenca del Bravo. http://www.colsan.edu. mx/investigacion/aguaysociedad/proyectofrontera/Documentos/LA% 20CUENCA%20DEL%20BRAVO.pdf

Windows to Universe (2017) Storm strength. https://www.windows2universe.org/ earth/Atmosphere/hurricane/intensity.html

Contents

Chapter 1
Precipitation in Mexico

José de Anda Sánchez

Abstract This chapter describes, in a general way, the main topics related to the distribution of precipitation in Mexico, while considering its climatic, hydrological, and geographical aspects. First, the distribution of precipitation within Mexico is described and how the resulting water is managed within the hydrologic administrative regions (HARs). In the following sections, the main climatic phenomena having a direct impact on the precipitation regime within the territory, such as the Mexican monsoon, El Niño, and tropical cyclones, are considered. Finally, main issues of concern related to precipitation, namely, droughts and the actual situation and future scenarios related to climate change in Mexico are addressed.

Keywords Mexico · Rainfall · Precipitation · Mexican monsoon · El Niño · Tropical cyclones · Droughts · Climate change

1.1 Introduction

There is a strong influence on Mexico's climate originating from the Northern Atlantic and Northeast Pacific Oceans' subtropical high-pressure systems (close to 30° North latitude). In addition, the location at the south of the intertropical convergence zone (ITCZ) influences the climate in this portion of Mexico. On the other hand, moist trade winds dominate the climate in the 6 months around summer. North America's polar air masses have a strong influence, mainly in winter and spring, causing a drastic decrease in the temperature in most of the country (Mendoza et al. 2005). "Northerners" is the popular name used to describe the intrusion of cold winds into the Northern and Central Mexico's highlands. This term is also used to describe the same phenomenon on the Gulf of Mexico's coastal plains and the Yucatan Peninsula. Occasionally, the northerners cause intense precipitation at the

J. de Anda Sánchez (✉)
Centro de Investigacion y Asistencia en Tecnologia y Diseño del Estado de Jalisco, A. C., Guadalajara, Jalisco, México
e-mail: janda@ciatej.mx

© Springer Nature Switzerland AG 2020
J. A. Raynal-Villasenor (ed.), *Water Resources of Mexico*, World Water Resources 6, https://doi.org/10.1007/978-3-030-40686-8_1

low-altitude zones and moderate precipitation in the high-altitude mountains in both Northern and Central Mexico. In Mexico, the rainfall season has two peaks, in June and September, which has more influence on Central and Southern Mexico. The arid climate dominating Northern Mexico receives impact during the rainfall season. A reduction in convective activity and precipitation is observed in July and August, and this phenomenon is the so-called midsummer drought (MSD) or "canícula", which usually influences agriculture and industrial activities (Mosiño and García 1966; Magaña et al. 1999; Perdigón-Morales et al. 2017). The variations in the rainfall regime during summer apparently result from low-level winds blowing over a region of warm water in July and August (Magaña et al. 2003). Probably, the variations of cloud cover during the season regulate the sea-surface temperature, keeping it warm. Magaña et al. (1999), noted that around the months of May–June, after the beginning of the summer's monsoon, the temperature of the surface of the sea in the eastern Pacific declines by about 1 °C. This phenomenon is produced by the interception of solar radiations stimulated by the increasing cloudiness and the increasing intensity of the easterly winds. Because of the variations in the sea-surface temperature in the region, the climate undergoes a significant reduction in deep convective activity in July–August (Mendoza et al. 2005).

The strong differences in the topography of the Mexican Plateau constitute another important factor that influences the variability of climate and precipitation in Mexico throughout the year (Wallén 1955; Mosiño and García 1974). A good understanding of the climate variations throughout the country' geography is very important to Mexican society since key decisions about several economic activities depend on this knowledge. However, the nonexistence of rainfall data in several parts of Mexico does not currently allow forming a complete sequence of the natural variability of climate (Therrell et al. 2002).

1.2 Distribution of Precipitation

Figure 1.1 shows the rainfall climatology observed in Mexico during 1998–2013 according to the product 3B42 dataset derived from the Tropical Rainfall Measuring Mission (TRMM) satellite. Its average intra-annual variability is displayed in the inserted box on the bottom left. In this box, it is shown that the precipitation regime in the country depends strongly on the annual seasons, so that 14.4% of the total precipitation occurs between December and April and 85.6% during the period May–November. In Northern Mexico, such a map shows an overall annual dry season (200 ± 100 mm/year); on the contrary, southeastern Mexico receives an average annual precipitation of greater magnitude (2000 ± 500 mm/year). On this map, one may see also that gradients of precipitation gradually increase from the high plateau of Mexico to the Eastern and Western Sierra Madre. Except for the Californian peninsula and the Northwestern Sonora, Mexico's coastal zones annually receive more than 1000 mm of rainfall. According to Fig. 1.1, Mexico's highest precipitation amounts occur in the region known as the Isthmus of Tehuantepec,

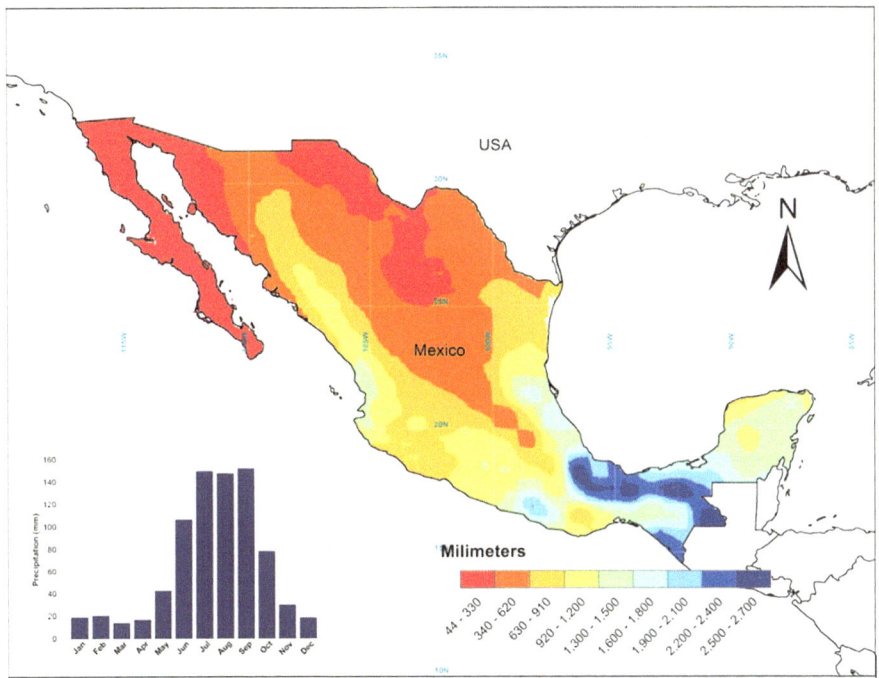

Fig. 1.1 TRMM 3B42 derived map of mean annual precipitation in Mexico between 1998 and 2013. The inset displays the monthly precipitation climatology averaged over Mexico (De Jesús et al. 2016)

which comprises the highlands of Oaxaca and Chiapas, and in the flood plains of Veracruz and Tabasco (De Jesús et al. 2016).

1.3 Hydrologic Administrative Regions

Since 1997, Mexico's water resources administration and conservation had been grouped into 13 hydrological-administrative regions (HARs)—the division had been done by integrating watersheds with a geographic-vicinity criterion, resulting in the basic units for water-resources management. There are significant variations among the regional characteristics. In Southeast Mexico (see Fig. 1.2), the HARs V, X, XI, and XII are grouped together because they comprise two-thirds of the country's renewable water resources, and such a region has 20% of the population with a 20% contribution to national GDP. In contrast, Northern, Central, and Northwestern Mexico have 33% of Mexico's renewable water resources, 80% of the population, and an 80% contribution to the national GDP. The available renewable water resources per capita in Southeast Mexico is seven times larger than that in the rest of Mexico (CONAGUA 2016).

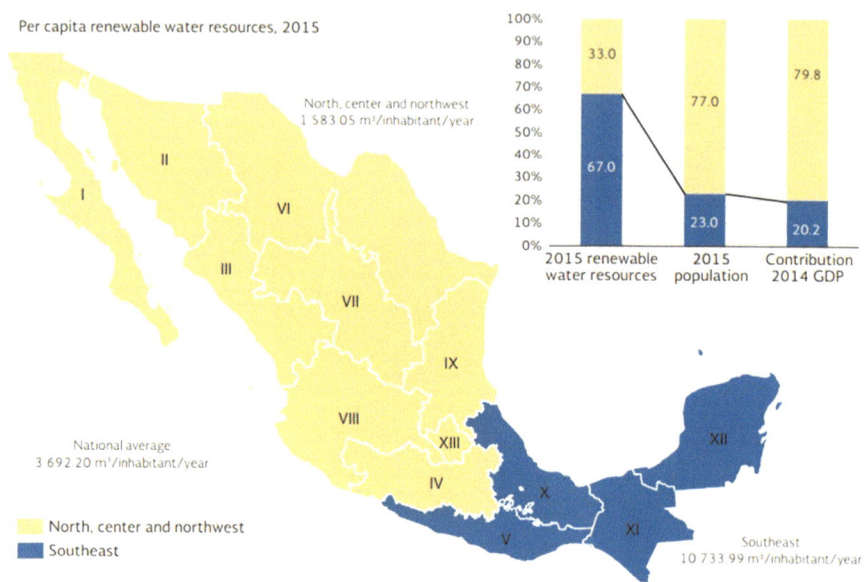

Fig. 1.2 Hydrologic Administrative Regions (HARs) of Mexico (CONAGUA 2016)

1.4 Distribution of Precipitation by HARs

In the period 1981–2010, the normal precipitation in Mexico was 740 mm. The World Meteorological Organization (WMO) has recommended that the normal values be obtained by averaging the measurements calculated for a continuous and relatively lengthy period, with a data record spanning at least 30 years considered to be a representative period for climatological data. The period must begin on the first day and end on the last day of the year's calendar. The geographical limits of any Hydrological Administrative Region (HARs) must coincide with municipal divisions to facilitate the integration of socioeconomic information. For 1981–2010, Table 1.1 shows the normal precipitation by HAR (CONAGUA 2016).

One can observe in Fig. 1.3 (CONAGUA 2016) that 68% of the normal monthly precipitation falls between June and September; at the same time, the monthly distribution of precipitation makes clear the availability of water resources. In the period 1981–2010 (see Table 1.1), the normal annual precipitation was 11 times greater during the period 1981–2010 in hydrological–administrative region XI, located in Southern Mexico, than the one registered in hydrological–administrative region I located in the Baja California Peninsula, which is the driest one. This regional variation in the normal precipitation is highlighted in Fig. 1.4 (CONAGUA 2016).

Table 1.1 Normal monthly precipitation by Hydrologic Administrative Region (HAR) in the period 1981–2010 (CONAGUA 2016)

HAR	J	F	M	A	M	J	J	A	S	O	N	D	Annual number
I	20	19	14	4	1	1	10	26	32	11	10	20	168
II	24	21	12	6	4	19	108	103	58	25	17	31	428
III	31	16	8	6	9	66	194	188	142	52	26	29	765
IV	12	8	6	11	48	179	199	197	194	84	15	6	962
V	8	8	6	15	71	230	200	219	242	113	20	7	1139
VI	19	11	11	17	28	40	63	61	64	32	12	15	372
VII	18	9	6	12	27	56	79	71	67	29	11	13	398
VIII	22	11	4	6	23	131	197	180	153	60	13	10	808
IX	26	20	19	38	67	120	137	119	166	89	30	23	855
X	51	40	30	43	84	222	261	264	293	179	97	64	1626
XI	65	54	36	49	135	276	223	265	331	224	109	76	1842
XII	45	35	31	39	90	167	153	173	208	147	72	49	1207
XIII	11	11	12	28	51	109	126	115	110	57	13	6	649
National Values	25	17	13	18	42	102	134	134	135	69	27	23	740

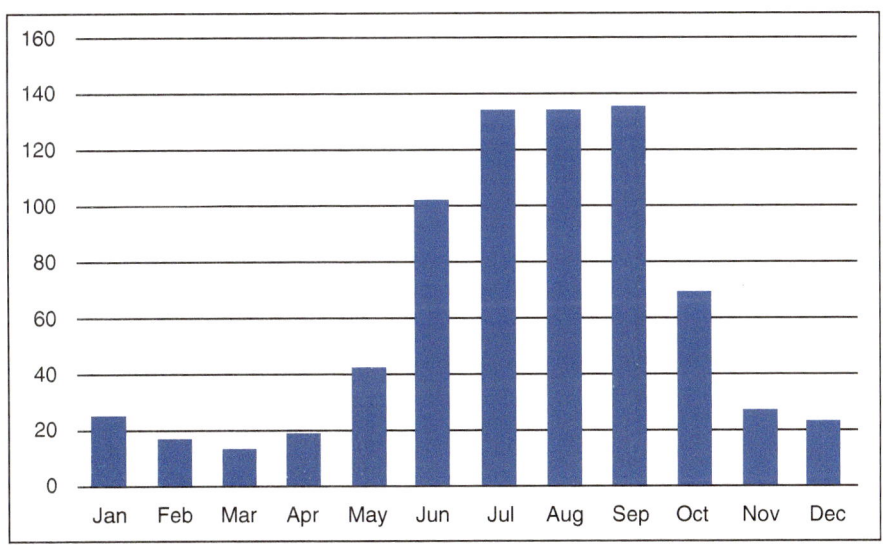

Fig. 1.3 Normal monthly national precipitations in Mexico in the period 1981–2010 (mm)

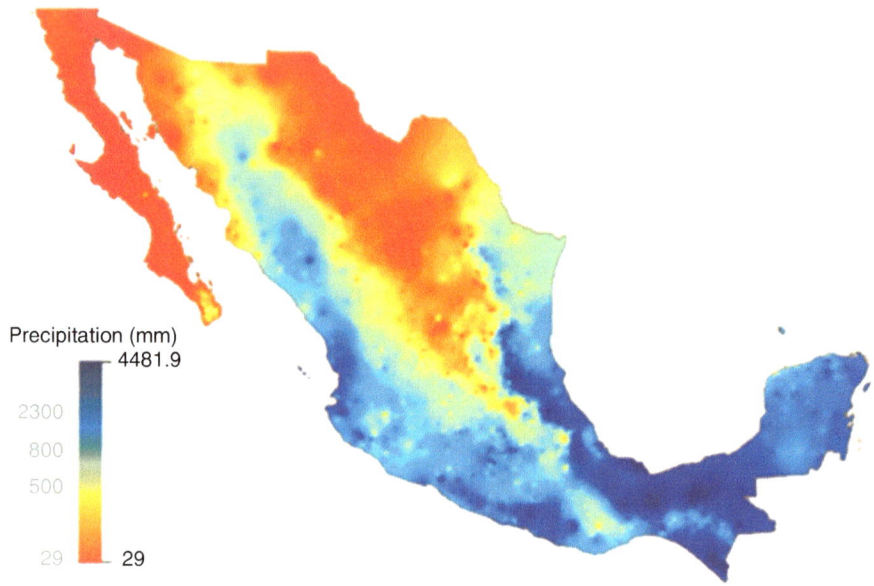

Precipitation (mm)

Fig. 1.4 Distribution of the precipitation in Mexico in the period 1981–2010 (CONAGUA 2016)

1.5   The Mexican Monsoon

The term "Mexican monsoon" is used in analogy with the better-known Asian monsoon for several reasons. Histograms of monthly mean rainfall and mean temperature from many of the stations in northwestern Mexico are similar to those in Southern Asia, with most of the annual rainfall taking place during a short period (2–4 months) and with the highest temperatures just prior to the onset of the rains (Douglas et al. 1993).

During summer, a significant feature of the climate in Northwestern Mexico and Southwestern US is the Mexican monsoon. The Mexican monsoon starts in early July and ceases at the end of September. In Northwestern Mexico, the amount of rainfall during July–September accounts for 60–80% of the annual rainfall, with rainfall totals along the Pacific piedmont slopes of the Sierra Madre Occidental Mountains exceeding 65 cm in some locations during the 3-month period cited earlier (Douglas et al. 1993; Stensrud et al. 1995).

Figure 1.5 shows practically the same trend display as in Fig. 1.1. In this case, the highest mean values of precipitation during summer are located in Northwestern Mexico, with values exceeding 5 mm day^{-1}. The precipitation increases toward Southern Mexico, where the mean of precipitation achieves values as great as 10 mm day^{-1}. The precipitation regime diminishes drastically over the highland

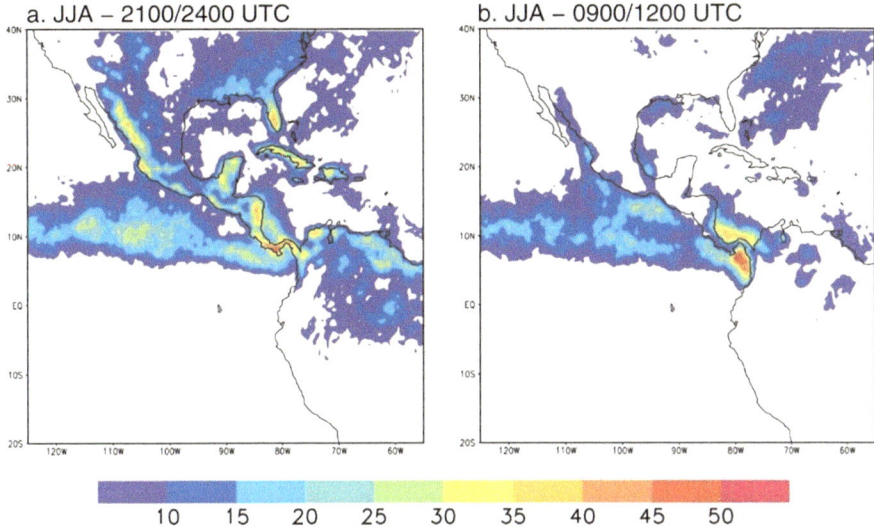

Fig. 1.5 Evening (2100–2400 UTC) (**a**) and night (0900–1200) (**b**) temporal frequency of cold clouds (infrared brightness temperature Tb < 235 K) during the boreal summer (JJA). The period of analysis extends from 1983 to 1991 (Vera et al. 2006)

plateau located in Northern and Central Mexico. However, the precipitation increases to amounts as large as 5 mm day^{-1} in the Gulf of Mexico coast, which includes the eastern Mexican seaboard. On the other hand, a maximum of precipitation during the summer season is also evident in Western Mexico. Until today, the interrelationships between the climatological variables describing the precipitation distribution during the monsoon period are not completely understood (Stensrud et al. 1995; Adams and Comrie 1997; Barlow et al. 1998). Recently, the fact was published that the interannual variability of the summer monsoon occurring in southern US and Northwestern Mexico is mainly controlled by several climatic factors in both in the ocean and on the continent (Higgins and Shi 2000). Thanks to international efforts carried out by the North American Monsoon Experiment (NAME) and the Monsoon Experiment South America (MESA), it is now possible to have a better understanding of this phenomenon. These international research programs have the following objectives (Vera et al. 2006):

1. To understand the key components of the American monsoon systems and their variability
2. To determine the role of these systems in the global water cycle
3. To improve observational datasets
4. To improve simulation and monthly-to-seasonal predictions of the monsoons and regional water resources

1.6   Impact of El Niño on Precipitation

NOAA (2017) has defined El Niño and the Southern Oscillation (ENSO) phenomena as "a periodic fluctuation in sea surface temperature (El Niño) and the air pressure of the overlying atmosphere (Southern Oscillation) across the equatorial zone of the Pacific Ocean." Also, NOAA (2017) commented that the name of El Niño is related to Jesus Christ's acknowledged birthday (well known as El Niño in Spanish) due to the fact that such a phenomenon attains its full strength at the end of the year, and the early Christian inhabitants of western equatorial South America associated the festivities celebrating the birth of Jesus Christ with the warm-water currents and their resulting climatic changes.

Normal climatic patterns around the world are affected by ENSO. The normal precipitation patterns are modified by El Niño, and its counterpart, La Niña. In Northwestern Mexico, the precipitation is modified during El Niño (it increases) and during La Niña (it decreases) winters; on the contrary, the precipitation decreases with El Niño and increases with La Niña in the region around the Isthmus of Tehuantepec. Over the Southern Gulf of Mexico, the number of *northerners* is increased by a southward shift in the position of the subtropical jet stream. Magaña et al. (2003) observed that during summer, El Niño may produce a precipitation deficit. Several phenomena associated with El Niño may cause negative precipitation anomalies over most of Mexico. Due to the enhanced subsidence associated with a southward shift of the Intertropical Convergence Zone (ITZC), severe droughts may be produced, and several additional phenomena such as more intense trade winds, fewer tropical cyclones over the Intra Americas Seas (IAS), and reduced relative humidity may be observed. In 1997, these phenomena caused major socioeconomic loses during the summer, and they could be directly related to El Niño. Climate conditions return to normal or result in enhanced precipitation during La Niña years (Magaña et al. 2003).

According to Bravo–Cabrera et al. (2010), the precipitation of Mexico is strongly influenced by the intensity of the monsoon, the position of the subtropical jet stream, and the frequency and intensity of hurricanes. The influence of each atmospheric phenomena on the precipitation regime is difficult to isolate and evaluate separately. On the other hand, in recent studies, it was found little variation, as a whole, in the precipitation data in Mexico due to the ENSO phenomena. The ENSO phenomenon has a statistically significant influence in the precipitation in some specific regions of the Mexico.

1.7   Tropical Cyclones and Rainfall

A key contributor to seasonal precipitation includes tropical cyclones (TCs), but they are not usually considered in seasonal climate forecasts. The tropical cyclones, in addition to the easterly waves (EWs), are major factors producing 50% of the summer rainfall in Mexico, Central America, and the Caribbean region. In semi-arid regions, this effect is easily observed in regions affected by TCs, such as in Northern Mexico. The tropical cyclones and the consequent increased precipitation may help in recovering the dam-storage levels and so reducing the effects of droughts. The low confidence in predicting the paths of tropical cyclones, with a seasonal timescale, is the sole source of uncertainty when using seasonal forecasts in water management and the infrastructure-planning process. The approach based mainly in the ENSO phases, by using statistical and dynamic techniques to predict seasonal TC activity, has not produced good results for regional forecasting of tropical cyclones. In this chapter, we try to analyze how to determine the likely paths of tropical cyclones and what are the chances of a year having higher than normal precipitation from the impact of a single event. In 1979–2009, the most likely TC tracks were determined, and they were linked to the probabilities of the landfalls of tropical cyclones, related to ENSO phases. In order to regionalize the tropical cyclone activity, several important factors were considered: the mean flow, **adiabatic** heating, sea-surface temperature, and atmospheric stability. The large-scale influence of seasonal forecasts of tropical cyclones, regarding the quantification of the effects, at regional scales, should be considered in order to enhance the seasonal forecasts of precipitation over tropical region of the American continent, (Dominguez and Magaña 2015).

In Mexico, the total annual rainfall occurring in the Baja California Peninsula is strongly influenced every year by the TCs, since it receives 10–80% of the total annual precipitation due to the impact of cyclones. The influence of cyclones in the rest of Mexico is estimated as 0–20%, without any specific distribution pattern. The year 2013 was atypical since the influence of cyclones in the total precipitation was estimated at 33%, the highest value since 1998. The analysis of climatic data shows that a TC's category correlates to the highest rainfall volumes, not only the ranking of TCs based solely on the highest wind speeds (Breña-Naranjo et al. 2015) (Fig. 1.6).

1.8 Droughts

Droughts, floods, and tropical storms are the most common natural hazards affecting agriculture, natural resources, and other economic sectors around the world (FAO 2015). In particular, droughts often cause damage to crop production, put at risk the survival of livestock, limit access to water for the population, and may cause animal and human deaths. FAO (2015) has defined a drought as a period of deficit of rainfall or lower than normal rainfall that reduces the water availability in the region and as a phenomenon that is highly variable in time and space (De Jesús et al. 2016; FAO 2015). According to several authors, the frequency of water deficits has become greater and they are more persistent, particularly in Central and Northern Mexico (Sancho y Cervera and Pérez-Gavilán Arias 1981; Velasco 1999; Méndez and Magaña 2010; De Jesús et al. 2016). The regions most affected by drought in Mexico have some common features: they have less annual precipitation than in

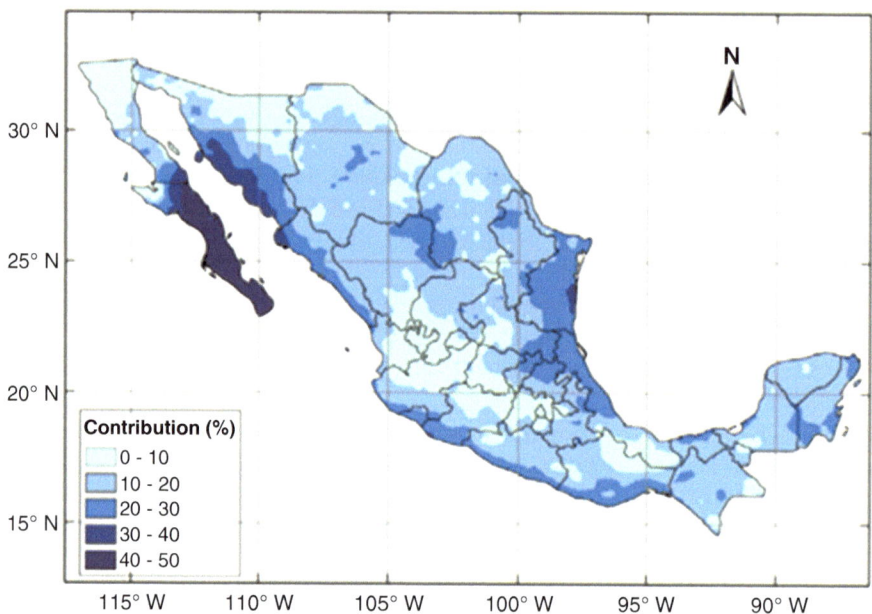

Fig. 1.6 Average contribution of TCs (obtained from the sum of tropical depressions, tropical storms, and hurricanes) to the mean annual rainfall (1998–2013) (Breña-Naranajo et al. 2015)

the rest of the country; they suffer frequent droughts; so, they have implemented the most extensive and intensive agriculture systems in largest irrigated areas using sprinkler and drip irrigation systems in Mexico and they also have the highest rate of economic growth and industrial development. As a consequence, they have the highest demand for water for agriculture and industry than any other regions. As depicted in Fig. 1.2, the regions of Northern Mexico produce 79.8% of the nation's GDP; while they contain 77% of the total population, they have only 33% of total renewable water for agriculture, industry, and municipal services. It is worth mentioning that agriculture in Mexico uses about 76.3% of the total available water because of lack of irrigation infrastructure; therefore, the nonirrigated areas depend on summer rains, causing them to be drastically affected by droughts (Velasco 1999; CONAGUA 2016).

Over Western North America, the prevailing long-lasting drought conditions since at least 1999 have affected snowpack, streamflow, precipitation, reservoir storage, and wildfire activity. The Palmer Drought Severity Indices (PDSI) show that severe and sustained drought began in 1994 in Mexico, and it continued with only short periods for the past 15 years. The period of drought in Mexico, which started at the end of the 1990s and continued in the beginning of the twenty-first century, is similar in some aspects to the drought of the 50s, which is considered as the most severe drought in Mexico's official climate records (1900–2008).

It is supposed that large-scale changes in ocean–atmospheric circulation patterns have contributed to the diminished rainfall regime in Mexico and have led to the current drought. But, in addition, global warming, changes in the land-use coverage, and misuse and over-exploitation of natural resources have also contributed to the early twenty-first-century drought in the country (Stahle et al. 2009). The long-lasting droughts have affected socioeconomic issues in Mexico and have also led to conflicts among its federal states and to transboundary water conflicts with the US. It is therefore of great importance for Mexican society to have information about the reasons that have resulted in the water scarcity in several regions of Mexico (Méndez and Magaña 2010).

However, it is important to mention that the scarce or abundant occurrence of water is commonly due to the natural climate variability in Central and Southern Mexico. Through analysis of sediment cores taken from the bottom of the sea and lakes and also of tree-ring data, it has been possible to reconstruct the main climatic events occurring during the period of 1500–1990 in Mexico. It was found that during the periods 1545–1600, 1752–1768, and 1859–1868, relatively dry conditions prevailed in Central and Southern Mexico. On the other hand, during the 1910s, 1930s, 1950s, and mid-1980s, the climatic records show dry conditions in Central and Southern Mexico, but relatively wet conditions in Northern Mexico (Méndez and Magaña 2010).

1.9 Climate Change and Precipitation

The latest published climate models predict that Mexican territory will be affected by warmer and dryer conditions. However, some models also predict an increase in precipitation, but, unfortunately, the increases do not offset the greater potential evaporation. According to these climatic scenarios in Mexico, Central and Northern Mexico will suffer a decrease in soil moisture; a reduction of water storage in surface-water bodies; and a decrease in the water tables in aquifers. This will cause severe damage to rain-fed and irrigated agriculture, the water-supply systems for urban and industrial uses and a reduction of water flow in hydropower facilities, as well as damage in several compartments to aquatic and terrestrial ecosystems (Eakin 2005; Karmalkar et al. 2011; INECC 2016). However, particularly for Mexico, there are still significant discrepancies among the predicted climatic variables and the observed variables when climate models are applied, especially regarding precipitation forecasts (Liverman and O'Brien 1991).

The compiled simulation data, produced by the Atmospheric General Circulation Model (AGCM), were utilized in developing the regional projections. Such a model was applied by dividing the world into several regions. The regions of Central America and the Caribbean were defined by the following geographical boundaries: from the Equator to 35° N latitude and from 60° to 120° W longitude. To assess future changes in precipitation regimes, it is necessary to apply the predictive climate models and to observe how far the model results represent the key climatic processes in the region. Some of the models usually applied in this case are the Coupled Model Intercomparison Project Phase 3 (CMIP3) and CMIP5 multi-model ensembles (IPCC 2013).

In Central America and the Caribbean region, during June–August, the CMIP5 models predicted the scenario with the highest temperatures by the end of the century. These models predict a warmer scenario over Central American than in the Caribbean during summer and winter. It is also observed that in October–March the predicted precipitation will decrease in the north of Central America including Southeast Mexico. For the Caribbean region, a scenario of decreased precipitation in the south is predicted, but with an increase of rainfall in the north. For the period April–September, the projections are for a reduction of the rainfall regime over the entire Central America and the Caribbean region. By using the CMIP3 and CMIP5 models along with a high-resolution model to predict the precipitation in Mexico by the end of the century, a similar reduction was observed in part of Mexico and also in the southern Caribbean during December–March and in Central America and the Caribbean during June–September (Fig. 1.7). The set of CMIP5 models shows better agreement in the rainfall regime during December–March, with an increase in the

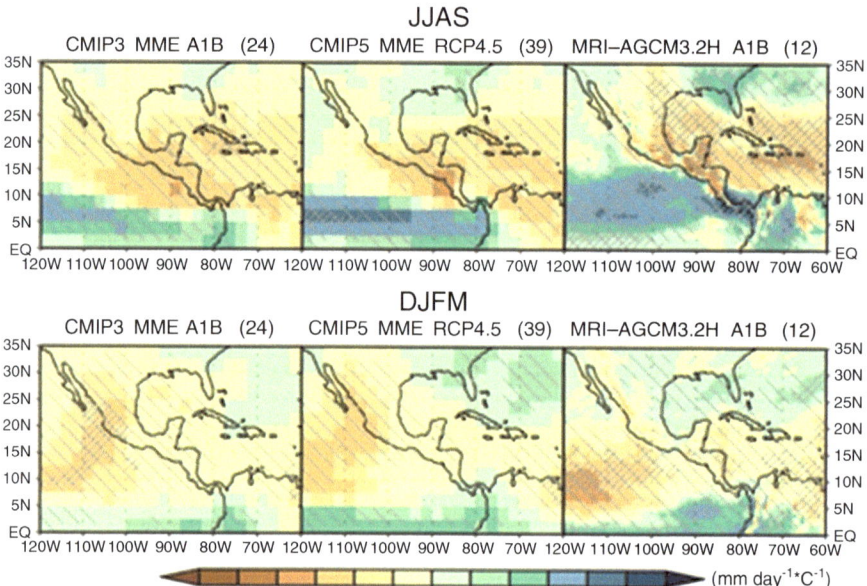

Fig. 1.7 Maps of precipitation changes for Central America and Caribbean in 2080–2099 with respect to 1986–2005 in June to September (above) and December to March (below) in the SRES A1B scenario with 24 CMIP3 models (left) and in the RCP4.5 scenario with 39 CMIP5 models (middle). Right figures are the precipitation changes in 2075–2099 with respect to 1979–2003 in the SRES A1B scenario with the 12-member 60-km mesh Meteorological Research Institute (MRI)-Atmospheric General Circulation Model 3.2 (AGCM3.2) multi-physics, multi-SST ensembles (Endo et al. 2012). Precipitation changes are normalized by the global annual mean surface-air temperature changes in each scenario. Light hatching denotes where more than 66% of models (or members) have the same sign with the ensemble mean changes, while dense hatching denotes where more than 90% of models (or members) have the same sign with the ensemble mean changes (IPCC 2013)

Northern Caribbean sector compared with the prediction of the CMIP3 models. Figure 1.7 also suggests an increase and a southern displacement of the East Pacific Intertropical Convergence Zone (ITCZ), which can influence the trend to drought in Southern Central America (Karmalkar et al. 2011; IPCC 2013).

References

Adams DK, Comrie AC (1997) The North American monsoon. Bull Am Meteorol Soc 78:2197–2213. https://doi.org/10.1175/1520-0477(1997)078<2197:TNAM>2.0.CO;2. Accessed 15 Dec 2017

Barlow M, Nigam S, Berbery EH (1998) Evolution of the North American monsoon system. J Clim 11:2238–2257. https://doi.org/10.1175/1520-0442(1998)011<2238:EOTNAM>2.0.CO;2. Accessed 17 Jan 2018

Bravo-Cabrera JL, Azpra Romero E, Zarraluqui Such V, Gay García C, Estrada Porrúa F (2010) Significance tests for the relationship between "El Niño" phenomenon and precipitation in Mexico. Geofis Int 49(4):245–261

Breña-Naranjo JA, Pedrozo-Acuña A, Pozos-Estrada O, Jiménez-López SA, López-López MR (2015) The contribution of tropical cyclones to rainfall in Mexico. Phys Chem Earth Parts A/B/C 83–84:111–122. https://doi.org/10.1016/j.pce.2015.05.011

CONAGUA (National Water Commission) (2016) Statistics on water in Mexico, 2016 edition. SEMARNAT (Ministry of the Environment and Natural Resources). CONAGUA (National Water Commission). Mexico City Mexico, 278 p. http://files.conagua.gob.mx/conagua/publicaciones/Publicaciones/EAMI2016.pdf. Accessed 2 Feb 2018

De Jesús A, Breña-Naranjo JA, Pedrozo-Acuña A, Alcocer-Yamanaka VH (2016) The use of TRMM 3B42 product for drought monitoring in Mexico. Water 8(8):325. https://doi.org/10.3390/w8080325

Dominguez CH, Magaña V (2015) The contribution of tropical cyclones to seasonal precipitation over Mexico. US CLIVAR climate variability and predictability program. Observing & modelling climate variability in the Intra-Americas Seas and impacts on the continental Americas and the Caribbean, September 9–11, 2015. https://usclivar.org/2015-iasclip-abstract/contribution-tropical-cyclones-seasonal-precipitation-over-mexico. Accessed 20 Nov 2017

Douglas MW, Maddox RA, Howard K, Reyes S (1993) The Mexican monsoon. J Clim 6:1665–1667. https://doi.org/10.1175/1520-0442(1993)006<1665:TMM>2.0.CO;2

Eakin H (2005) Institutional change, climate risk, and rural vulnerability: cases from Central Mexico. World Dev 33(11):1923–1938. https://doi.org/10.1016/j.worlddev.2005.06.005

Endo H, Kitoh A, Ose T, Mizuta R, Kusunoki S (2012) Future changes and uncertainties in Asian precipitation simulated by multiphysics and multi–sea surface temperature ensemble experiments with high-resolution Meteorological Research Institute atmospheric general circulation models (MRI-AGCMs). J Geophys Res 117:D16118

FAO (2015) The impact of disasters on agriculture and food security. Food and Agriculture Organization of the United Nations, Rome, 77 p. ISBN 978-92-5-108962-0. http://www.fao.org/resilience/resources/resources-detail/en/c/346258/. Accessed 14 Nov 2017

Higgins RW, Shi W (2000) Dominant factors responsible for interannual variability of the summer monsoon in the Southwestern United States. J Clim 13:759–776. https://doi.org/10.1175/1520-0442(2000)013<0759:DFRFIV>2.0.CO;2. Accessed 21 Jan 2018

INECC (2016) Atlas Nacional de Vulnerabilidad al Cambio Climático (ANVCC). Instituto Nacional de Ecología y Cambio Climático México. https://wwwgobmx/inecc/acciones-y-programas/atlas-nacional-de-vulnerabilidad-ante-el-cambio-climatico-anvcc-80137. Accessed 30 Jan 2018

IPCC (2013) Chapter 14: climate phenomena and their relevance for future regional climate change. In: Stocker TF, Qin D, Plattner G-K, Tignor M, Allen SK, Boschung J, Nauels A, Xia Y, Bex V, Midgley PM (eds) Climate change 2013: the physical science basis, Contribution of Working Group I to the Fifth Assessment Report of the Intergovernmental Panel on Climate Change. Cambridge University Press, Cambridge/New York, pp 1217–1308. https://doi.org/10.1017/CBO9781107415324.028

Karmalkar AV, Bradley RS, Diaz HF (2011) Climate change in Central America and Mexico: regional climate model validation and climate change projections. Clim Dyn 37:605. https://doi.org/10.1007/s00382-011-1099-9

Liverman D, O'Brien KL (1991) Global warming and climate change in Mexico. Glob Environ Chang 1(5):351–364. https://doi.org/10.1016/0959-3780(91)90002-B

Magaña V, Amador JA, Medina S (1999) The midsummer drought over Mexico and Central America. J Clim 12:1577–1588. https://doi.org/10.1175/1520-0442(1999)012<1577:TMDOMA>2.0.CO;2

Magaña VO, Vázquez JL, Pérez JL, Pérez JB (2003) Impact of El Niño on precipitation in Mexico. Geofis Int 42(3):313–330

Méndez M, Magaña V (2010) Regional aspects of prolonged meteorological droughts over Mexico and Central America. J Clim 23(5):1175–1188. https://doi.org/10.1175/2009jcli3080.1

Mendoza B, Jáuregui E, Diaz-Sandoval R, García-Acosta V, Velasco V, Guadalupe Cordero G (2005) Historical droughts in central Mexico and their relation with El Niño. Am Meteorol Soc 44:709–716. https://doi.org/10.1175/JAM2210.1

Mosiño P, García E (1966) The midsummer droughts in Mexico. In: Proceedings of the regional Latin American conference, vol 3. International Geophysical Union, Mexico City, pp 500–516

Mosiño P, García E (1974) The climate of Mexico. In: Climates of North America, Bryson RA, Hare FK (eds) World survey of climatology, Landsberg HE (ed. in Chief), vol 11. Elsevier, Amsterdam, pp 365–404

NOAA (2017) El Niño/Southern Oscillation (ENSO) technical discussion. National Centers for Environmental Information. National Oceanic and Atmospheric Administration. https://www.ncdc.noaa.gov/teleconnections/enso/enso-tech.php. Accessed 24 Feb 2018

Perdigón-Morales J, Romero-Centeno R, Pérez PO, Barrett BS (2017) The midsummer drought in Mexico: perspectives on duration and intensity from the CHIRPS precipitation database. Int J Climatol 38:2174. https://doi.org/10.1002/joc.5322

Sancho y Cervera J, Pérez-Gavilán Arias D (1981) A perspective study of droughts in Mexico. J Hydrol 51(1–4):41–55. https://doi.org/10.1016/0022-1694(81)90114-1

Stahle DW, Cook ER, Villanueva-Díaz J, Fye FK, Burnette DJ, Griffin D, Acuña-Soto R, Seager R, Heim RR Jr (2009) Early 21st-century drought in Mexico. Eos 90(11):89–100

Stensrud DJ, Gall RL, Mullen SL, Howard KW (1995) Model climatology of the Mexican monsoon. J Clim 8:1775–1794. https://doi.org/10.1175/1520-0442(1995)008<1775:MCOTMM>2.0.CO;2

Therrell MW, Stahle DW, Cleaveland MK, Villanueva-Diaz J (2002) Warm season tree growth and precipitation over Mexico. J Geophys Res Atmos 107(D14):ACL6-1–ACL6-8. https://doi.org/10.1029/2001JD000851

Velasco I (1999) Severe droughts becoming recurrent, more persistent in Mexico. Drought Network News (1994–2001), 88. http://digitalcommons.unl.edu/droughtnetnews/88. Accessed 11 Dec 2017

Vera C, Higgins W, Amador J, Ambrizzi T, Garreaud R, Gochis D, Gutzler D, Lettenmaier D, Marengo J, Mechoso CR, Nogues-Paegle J, Silva Dias PL, Zhang C (2006) Toward a unified view of the American monsoon systems. J Clim 19:4977–5000. https://doi.org/10.1175/JCLI3896.1

Wallén CC (1955) Some characteristics of precipitation in Mexico. Geogr Ann 37(1/2):51–85

Chapter 2
Groundwater Resources of Mexico

Carlos Gutiérrez-Ojeda and Oscar A. Escolero-Fuentes

Abstract Groundwater constitutes an essential reserve for Mexico's socioeconomic development because more than half of its territory is arid or semi-arid, and the water supply depends heavily on groundwater resources, in fact, about one third of the total consumption comes from aquifers. It is used to irrigate more than two million hectares of land (i.e., one third of the total irrigated land), it supplies about 70% of water requirements for the public-urban sector (60 million people living in urban centers), it provides water for most of the industrial facilities and almost the entire water demand of the rural population (20 million people). About 105 regional aquifers are subject to intensive exploitation, which has caused a serious environmental impact during the past four decades and the mining of 11 km^3/year of the underground water reserve. In response, programs have been initiated for managed aquifer recharge, extraction of brackish groundwater for desalination, and exploration of aquifer strata at great depths (more than 2000 m).

Keywords Groundwater · Aquifers · Brackish groundwater · Legal tools · Managed aquifer recharge

2.1 Hydrogeological Regions of Mexico

The increase in water demand to sustain socioeconomic development, due to the increase in population, production, and agricultural irrigation, is a permanent condition that has impacts throughout the country. This has occurred with greater intensity in the arid or densely populated regions, and especially in the megalopolises. To address these problems arising from socioeconomic

C. Gutiérrez-Ojeda (✉)
Instituto Mexicano de Tecnologia del Agua, Jiutepec, Morelos, Mexico
e-mail: cgutierr@tlaloc.imta.mx

O. A. Escolero-Fuentes
Instituto de Geologia, Universidad Nacional Autonoma de Mexico, Ciudad de Mexico, Mexico

© Springer Nature Switzerland AG 2020
J. A. Raynal-Villasenor (ed.), *Water Resources of Mexico*, World Water Resources 6,
https://doi.org/10.1007/978-3-030-40686-8_2

15

development, information is needed on the characteristics of the aquifer strata and the functioning of the groundwater-flow systems. In the past, most groundwater studies have been conducted within the boundaries of the exploitation zones or administrative boundaries.

One of the first works on the hydrogeological regionalization of Mexico was carried out by Alfonso De la O Carreño (1951, 1954). In this work, Volume I (1951) presents a synthesis of the surface hydrology of the republic and a hydrographic division of the country into 182 basins. In Volume II (1954), a regionalization scheme called "The Geohydrological Provinces of Mexico" is presented, from the distribution of the 182 basins in eleven Geohydrological Provinces, based on a purely hydrographic criterion.

Subsequently, in 1978, an integration effort was carried out at the national level by the now extinct Secretary of Agriculture and Hydraulic Resources (SARH 1978), with the formation of the "Geohydrological Atlas: National Geohydrological Information Bank", Volume 1. This document integrated the information on 140 known groundwater exploitation zones at that date and the level of progress in the studies carried out in each region. An important aspect was that they proposed the regionalization of aquifers within the national territory, grouping the exploitation zones into regional aquifers in Quaternary alluvions, regional aquifers in Tertiary basins, and regional aquifers in limestones.

Several years later, Velazquez-Aguirre and Ordaz-Ayala (1992, 1993) published almost simultaneously the work called "Provincias Hidrogeologicas de Mexico"; both articles have the same content and are based on the work edited by Back et al. (1988). They considered the Hydrological Regions of Mexico, the Physiographic and Geological Provinces and the Tectostratigraphic Terrains of Mexico and proposed a regionalization of the country, dividing it into eleven Hydrogeological Provinces, as shown in Fig. 2.1.

2.2   Groundwater Resources Evaluation

2.2.1   Aquifer Delimitation

Aquifer delimitation is a complex task due to inaccessibility, irregular geometry, heterogeneous lithology, and the dimensions involved (Chavez et al. 2006): delineation of their lateral and vertical borders—even with modern prospecting equipment—can only be carried out in a simplified and presumptive way (Alcocer and Chavez 2016).

Current aquifer knowledge was obtained through studies performed by CONAGUA, academic institutions, research centers, and private companies. Aquifer dimensions, stratigraphy and boundary conditions were defined, as well as the flow-system components (Brassington 2007).

For administrative-facility and water-management purposes, the country has been divided into 653 aquifers (Fig. 2.2). The lateral and vertical borders are defined in a simplified form: in many cases, they are not strictly physical borders, but rather

Fig. 2.1 Hydrogeological Provinces of Mexico. (Modified from Velazquez-Aguirre and Ordaz-Ayala 1992, 1993)

conventional boundaries that coincide with the watershed divides, and in the low-lands with political (municipal, state) or hydrological (river) boundaries (Chavez et al. 2006).

The administrative legal reference framework that defines the official names and limits of the Mexican aquifers as management units was published in the Official Gazette of the Federation (DOF 2001): all groundwater extraction sources and the respective water rights are associated with their management units (Alcocer and Chavez 2016).

Each aquifer is geographically limited by a simplified georeferenced polygon, and they have a lower limit. In some basins, there are deeper aquifers that are hydraulically separated from those officially defined, although the knowledge about them is incipient (Alcocer and Chavez 2016). However, most of the available groundwater is stored within the first 400 m below the ground surface, where most permeable aquifers are found, and the recharge is more dynamic (Arreguin and Lopez 1995).

Fig. 2.2 Mexican aquifers. (Modified from CONAGUA 2015a)

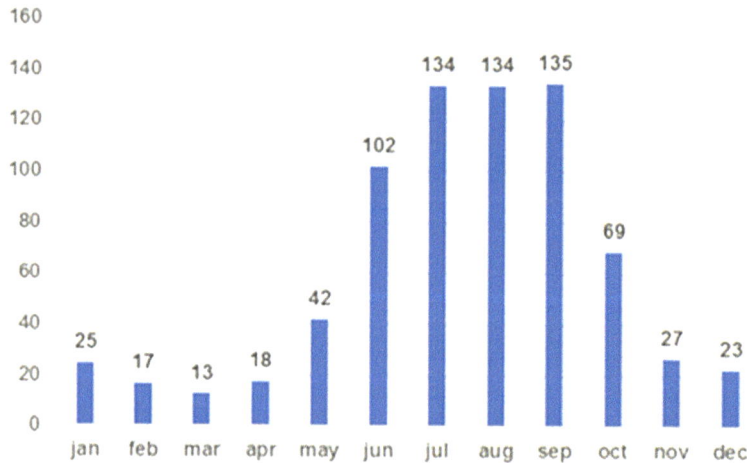

Fig. 2.3 Average monthly rainfall in Mexico. (CONAGUA 2015a)

2.2.2   Recharge

Mexico has a total area of 1964 Mkm^2 and had an estimated population of 119.9 million inhabitants in 2015 (INEGI 2018). The average annual precipitation is 740 mm, 77% of which occurs between June and October (Fig. 2.3). The spatial

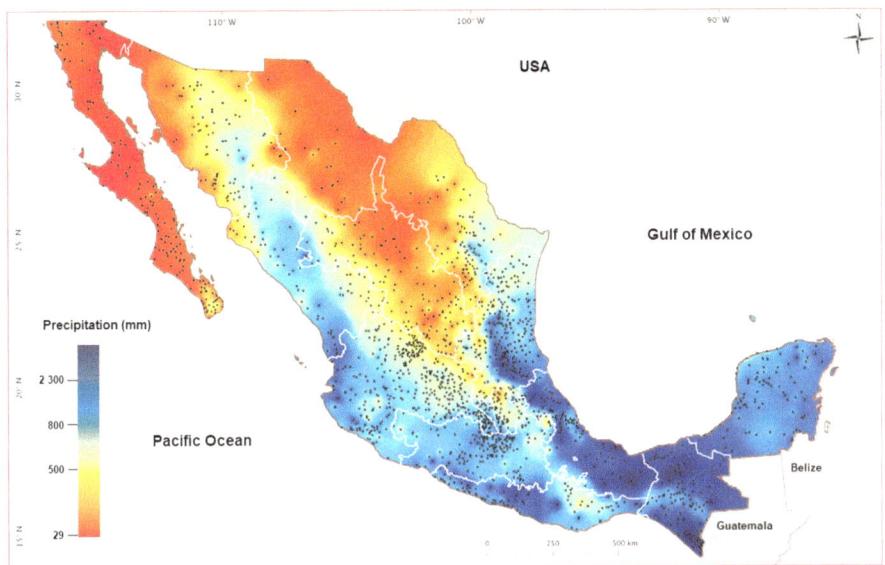

Fig. 2.4 Spatial distribution of annual precipitation. (Modified from CONAGUA 2015b)

distribution of rainfall is uneven, with great differences between the north and south of the country (Fig. 2.4).

Mexico receives about 1449.5 km^3/year of water from precipitation, of which 72.5% returns to the atmosphere through evapo-transpiration, 21.2% is drained by rivers or streams and the remaining 6.3% is infiltrated into the soil and becomes aquifer recharge (CONAGUA 2015a). The total annual volume of aquifer recharge is estimated at 91.8 km^3/year, while 33.3 km^3/year are extracted from the subsoil (Fig. 2.5).

An additional input of water for the aquifers comes from deep percolation losses from excess irrigation, leaks in the urban water-distribution systems, artificial and induced recharge, and upward flows from deeper aquifers (Alcocer and Chavez 2016). This amount of water recharge is estimated to be 11 km^3/year (CONAGUA 2010).

2.2.3   Water Quality

Figure 2.6 shows the groundwater salinity expressed in terms of total dissolved solids (TDS) content. Most of the 653 aquifers contain water with a salinity less than 1000 mg/L of TDS, which is the maximum permissible limit for human consumption (Chavez et al. 2006).

Soils salinization and brackish groundwater occur in arid areas with low rainfall and high evaporation rates, shallow groundwater levels, dissolution of evaporitic minerals, and the presence of highly saline congenital water. By the end of 2015,

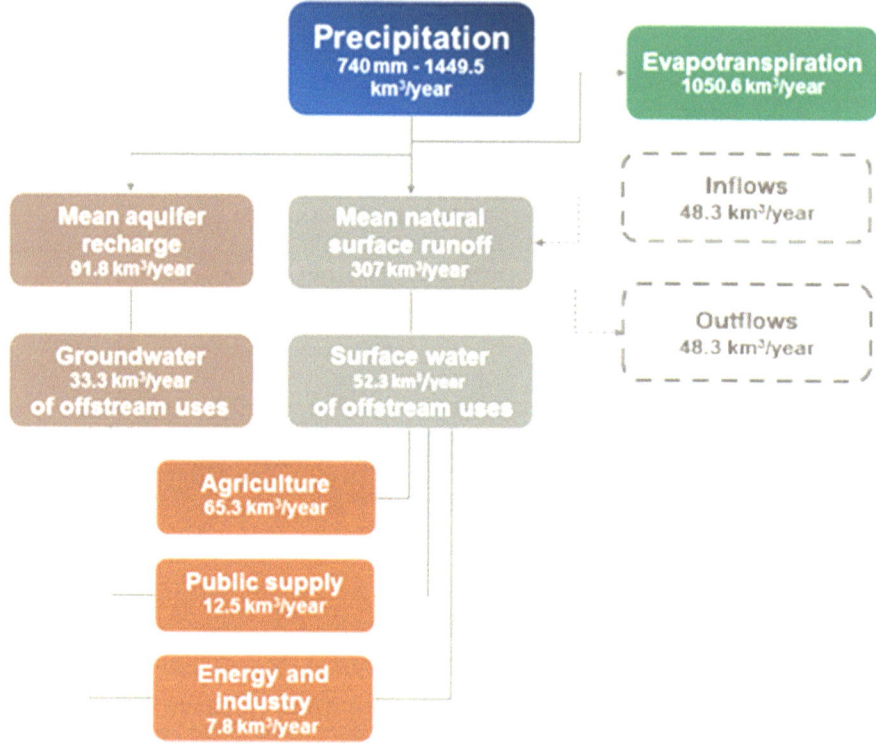

Fig. 2.5 Water cycle at the national level. Proprietary development, with data from CONAGUA (2017). (Arreguin-Cortes and Cervantes-Jaimes 2018)

32 aquifers had been identified with these effects, mainly in the Baja California Peninsula and Mexican highlands (CONAGUA 2016b).

Several aquifers in the north and center of the country experience the natural presence of chemical elements in concentrations quite above the Mexican water quality standard for human consumption (Chavez et al. 2006). In particular, hydrogeological studies have documented the origin and distribution of fluorine and arsenic in aquifers of Durango, Zacatecas, Chihuahua, San Luis Potosi, Guana-juato, and Coahuila states.

2.3 Transboundary Aquifers

Mexico participates in the International Boundary and Water Commissions (IBWC) with the USA (northern border), Guatemala, and Belize (southern border). They act independently of each other and make recommendations to the governments of each country for the solution of bilateral issues regarding limits and waters (SRE 2017).

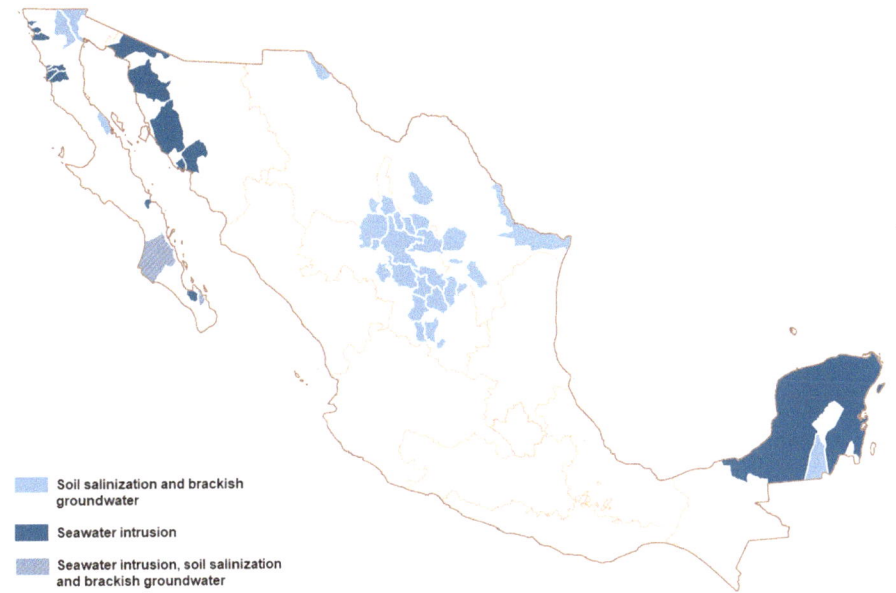

Fig. 2.6 Mexican aquifers with seawater intrusion, soil salinization, and brackish groundwater. (Modified from CONAGUA 2016b)

Table 2.1 Transboundary aquifers of Mexico (IGRAC 2014; UNESCO 2009)

Mexico–USA		Mexico–Guatemala	
No	Description	No	Description
8 N	San Diego–Tijuana	1C	Soconusco–Suchiate/Coatan
9 N	Cuenca Baja del Rio Colorado	2C	Chicomuselo–Cuilco/Selegua
10 N	Sonoyta–Papagos	3C	Ocosingo–Usumacinta–Pocom-Ixcan
11 N	Nogales	4C	Marquez de Comillas–Chixoy/Xaclbal
12 N	Santa Cruz	5C	Boca del Cerro–San Pedro
13 N	San Pedro	6C	Trinitaria–Nenton
14N	Conejos Medanos–Bolson de la Mesilla		
15N	Bolson del Hueco–Valle de Juarez	**Mexico–Guatemala–Belize**	
16 N	Edwards–Trinity–El Burro	7C	Peninsula de Yucatan–Candelaria-Hondo
17 N	Cuenca Baja del Rio Bravo/Grande		

While the 1944 treaty regulates the surface-water resources between Mexico and the USA, there is no groundwater treaty with any bordering country (Hatch-Kuri and Ibarra-Garcia 2015; Sanchez and Eckstein 2017). Joint efforts carried out in the framework of the IBWC include the following: exchange of information, reciprocal consultation, conflict management and preparation of reports, studies and integrated models (SRE 2017).

Mexico recognizes ten major transboundary aquifers with the USA, six with Guatemala, and one with Guatemala-Belize (Table 2.1; Fig. 2.7).

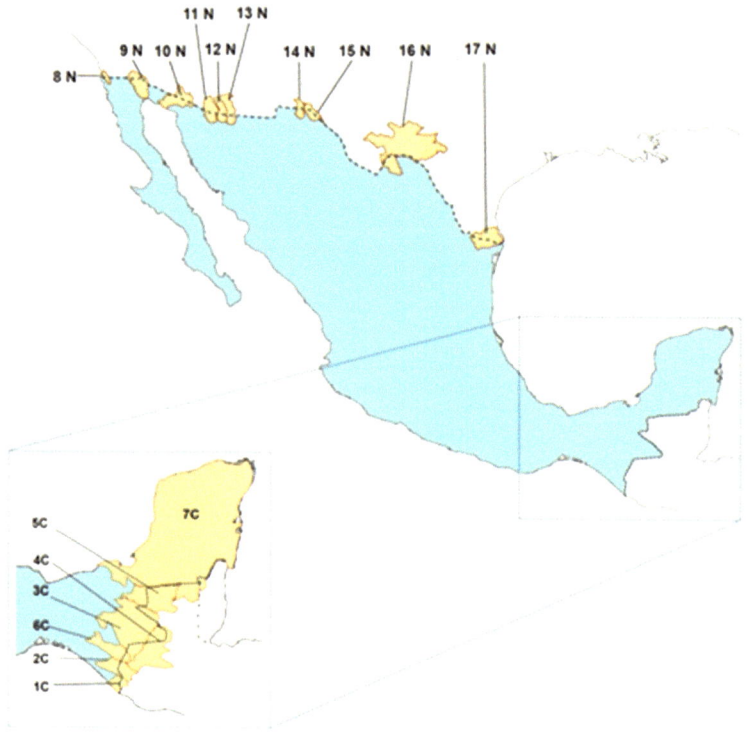

Fig. 2.7 Transboundary aquifers of Mexico. (Modified from UNESCO 2009)

Cross-border issues related to groundwater between Mexico and the USA are dealt with on a case-by-case basis by the IBWC, although there are differences between both countries regarding legislation and jurisdiction (federal in Mexico, not federal in the USA), data collection, available data, aquifer boundaries and methodologies (Sanchez et al. 2016). The official documents to manage binational groundwater include Minute 242 (limits extractions in San Luis Rio Colorado, Sonora–Yuma, Arizona area), Minute 289 (monitoring of groundwater quality), and the 2009 Agreement (four hydrogeological studies on transboundary aquifers of mutual interest: Hueco Bolson, Mesilla Bolson, San Pedro and Santa Cruz aquifers) (SRE 2017).

2.4   Legal Tools for Water Management: Water Rights, Closures, Reservations, and Regulations

The agrarian laws that emanated from the Mexican Revolution established in the Constitution of 1917 that the nation is the "original owner" of the lands, the waters, and the resources of the subsoil: Article 27 of the constitution and its reform of 1992 established the ways for the state to restitute or endow farmers with land through the creation of "ejidos", a form of possession of land owned by the nation, which benefit the "ejidatarios" to work collectively.

The 1917 Mexican Constitution (Paragraph 5, article 27) enables landowners to extract the underlying water through artificial means. However, the federal government may intervene when public interests or third parties are affected. The forms of intervention were established in the 1926 Federal Water Law and include pumping prohibitions in restricted areas, regulations, and reserves. Between 1958 and 1980, 85 decrees of prohibition were issued, in which the authorization for new wells was restricted in more than 50% of the national territory, and regulations for the operation and the use of water in the irrigation districts were issued (Fig. 2.8).

In 1986, the modifications of the Federal Law of Rights established water-use fees. It divided the country into zones according to the degree of water availability and set differentiated tariffs that reflected the degree of water scarcity in each zone. Technical standards were later issued to regulate maximum permissible limits in wastewater discharges and for the construction of drinking-water wells.

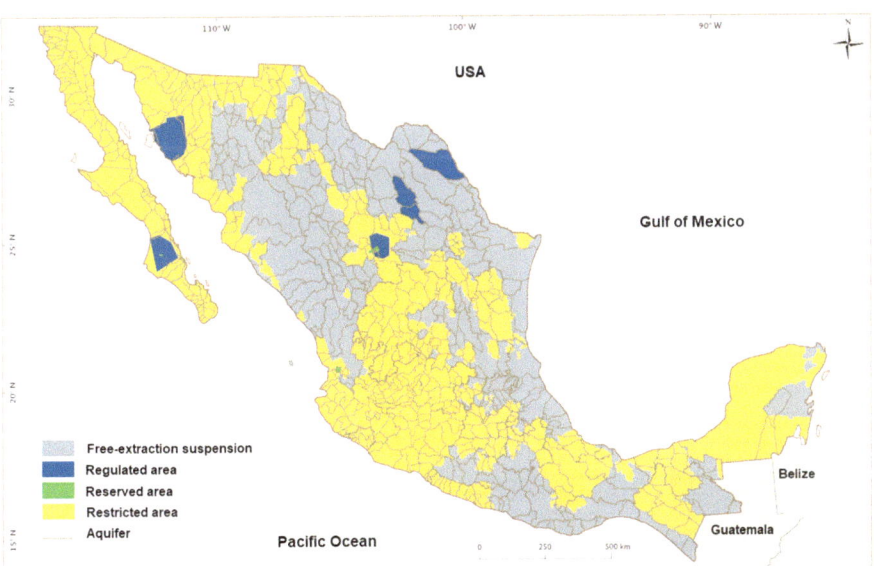

Fig. 2.8 Groundwater management areas in 2015. (Modified from CONAGUA 2016b)

The 1992 National Water Law established the cases requiring the federal government's intervention:

- Prevention of aquifer overuse
- Protection and recovery of ecosystems
- Protection of drinking-water supplies
- Preservation and control over water quality
- Shortage for extraordinary drought

The 1992 law authorizes the federal government to grant concession titles for the extraction of water or permits for the discharge of wastewater into nationally owned water bodies (lakes, rivers, aquifers). These water rights are independent of the ownership of the land, and each concession title authorizes the user to extract a certain annual volume of water from a specific aquifer. The concession titles can be granted for periods from 1 to 50 years, and water rights and discharge permits must be registered in the Public Registry of Water Rights (REPDA). CONAGUA can authorize the extension of water rights and/or the transfer of these rights between users, and it is necessary to register in the REPDA the change in ownership of the concession title.

The zones where it is forbidden to increase the extraction of groundwater are those regions in which, due to their hydrogeological conditions or considerations of public interest, the state has decreed or decreed the temporary or conditional prohibition to carry out new works to extract the waters of the subsoil (Article 9). They are classified in (Article 11):

I. Closed or restricted areas where it is not possible to increase the extraction of groundwater due to the risk of dangerously collapsing or depleting aquifers
II. Closed or restricted areas where only extractions for domestic purposes are allowed
III. Closed or restricted areas in which the capacity of aquifers allows limited extractions for domestic, industrial, irrigation, and other uses

The regulated zones are those in which the presidential decree establishes rules and limitations for the extraction of groundwater, in a specific aquifer. These regulations may impose limits on the duration of concession titles, criteria for the transmission of water rights among users, zoning to protect certain uses of water, additional obligations for water users, etc.

The reserve areas are those in which a presidential decree establishes the delimitation of a territory or a volume of water that is set aside for a specific use of water extracted from the subsoil, such as the strategic reserve of water for potable use in cities that have limitations to find new sources of water or that are estimated to experience great growth in water demand in the future.

As of the year 2000, a new strategy for groundwater management was established, based on the establishment of official standards of mandatory compliance, and in which the technical procedures for making decisions regarding groundwater management were established. As part of this approach, administrative limits of the main country's aquifers have been published in official decrees.

In 2002, the Official Mexican Standard NOM-011-CONAGUA-2000 was published (DOF 2002) and was later modified in 2015 (DOF 2015); the standard establishes the technical procedure to determine the availability of water in a specific aquifer, based on the estimated recharge of the aquifer and the volume of water authorized for extraction and registered in the Public Registry of Water Rights.

This is the criterion currently used to grant new water concessions in those aquifers in which the availability of water has a positive value.

In 2009, the norms that establish the technical procedure to carry out the artificial recharge of aquifers with treated wastewater NOM-014-CONAGUA-2003 and with surface water, rainwater, or storm water NOM-015-CONAGUA-2007 were published (DOF 2009a, b, respectively). Altogether, these standards constitute the legal instruments that are currently used to make decisions regarding groundwater management.

2.5 Participation of Users: Basin Councils and COTAS

In 1987, the Mexican government launched a strategy for the participation of organized groundwater users in water management as an institutional response to the problems arising from intensive groundwater use. Through the organization of groundwater user groups, the goal was to obtain a consensus for the approval of "Aquifer Regulations" that would enforce a reduction in groundwater extraction. The first water-user association, called "Water Group", was formed in August 1990 in the Comarca Lagunera region, leading to the approval of the "Regulation Decree for the Aquifer of the Comarca Lagunera". Following this same scheme, another water group was constituted in 1991 in the Valley of Santo Domingo preceding the approval of the "Regulations Decree" for this aquifer. Other user groups were also organized in the north and central parts of the country, although their decree was never approved (Escolero and Martinez 2007).

Following a stagnation period between 1993 and 1995 due to changes in the federal government, the formation of new user groups started again by the end of 1995. These new groups known as Technical Committees for Groundwater or *Comites Tecnicos de Aguas Subterraneas* (COTAS) were first launched in the Queretaro Valley aquifer. These committees were then established in various aquifers in the center and north of the country. Two years later, in the state of Guanajuato, local authorities took the initiative of promoting the creation of such committees for all the state's aquifers by supporting them financially through a trusteeship established for this purpose.

To support groundwater management, the 2004 law considers the conformation of COTAS as collegiate bodies of mixed integration and auxiliary bodies not subordinated to CONAGUA or basin organizations (Fig. 2.9).

The COTAS are bodies that support the basin councils so that they can carry out their coordination and concentration of tasks among the CONAGUA, the federal, state or municipal government agencies and the representatives of the users of the

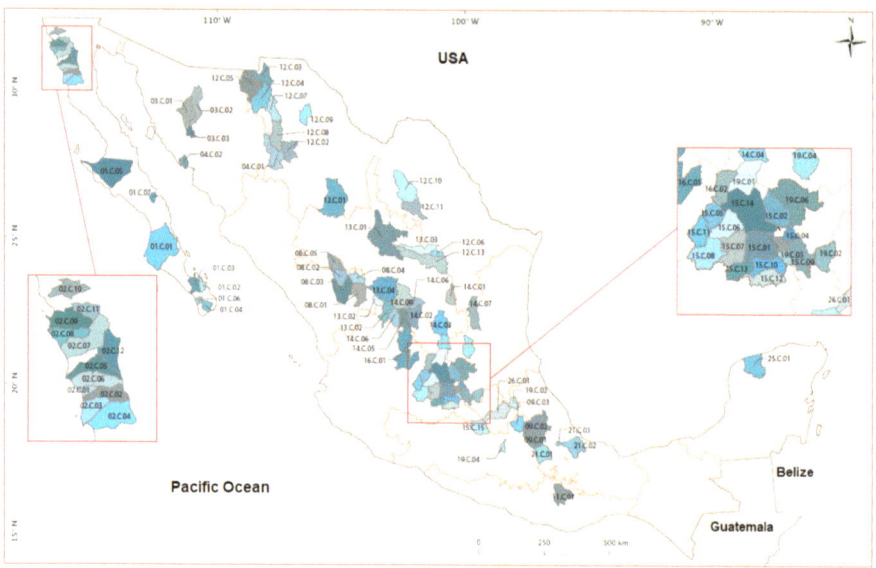

Fig. 2.9 Technical Committees for Groundwater (COTAS) in 2015. (Modified from CONAGUA 2016b)

respective (superficial) hydrological basin all for the purpose of formulating, executing programs and actions for the better administration of the so-called aquifers within their jurisdiction.

2.6 Groundwater Use

2.6.1 Water Balance

Groundwater constitutes an essential reserve for Mexico's socioeconomic development since more than half of its territory is arid or semi-arid and water supply depends heavily on groundwater resources: about one third of the total consumptive use comes from the aquifers. It is used to irrigate more than two million hectares of land (i.e., one third of the total irrigated land, mainly in the northern, northwestern, and central parts of Mexico), it supplies about 70% of water requirements for public-urban sector (60 million people living in urban centers), it provides water for most of the industrial facilities (in northern and central Mexico) and almost the entire water demand of rural population (20 million people) (Chavez et al. 2006).

Of the 33.3 km^3 of groundwater used every year, agriculture consumes 70.4%; the public uses 22.2% (2% for the rural population), industries 6.1%, and electricity generation 1.4%. There are about 130,000 registered wells, 80,000 of which are used for irrigation purposes and 15,000 for municipal and industrial supply (Arreguin and Lopez 1995).

Global national groundwater balance is positive since total extraction only represents 36.3% of the total recharge. However, it does not reflect the critical situation of the vast arid regions of central and northern Mexico, where critical aquifer overexploitation occurs, and underground storage is being depleted. The agricultural sector and urban population are growing at an accelerated rate, with the consequent increase in water demand.

2.6.2 Water Availability

In 2002, CONAGUA established a mandatory national standard (DOF 2015) to determine the water availability in basin and aquifer units, which establishes the minimum requirements to calculate the mean annual groundwater availability. To date, there are 408 aquifers with groundwater availability and 245 without availability (Fig. 2.10; DOF 2018).

Aquifers in the central and northern portions of the country—where Mexico's population and economic activities are concentrated—have greater water extraction rates than the average rate of natural recharge (Arreguin et al. 2011), as well as lower annual rainfall (50–500 m); in contrast, there is greater groundwater availability at the south and southeast, where average annual rainfall ranges between 1000 and 2000 mm (Aboites et al. 2008). Water stress in high-demand zones and with low groundwater availability has increased aquifer overexploitation (CONAGUA 2006).

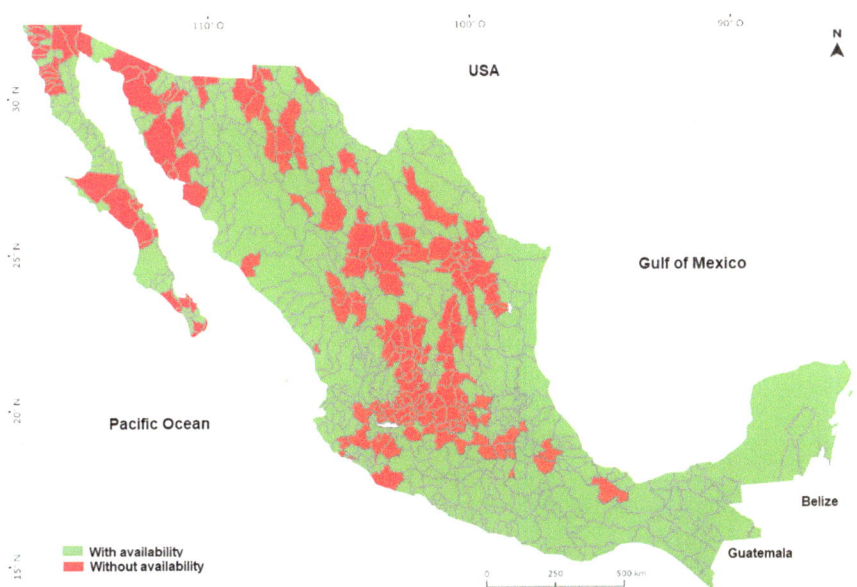

Fig. 2.10 Groundwater availability. (Modified from CONAGUA 2015a)

2.7 Impacts of Groundwater Extraction

2.7.1 Aquifer Overexploitation

Considering that there is no a unique and single definition of the term overexploitation, the criterion used in practice by CONAGUA is the following: an aquifer is overexploited when the permanent volume of water extracted surpasses the recharge, in such a way that the natural discharge is eliminated and the difference between extraction and recharge derives from the storage (Alcocer and Chavez 2016).

According to this criterion, 105 of the 653 aquifers are overexploited (Fig. 2.11; CONAGUA 2016a). This number has increased substantially in the past five decades: 32 in 1975, 36 in 1981, 80 in 1985, 97 in 2001, and 102 in 2003 (Chavez et al. 2006). It is estimated that the rate of overexploitation has reached 6 km^3/year, with an accumulated total loss of groundwater storage of 110 km^3 during the period 1950–2015 (Alcocer and Chavez 2016).

About 42 million inhabitants depend partially on overexploited aquifers for their various activities. The overexploited aquifers are directly related to the most important industrial and urban centers of the country, turning water-resources availability into a limiting factor for the sustainable development of the region. The most severe cases are found at the northwestern, northern, and central regions of the country, due mainly to agricultural development, as well as in the Valley of Mexico and Lerma

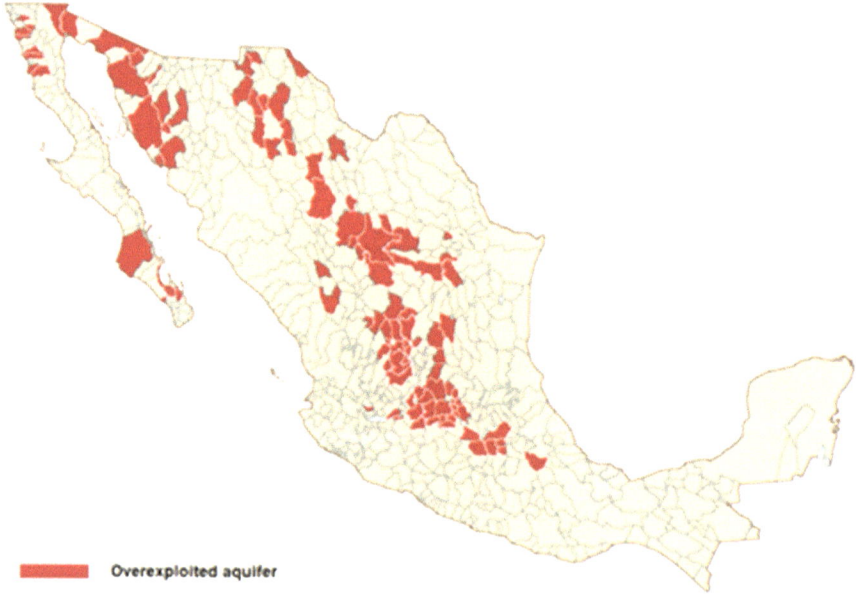

Overexploited aquifer

Fig. 2.11 Overexploited aquifers of Mexico. (Modified from CONAGUA 2015a)

River Basin (Alcocer and Chavez 2016). The overexploitation effects observed in the environment of such regions are: groundwater drawdown of several tens of meters, springs' depletion, disappearance of lakes and wetlands, base-flow reductions, elimination of native vegetation, loss of ecosystems, decreased well performance, increased costs of water extraction, land subsidence, groundwater contamination, and saline intrusion (Chavez et al. 2006).

Despite all the negative effects, aquifer overexploitation should be a feasible and even recommendable measure to compensate surface-water scarcity during prolonged droughts, if it is a temporary and controlled activity that makes it possible to reach aquifer equilibrium in the long term (Tuinhof et al. 2006)

2.7.2   Land Subsidence

Overexploitation of groundwater resources in Mexico has caused a differential land subsidence phenomenon (Murck et al. 1997) in or nearby urban areas settled over compressible deposits, such as Mexico City, Celaya, Aguascalientes, Queretaro, Morelia, Salamanca, Toluca, and Irapuato (Carreon-Freyre and Cerca 2006; Calderhead et al. 2011; Huizar-Alvarez et al. 2011), and in some agricultural regions of San Luis Potosi and Aguascalientes states (Chaussard et al. 2014). Indirect consequences of subsidence include: infrastructure damage in urbanized areas; differential changes in elevation and gradient of stream channels, drains, and water-transport structures; failure of water-well casings; groundwater contamination; and flood risk increase (UNESCO 1984).

Subsidence effects are observed in areas overlying aquifer systems composed of compressible deposits of variable thickness (Huizar-Alvarez et al. 2011; Bouwer 1977; Jachen and Holzer 1979; Holzer and Pompeyan 1981). Vertical subsidence rates registered in Mexico City are over 30 cm/year and 5–10 cm/year in the other areas (Chaussard et al. 2014).

2.7.3   Seawater Intrusion

Because of the intensive groundwater exploitation, 18 coastal aquifers have experienced seawater intrusion (CONAGUA 2016b). The most critical cases are in Baja California Peninsula and the state of Sonora. The saline water that migrated inland has disabled many wells and important agricultural areas and increased the groundwater salinity until reaching concentrations unfit for most common uses (Chavez et al. 2006).

2.7.4 Hydrocarbon Contamination

Spills or leaks of gasoline and diesel represent another relevant groundwater contamination risk, especially for aquifers that supply water to large cities, although their impact is generally at local and temporal scales (Chavez et al. 2006). More than 50 cases have been documented, mainly in states with activities linked to extraction, conduction, and distribution of hydrocarbons.

2.8 New Approaches for the Development of Groundwater Resource

The traditional sources of groundwater are being depleted or are totally committed to current users and the growing demand for water for public-urban, agricultural, and industrial use; as a result, new strategies have been developed to search for new and alternative sources of groundwater.

One of the strategies commonly applied in Mexico is the managed aquifer recharge, for which official norms NOM-014-CONAGUA-2003 and NOM-015-CONAGUA-2007 establish the criteria and requirements to carry out such activities (DOF 2009a, b, respectively). Based on these regulations, numerous studies, projects, and aquifer recharge tests have been carried out, driven mainly by municipal water companies. Escolero et al. (2017) present a useful compilation of aquifer recharge works carried out in Mexico.

In the search for new sources of groundwater, exploratory wells are being drilled more than 2000-m deep in Mexico City, which has made it possible to locate aquifer strata not previously explored at depths between 1000 and 2000 m, containing water of good quality and high yield (Morales-Casique et al. 2014). These results have generated positive expectations, and it is planned to drill more deep wells in the city to achieve the extraction rate of 1 m^3/s. This water source will replace shallow wells located in areas of intensive exploitation, related to land subsidence in the urban area, or replace imported volumes from sources external to the basin of Mexico. Based on these results, studies are being conducted in other regions of the country to search for other deep aquifer strata.

More recently, the extraction of brackish water in coastal aquifers has been promoted to supply desalinization plants and for use in water supply to urban areas, real estate and tourism complexes and in greenhouses producing agricultural products for export. The national water authority has established criteria and guidelines for authorizing the construction of wells and the extraction of brackish water in coastal areas to support this type of work. Several related developments are currently under construction in northern Mexico.

References

Aboites L, Cifuentes E, Jimenez B, Torregrosa ML (2008) Pendientes nacionales del agua – Agenda del agua. Academia Mexicana de Ciencias, Red del Agua. Primera edicion 2008, IS BN 978-607-95166-0-4

Alcocer YVH, Chavez GR (2016) Aprovechamiento del agua subterranea. H2O Gestion del Agua, Revista del Sistema de Aguas de la Ciudad de Mexico, abril-junio 2016, No 10

Arreguin CF, Lopez PM (1995) An overview of Mexico's water regime and the role of groundwater. In: National Research Council (ed) Mexico City's water supply: improving the outlook for sustainability. National Academies Press, Washington, DC

Arreguin CF, Lopez PM, Marengo MH (2011) Mexico's water challenges for the 21st century. In: Water resources in Mexico: scarcity, degradation, stress, conflicts, management, and policy, Hexagon series on human and environmental security and peace, vol 7. Springer-Verlag, Berlin/ Heidelberg. https://doi.org/10.1007/978-3-642-05432-7_2

Arreguin-Cortes FI, Cervantes-Jaimes CE (2018) Water security and sustainability in Mexico. In: Raynal-Villaseñor JA (ed) Water resources of Mexico. Springer International Publishing A G, Cham

Back W, Rosenshein JS, Seaber PR (1988) Hydrogeology. The geology of North America, vol 0–2. The Geological Society of America, Boulder

Bouwer H (1977) Land subsidence and cracking due to groundwater depletion. Groundwater 15:358–364

Brassington R (2007) Field hydrogeology. Wiley, Chichester

Calderhead AI, Therrien R, Rivera A, Martel R, Garfias J (2011) Simulating pumping-induced regional land subsidence with the use of InSAR and field data in the Toluca Valley, Mexico. Adv Water Resour 34(2011):83–97

Carreon-Freyre DC, Cerca M (2006) Integration of geological properties in the study of the subsidence and fracturing phenomena in two urban areas of Mexico. IAEG2006 Paper number 291. The Geological Society of London

Chaussard E, Wdowinski S, Cabral-Cano E, Amelung F (2014) Land subsidence in Central Mexico detected by ALOS InSAR time-series. Remote Sens Environ 140(2014):94–106

Chavez R, Lara F, Sencion R (2006) El agua subterranea en Mexico: condicion actual y retos para un manejo sostenible. Bol Geol Min 117(1):115–126. ISSN: 0366-0176

CONAGUA (2006) Water in Mexico. Comision Nacional del Agua, First English edition, March, 2006, ISBN 968-817-730-X

CONAGUA (2010) Estadisticas del Agua en Mexico, edicion 2010. Comision Nacional del Agua, México

CONAGUA (2015a) Estadisticas del Agua en Mexico, edicion 2015. Comision Nacional del Agua, diciembre de 2015, México

CONAGUA (2015b) Atlas del Agua en Mexico 2015. Comision Nacional del Agua, diciembre 2015, México

CONAGUA (2016a) NUM3RAGUA. Comision Nacional del Agua, octubre 2016

CONAGUA (2016b) Atlas del Agua en Mexico 2016. Comision Nacional del Agua, octubre 2016

CONAGUA (2017) Estadisticas del Agua en Mexico, Edicion 2016. Comision Nacional del Agua, http://201.116.60.25/publicaciones/EAM_2016.pdf Accessed 7 Mar 2018

De la OCA (1951) Provincias Geohidrologicas de Mexico. Tomo I, Boletin num. 56. Instituto de Geologia, UNAM, Mexico. 137 p

De la OCA (1954) Provincias Geohidrologicas de Mexico. Tomo II, Boletin num. 56. Instituto de Geologia, UNAM, Mexico. 166 p

DOF (2001) Acuerdo por el que se establece y da a conocer al publico en general la denominacion unica de los acuiferos reconocidos en el territorio de los Estados Unidos Mexicano. Diario Oficial de la Federacion, 5 diciembre 2001, Segunda Seccion

DOF (2002) NORMA Oficial Mexicana NOM-011-CNA-2000, Conservacion del recurso agua-Que establece las especificaciones y el metodo para determinar la disponibilidad media anual de las aguas nacionales. Diario Oficial de la Federacion. 17 abril 2002

DOF (2009a) NORMA Oficial Mexicana NOM-014-CONAGUA-2003, Requisitos para la recarga artificial de acuiferos con agua residual tratada. Diario Oficial de la Federacion. 18 agosto 2009

DOF (2009b) NORMA Oficial Mexicana NOM-015-CONAGUA-2007, Infiltracion artificial de agua a los acuiferos. – Caracteristicas y especificaciones de las obras y del agua. Diario Oficial de la Federacion. 18 agosto 2009

DOF (2015) NORMA Oficial Mexicana NOM-011-CONAGUA-2015, Conservacion del recurso agua-Que establece las especificaciones y el metodo para determinar la disponibilidad media anual de las aguas nacionales. Secretaria de Medio Ambiente y Recursos Naturales, Diario Oficial de la Federacion, 27 marzo 2015, Primera Seccion

DOF (2018) ACUERDO por el que se actualiza la disponibilidad media anual de agua subterranea de los 653 acuiferos de los Estados Unidos Mexicanos, mismos que forman parte de las Regiones Hidrologico-Administrativas que se indican. Secretaria de Medio Ambiente y Recursos Naturales, Diario Oficial de la Federacion, 4 enero 2018, Primera Seccion

Escolero O, Martinez S (2007) The Mexican experience with groundwater management. In: The global importance of groundwater in the 21st century: Proceedings of the International Symposium on Groundwater Sustainability. National Ground Water Association, NGWA Press, ISBN 1-56034-131-9. p. 97–103

Escolero O, Gutierrez-Ojeda C, Mendoza-Cazares EY (2017) Manejo de la recarga de acuiferos: Un enfoque hacia Latinoamerica, Publicado por Instituto Mexicano de Tecnologia del Agua (IMTA), Mexico. http://www.geologia.unam.mx/contenido/libro-recarga-acuiferos-latinoamerica. Accessed 7 Mar 2018

Hatch-Kuri G, Ibarra-Garcia V (2015) Las aguas subterraneas transfronterizas Mexico-Estados Unidos: importancia e invisibilidad dentro del contexto del TLCAN. Transboundary groundwater US–Mexico: invisibility and importance in the context of NAFTA. America Latina Hoy 69(2015):75–93

Holzer TL, Pompeyan EH (1981) Earth fissures and localized differential subsidence. Water Resour Res 17:223–227

Huizar-Alvarez R, Mitre-Salazar LM, Marin-Cordova S, Trujillo-Candelaria J, Martinez-Reyes J (2011), Subsidence in Celaya, Guanajuato, Central Mexico: implications for groundwater extraction and the neotectonic regime. Geofis Int 50 3: 255–270

IGRAC (2014) Transboundary aquifers of the world. International Groundwater Resources Assessment Centre, Paris

INEGI (2018) Poblacion, Instituto Nacional de Estadistica, Geografia e Informatica, http://www.beta.inegi.org.mx/temas/estructura/. Accessed 11 Mar 2018

Jachen RC, Holzer TL (1979) Geophysical investigations of ground failure related to groundwater extraction in Picacho basin. Arizona. Groundwater 17:574–585

Morales-Casique E, Escolero O, Arce JL (2014) Resultados del pozo San Lorenzo Tezonco y sus implicaciones en el entendimiento de la hidrogeologia regional de la Cuenca de Mexico. Rev Mex Cienc Geol 31(1):64–75

Murck B, Skiner G, Porter S (1997) Subsidence in dangerous earth: an introduction to geologic hazards. Wiley, New York, pp 173–189

Sanchez R, Eckstein G (2017) Aquifers shared between Mexico and the United States: management perspectives and their transboundary nature. Groundwater 55(2)

Sanchez R, Lopez V, Eckstein G (2016) Identifying and characterizing transboundary aquifers along the Mexico–US border: an initial assessment. J Hydrol 535(2016):101–119

SARH (1978) Atlas Geohidrologico: Banco Nacional de Informacion Geohidrologica. Editado e impreso bajo contrato con la empresa GEORAMA, S. A, de C.V. 282 p

SRE (2017) Aguas Subterraneas en la frontera entre Mexico y Estados Unidos. Secretaria de Relaciones Exteriores, Comision Internacional de Limites y Aguas (International Boundary and Water Commission), February 2017, https://cila.sre.gob.mx/cilanorte/index.php/boletin/11-doctos/76-aguas-subterraneas. Accessed 7 Mar 2018

Tuinhof A, Dumars C, Foster S, Kemper K, Garduno H, Nanni M (eds) (2006) Groundwater resource management: an introduction to its scope and practice (English), GW mate briefing note series; no. 1. World Bank, Washington, DC

UNESCO (1984) Guidebook to studies of land subsidence due to ground-water withdrawal. Prepared for the International Hydrological Programme, Working Group 8.4, Joseph F. Poland, Chairman and Editor

UNESCO (2009) Atlas of transboundary aquifers. UNESCO-International Hydrological Programme: Division of Water Science, ISARM Programme Paris, S. Puri and A. Aureli

Velazquez-Aguirre L, Ordaz-Ayala A (1992) Provincias Hidrogeologicas de Mexico. Ingenieria Hidraulica en Mexico, Vol. I, enero-abril

Velazquez-Aguirre L, Ordaz-Ayala A (1993) Provincias Hidrogeologicas de Mexico. Bol Soc Geol Mex 52(1):15–33

Chapter 3
Hydrogeology of Mexico

Ignacio Reyes-Cortes and Abundio Osuna-Vizcarra

Abstract Hydrogeology of Mexico includes treatment of hydrometeorology, which relates the presence of groundwater within the hydrological cycle and the direction of water flow. Geology and hydrogeology relate to the quantification of available water in the aquifer system because hydrogeological modeling can make predictions and facilitate operation of the aquifer system according to aquifer classification. Rock characterization and the distribution of their water-bearing capacity in Mexico are defined by parameters such soil moisture, infiltration, subsurface water, fissure and vein water, and hydrothermal activity. To know the effect of groundwater sources and their potential contamination, it is necessary to create detailed cartography for each aquifer.

Keywords Hydrometeorology · Geology · Hydrothermal activity

3.1 Introduction

The hydrogeology of Mexico has very particular features that define its climate, precipitation pattern and stream densities. The territorial extension of Mexico includes 1964 million km^2 of which 1959 million km^2 correspond to the continental surface and the rest to island areas. There are different factors that determine Mexico's climate. The southern part of the country is in the intertropical zone of the globe, whereas the northern part is in the temperate zone. Mexico is located at the same latitude as the Saharan and Arabian deserts. Mexico sits astride the latitude of the Tropic of Cancer, and it is between the two oceans, the Atlantic and Pacific, east and west, respectively.

The second factor is the geographical accident that characterizes Mexico's relief (Fig. 3.1). Mexico has two large ranges, which parallel both coasts. Coastal plains

I. Reyes-Cortes (✉) · A. Osuna-Vizcarra
Facultad de Ingenieria, Universidad Autonoma de Chihuahua, Chihuahua, Chihuahua, Mexico
e-mail: ireyes@uach.mx; aosuna@uach.mx

© Springer Nature Switzerland AG 2020
J. A. Raynal-Villasenor (ed.), *Water Resources of Mexico*, World Water Resources 6,
https://doi.org/10.1007/978-3-030-40686-8_3

1 Península de Baja California
2 Llanura Sonorense
3 Sierra Madre Occidental
4 Sierrasy Llanuras del Norte
5 Sierra Madre Oriental
6 Grandes Llanuras de Norteamérica
7 Llanura Costera del Pacífico
8 Llanura Costera del Golfo Norte
9 Mesa del Centro
10 Eje Neovolcánico
11 Península de Yucatán
12 Sierra Madre del Sur
13 Llanura Costera del Golfo del Sur
14 Sierras de Chiapas y Guatemala
15 Cordillera Centroamericana

FUENTE: Instituto Nacional de Estadística Geografía e Informática, Dirección General de Geografía
         Cartas Fisiográficas escala 1:1 000 000, México

Fig. 3.1 Map presenting each of the physiographic provinces in a schematic way

develop their widths because of the large amount of sediments eroded from the mountain ranges (the Sierras Madres) and the interior of the country. The wedge clastic sediments accumulate against the ranges, forming large talus, but most of them are carried to the coast. The gulf and ocean currents allocate them along the coastal beaches forming bars and barriers. The geographical location and their relief have a direct impact on the availability of water resources.

The urban areas extract most of their water supply from groundwater reserves. So, groundwater supply is growing as a problem in Mexico, but quality is also becoming a concern nowadays. The problem implicates the management and protection of this renewable resource in Mexico and its significant value for maintaining and developing the hydrogeological systems for the future.

Hydrogeology involves many factors of science and engineering, including geology, geophysics, mathematics, computer applications, chemistry, physics, biology, and soil science. Hydrogeology is fundamental in the planning, designing, and implementation of any kind of water project. Such projects include the study of groundwater resource conservation, up to their development and restoration for domestic, agricultural, and industrial purposes.

Groundwater engineering projects in many everyday activities, such as mining, foundation, and road building, can be involved to facilitate dewatering of slopes, tunnels, and open excavations, as well as delineation and remediation of contaminated aquifers. It is important to understand the groundwater flow in porous and fractured media, recharge/discharge relationships, and the contaminant transport mechanisms, as well as hydraulics.

Hydrogeological studies of Mexico begin with the overview of relief forms as identified and defined based on the comprehensive analysis of topographic, geological, hydrological, and pedological or soil information. These constitute relatively homogeneous units, representing the various physiographic provinces in which the country can be divided, according to their geology and topography.

The relief is the way in which the Earth's surface exhibits itself; in Mexico, there is an extraordinary variety of relief. It varies from large mountain ranges to great coastal plains passing through wide valleys, canyons, high plateaus, and depressions, among yet other physiographic features.

The Sierra Madre Occidental, with its plateaus and canyons, is one of the most outstanding features of the relief of Mexico. Other features are the Sierra Madre Oriental formed mainly by folded sedimentary rocks and the Neo-Volcanic Axis, which crosses the country from east to west. It is where we can find the highest volcanic mountains, some more than 5000 m in altitude. However, the physiographic knowledge of a region implies, in addition to the identification of the main relief features, an explanation of the processes that model it and that create their current appearance. The physiographic provinces are regions in which the relief is the result of the action of the same set of soil modeling agents and of the same geological origin. In addition, they sustain the same or very similar types of soil and vegetation.

The 20° N parallel marks change in the direction of the main features of the country's relief since, toward the north of this parallel, the main relief features are oriented almost from northwest to southeast. This difference in orientation indicates that, in the formation of the relief of Mexico, there intervened two different orogenic events, so that structurally the northern part belongs to the North America basement, while the southern belongs to Central America. This explains the enormous complexity of relief in the Central part of the country.

For a better understanding of this structural diversity, the country is divided into 15 physiographic regions (Fig. 3.1). The physiographic provinces are as follows: 1, Peninsula of Baja California; 2, Sonoran Plain; 3, Sierra Madre Occidental; 4, Sierras and Plains of the North; 5, Sierra Madre Oriental; 6, Great Plains of North America; 7, Pacific Coastal Plain; 8, North Gulf Coastal Plain; 9, Mesa del Centro; 10, Neo-Volcanic Axis; 11, Yucatan Peninsula; 12, Sierra Madre del Sur; 13, South Gulf Coastal Plain; 14, Sierras de Chiapas and Guatemala; and 15, Central American Mountain Range.

Water is used in so many ways to meet essential human needs and to produce goods and services. The water volumes (in the case of those for public urban or domestic uses) are allocated to the registered users of the Nation Water Record in the Public Register of Water Rights (PRWR or REPDA "Registro Público de Derechos de Agua" in Spanish).

The REPDA records the water uses, which are classified into various categories. It stipulates a difference between offstream and instream uses. Offstream use is that one that is consumed during a specific activity. It is defined as the difference between the volumes of a given quality extracted minus the volume of a given quality discharged. It appears in the respective norm or policy as the National Water Law. In 2014, the regulation added a new instream use, the "ecological conservation," which was assigned a volume of 9.46 hm^3/year. The data are from December 31, 2015. The regionalization of volumes is carried out based on the location of the water use as registered in the REPDA, rather than the jurisdictional area.

The evolution in the volume assigned for offstream uses, in the period 2006–2015, shows that 61.1% of the water used for offstream uses comes from surface-water sources such as rivers, streams, and lakes. The remaining 38.9% comes from groundwater sources like aquifers. There are variations in the volumes allocated over time. Compared to 2006, in 2015, the volume of surface water assigned was 6.9% greater, meanwhile the groundwater assigned was 17.5% greater in the same years.

The greatest allocated volume is for agricultural irrigation, and this reflects that Mexico has one of the most considerable irrigation infrastructures in the world (INEGI 2016).

Hydropower is considered an instream use of water resources. In 2015, 138.7 billion m^3 of water was used nationwide. While the water is used several times in all power plants, the water uses can be offstream or instream sources, and it could be surface or groundwater. There is no predominant source when the difference between surface and groundwater sources is less than 5%.

The agriculture and public-supply water uses account for 90.9% of the water volume used nationwide in 2015. The evolution of distribution of uses and the evolution of volumes can be traced over time. The hydrological–administrative regions (HARs) that have the biggest consumption of water volumes are as follows: VIII Lerma-Santiago-Pacific, IV Balsas, III Northern Pacific, and VI Rio Bravo. That agriculture water use accounts for over 80% of the total water volumes in those HARs. The exception is IV Balsas due to the Petacalco power plant. It is located near the estuary of the Balsas River and uses a significant volume of water.

The largest water use in Mexico is for agriculture. According to the "VII Agricultural, Livestock and Forest Census" of 2007, the surface extent of agricultural use was 30.2 million hectares, of which 18% was for irrigation and the remainder was rain fed ("temporary watering").

Mexico is the seventh largest country worldwide in regard to irrigation infrastructure about agricultural areas. There are 6.5 million hectares, of which about half corresponds to 86 irrigation districts and the rest corresponds to 40,000 irrigation units. Of the water volume, 35.9% of that used for agriculture comes from groundwater. Considering that there are annual variations, the increased volume of groundwater for this grouped use is 19.3% higher than in 2006.

The grouped use for public supply is the one that delivers water to drinking-water networks, which supply domestic users (homes), as well as various industries and services. An essential need of the population is to have access to water of enough quantity and quality for human consumption, given that it has a direct influence on their health and general well-being. In the grouped use for public supply, groundwater is the main source, with 58.6% of the volume. The water demand for public supplies increased by 32.3% between 2006 and 2015. In Mexico, drinking-water services, sanitation, sewerage and wastewater treatment, and disposal are regulated by municipalities, through their water utilities.

A grouped water use includes the industry that takes its water directly from rivers, streams, lakes, or aquifers. The secondary sector, known as industry, is formed by mining, electricity, water, and piped-gas supply to end users, as well as the construction and manufacturing industries (INEGI 2013a).

Although it only represents 4.3% of the total offstream use of water, the grouped water use for the water self-supplying industry shows the largest growth trend. In the 2006–2015 period, the used water volume from groundwater sources increased 51.4%.

3.2 Hydrometeorology

In the hydrological cycle, a significant proportion of the rainfall returns to the atmosphere in the form of evapotranspiration, while the rest of the water flows downstream through streams and water bodies along the surface. The latter constitutes the runoff waters; otherwise, it infiltrates into the subsurface as groundwater.

The basins are natural units of the land, limited by the water divides that conform the relief. For management purposes of national waters, there were 731 hydrological documented basins by December 31, 2015, of which 627 were in a situation of availability. The basins of Mexico are organized into 37 hydrological regions and grouped into 13 hydrological–administrative regions (RHA) (CONAGUA 2016).

The groundwater in Mexico is divided into 653 aquifers. While their geographical limits were published in the period 2003–2009, the publication of water availability and updates have been carried out since 2003 until today. The federal government has 3160 monitoring stations to measure climate variables, including temperature, rainfall, evaporation, and the velocity and direction of the wind. Of these stations, 88 are observatories that transmit weather information in real time.

The hydrometric stations measure the water flow of the rivers, as well as the extraction from dams. There are 861 hydrometric stations in Mexico, including some automatic ones. The hydroclimatological stations measure both climatological and hydrometric parameters, and the measurement infrastructure of Mexico enables analysis of the hydrological cycle.

Renewable water is defined as the maximum amount of water that is feasible to exploit annually in a region. It is an amount of water that is replenished by rain and water from other regions or countries (imports). It is calculated as the annual mean natural runoff, plus the total annual recharge of the aquifers, plus the entrance flows minus the flows of output water to other regions (Gleick 2002). Annually, Mexico receives approximately 1,449,471 million m^3 of water in the form of precipitation. However, it is estimated that 72.5% evaporates and returns to the atmosphere, 21.2% flows through rivers downstream, and the remaining 6.3% infiltrates the subsurface naturally to recharge aquifers. Aquifers have renewal periods, understood as the ratio of estimated storage divided by the annual recharge. Sometimes, they are exceptionally long periods. These aquifers are considered as nonrenewable waters.

Considering the outflows (exports) and inflows (imports) of water with neighbor countries, Mexico has annually 446,777 million m^3 of renewable freshwater. The inflows represent the volume of water that drains into the country, generated in the transboundary basins that Mexico shares with its neighbor countries (the United States, Guatemala, and Belize). The outflows represent the volume of water that Mexico must deliver to the United States in accordance with the "Water Treaty" of 1944, formally the "Treaty between the Government of Mexico and the Government of the United States of America of the distribution of the international waters of the Colorado, Tijuana, and Bravo rivers."

Renewable water must be analyzed from the three perspectives:

Temporal distribution: There are large variations of renewable water throughout the year in Mexico. Most of rain occurs during summer, while the rest of the year is relatively dry.

Spatial distribution: Abundant precipitation occurs in some regions of the country that have a low population density, while in other regions, the opposite occurs.

Analysis area: The water problem and its attention are predominantly of a local nature. The indicators calculated on a large scale hide the strong variations that exist throughout the country.

Some hydrological-administrative regions (HAR) have limited renewable water, such as in the HAR I Peninsula of Baja California (renewable water of 1115 m^3/inhab/year); the HAR VI Rio Bravo (renewable water of 1004 m^3/inhab/year); and the HAR VIII Lerma-Santiago-Pacific (renewable water of 1451 m^3/inhab/year). Besides, as in the HAR XIII Aguas del Valle de Mexico (renewable water of 148 m^3/inhab/year), this last value of renewable water per capita is very low (EAM 2016).

The normal precipitation of the country in the 1981–2010 period was 740 mm/year. It is important to point out that the monthly distribution of precipitation accentuates the problems related to the availability of the resource because 68% of the total annual mean rainfall is concentrated between the months of June and September.

The hydrological–administrative region XI Frontera Sur receives the largest amount of rainfall as renewable water. The annual mean rainfall from 1981 to 2010 was 11 times higher than in the hydrological–administrative region I Peninsula of Baja California, the driest. The above-normal precipitation occurred mainly in Tabasco, the Papaloapan River Basin between Oaxaca and Veracruz, and in the northcentral basins. The cumulative precipitation that occurred in Mexico from 1 January to 31 December 2015 reached 872 mm, which was 18% higher than the normal average of the 1981–2010 period (740 mm).

Tropical cyclones are natural phenomena which generate most of the moisture transported from the sea into the interior territory. Hurricane rains account for the largest part of the annual rainfall in various regions of the country. Between 1970 and 2015, 224 tropical hurricanes hit the coasts of Mexico. The greatest number of hurricanes, 140, hits the Pacific coast.

Droughts occur when rainfall is significantly lower than the normal average recorded amounts, which causes serious hydrological disequilibrium that affects agricultural production systems. When rain is scarce, temperatures increase, and vegetation develops only with difficulty. Droughts are the most expensive natural disasters because they affect more people than other forms of natural disasters. In addition, droughts can be linked to the phenomena of soil degradation and deforestation. In the dry season, the risk of forest fires increases significantly (INEGI 2013b).

Mexico, as a partner with the United States and Canada, participates in the "North American Drought Monitor" (NADM), which analyzes weather conditions to monitor droughts in North America, continuously and on a large scale. May is the month of interest in the year because is when the dry season usually ends and the rainy season begins. Another interesting point to review the evolution of the drought is the month of November, when the rainy season usually ends and the dry season begins.

Both drought and intense rainfall, coupled with factors such as topography, land use, and vegetation coverage, can cause damage to social and economic activities. The vulnerability of Mexico is assessed based on a three-component model: the degree of drought–intense rainfall exposure; the sensitivity or estimation of the damage to commercial and industrial economic activities, as well as the agriculture impact; and the adaptability or degree of aquifer exploitation.

The country's rivers and streams constitute a hydrographic network 633,000 km in length, in which 51 major rivers carry 87% of the country runoff, and their basins cover 65% of the country surface. The basins of the Bravo and Balsas rivers stand out for their surface extension, and the Bravo and Grijalva–Usumacinta rivers stand out for their length. The rivers Lerma and Nazas–Aguanaval belong to the interior drainage.

3.3 Geology

There does not yet exist a classification of the territory into natural units or provinces, as a function of its geographical, physiographic, metallogenic, geophysical, tectonic, and geological characteristics. However, some schematic maps have been generated, for example, for the Sierra Madre del Sur that generally characterize it based on physiographic province (Raisz 1959), based on morphotectonics (Guzmán and de Cserna 1963), metallogenic (Salas 1975), or geological data (López-Ramos 2018).

Here is the definition of the geological province by Ortega-Gutiérrez et al. (1991, 1992): "The geological province is a cartographic part of the solid surface of the planet, from hundreds to millions of square kilometers of extension. Which characterized by its rocks, by its structure and by a sequence of events, that it integrates a singular evolutionary history different from that of the adjacent areas, of which it is separated by stratigraphic, tectonic or both limits." It enables description of 35 provinces by their geological nature.

The geological age stands out as one of the most distinctive features of the territory, since almost three quarters of the geological provinces of the country exhibit, mostly, rocks of Mesozoic or Cenozoic age, that is, of the last 225 million years, while that the Precambrian rocks appear only in limited quantities in about 12% of them.

Another distinctive feature of the geology of Mexico is the asymmetric distribution of the provinces in terms of their origin because the sedimentary provinces (of superficial origin with granular pores) are concentrated in the eastern half of the country, while in the western half, the magmatic and metamorphic provinces (of deep origin with fissure pores) are located. This clear division of Mexican geology is explained by the passive nature of the eastern margin of Mexico since the Jurassic, in contrast to the convergent and active western margin, which has existed since the same time.

The result of this contrasting geological history is that concentration of metallic ore bodies is in the western half, meanwhile the energetic ore bodies, of superficial origin (oil, gas, and coal), are located in the eastern half.

Because of its extension, the largest geological province of Mexico is the ignimbrite belt, which occupies the Sierra Madre Occidental. It has an area of approximately 300,000 km^2, is 1600 km long, and has an average width of 250 km. Meanwhile, the smallest province is Juchiteca, in southern Mexico, just a few hundred km^2 in extent. Finally, it should be noted that the geological provinces of Mexico describe only the geological units exposed on the surface or very close to it, but their distribution in depth, revealed by drilling or geophysical methods, is not considered.

Mexico is divided into 35 geological provinces, indicating their age, origin, and corresponding geotectonic environment (Fig. 3.2). The provinces, by their characteristic age, have the following distribution: 2 of the Precambrian, 3 of the Paleozoic, 13 of the Mesozoic, 16 of the Cenozoic, and 1 of the Cenozoic-Mesozoic. Regarding their predominant origin, the distribution is as follows: six plutonic, five volcanic, ten sedimentary-marine, four sedimentary-continental, five volcano-sedimentary, and five of complex origin.

This classification of the geological provinces of Mexico is presented as a general guide that facilitates extracting information in a prompt manner and understanding more easily the complex geological constitution of Mexico. It can also serve as a basis for more detailed and specific hydrogeological investigations and compilations by developing, for example, the integral study of the hydrogeology of each province, including as a fundamental goal the content and evolution of aquifers.

3.4 Hydrogeology

Hydrogeology, as a part of geology, studies rocks and structures formed during remote periods of time. This spans the way sediments are deposited to create various rock beds beneath the surface and the way that rocks obtain granular and fissured porosity over the time to form complex hydraulic continuity.

Fig. 3.2 Geological provinces of Mexico

1, Plataforma de Yucatán (C, sm, p); 2, Cuenca Deltaica de Tabasco (C, sc, g); 3, Cinturón Chiapaneco de Pliegues y Fallas (C, sm, or); 4, Batolito de Chiapas (P, p, ra); 5, Macizo Ígneo de Sononusco (C, p, ra); 6, Cuenca de Tehuantepec (C, sm, g); 7, Cuenca Deltaica de Veracruz (C, sc, g); 8, Macizo Volcánico de los Tuxtlas (C, v, ac); 9, Cuicateca (M, vs, as); 10, Zapoteca (pE, c, co); 11, Mixteca (P, c, co); 12, Chatina (M, p, ra); 13, Juchiteca (P, vs, as); 14, Plataforma de Morelos (M, sm, p); 15, Faja Volcánica Transmexicana (C, v, ac); 16, Complejo Orogénico de Guerrero-Colima (M, vs, as); 17, Batolito de Jalisco (M, p, ra); 18, Macizo Ígneo de Palma Sola (C, v, ac); 19, Miogeoclinal del Golfo de México (C, sm, g); 20, Cinturón Mexicano de Pliegues y Fallas (M, sm, or); 21, Plataforma de Coahuila (M, sm, p); 22, Zacatecana (M, c, co); 23, Plataforma de Valles-San Luis Potosí (M, sm, p); 24, Faja Ignimbrítica Mexicana (C, v, ac); 25, Cinturón Orogénico Sinaloense (M, vs, as); 26, Chihuahuense (C-M, c, co); 27, Cuenca de Nayarit (C, sm, g); 28, Cuenca Deltaica de Sonora- Sinaloa (C, sc, g); 29, Sonorense (pE, c, co); 30, Delta del Colorado (C, sc, g); 31, Batolito de Juárez San Pedro Mártir (M, p, ra); 32, Cuenca de Vizcaíno-Purísima (C, sm, g); 33, Cinturón Orogénico de Cedros Margarita (M, vs, cs); 34, Faja Volcánica de La Giganta (C, v, ac); 35, Complejo Plutónico de La Paz (M, p, ra). Explicación: Edad.- pE-precámbrico; P-paleozoico; M-mesozoico; C-cenozoico; Origen.- m-metamórfico; p-plutónico; v-volcánico; vs-volcano sedimentario; sm-sedimentario marino; sc-sedimentario continental; c-complejo; Ambiente geotectónico.- cs-complejo de subducción; ra-raíz de arco; as- arco submarino; ac-arco continental; g-geoclinal; or-orógeno; p-plataforma; co-compuesto

Hydrogeology studies interactions within geological systems of water, as a natural resource to supply people in Mexico, whether it is piped to homes or drawn from wells. Hydrogeology reports where the water is and how it moves within the aquifer. It is important to protect it as a natural resource. Hydrogeologists need geological maps, geophysical survey, and analyses of water samples from underground and the surface to build detailed pictures of how water flows through the aquifer (Klimentov and Bogdanov 1977).

Southeastern Mexico has abundant rainfall and plenty of surface-water resources. However, most of Mexico uses water stored underground to supply multiple needs. In this respect, hydrogeology studies aid the solution of water-supply problems, by locating suitable reserves and assessing the amount of water that can be extracted, without damaging the underground aquifers or surrounding ecosystems. Hydrogeology also studies how groundwater flows through rock to identify where and when water pollution risks exist and how to protect the aquifers.

Natural freshwater in Mexico consists of rivers, lakes, and some reservoirs. However, over 97% of the planet's freshwater is located under the surface in the form of groundwater, and México is not an exception as it has large volumes of groundwater.

In most towns in northern Mexico, groundwater is the main source of water supply for everyday use. Boreholes drilled into the underground water-saturated layer yield water for irrigation and urban municipality uses. At least 50% and up to 85% of water requirements come from urban municipal water and sanitation boards (juntas municipales de agua) that supply water from groundwater reserves. The same picture occurs in rural areas because of the low-cost rural water supply. Groundwater is cheap to develop because of its naturally good quality and widespread occurrence. It is available even in times of extreme drought due to the large amounts stored under the surface. Aquifer also protects water from catastrophic events like natural disasters such as hurricanes or spill events: then groundwater reserves can easily be developed to satisfy emergency needs of the population.

Some areas may not have undergrounds saturated with water. People there have limited amounts of intermittent water supply from the runoff water. However, most of the time, deeper under the surface the material is saturated with water, which has accumulated over hundreds or even thousands of years. These areas, known as aquifers, supply permanent springs and can tapped with boreholes to satisfy water needs. When groundwater is drawn from deep aquifer supplies, it is tapping into water down far below the surface. To reach that deep, water has infiltrated through rock layers down to the aquifer. The water in these deep areas may be protected from pollution caused by recent surface spills. The rock layers work as a filter that provides high-quality water where none was previously available (Pinneker 1977).

Currently, there are more environmental threats than before. Many areas within Mexico, especially the more developed economically, have, in addition to the federal law, local regulations to help protect the environment. Economic development brings industry and larger cities. Both demand more services and utilities. These places become potential sources of environmental pollution. Water brought to homes in sealed pipe systems keep water away from sources of pollution. However, the danger of pollutants seeping through the surface into wells must potentially exist, especially within urban areas. Therefore, groundwater is still vulnerable to pollution.

The same features of groundwater that make it useful also make it a fragile resource. Leaks from rusting storage tanks in a closed gasoline station might eventually enter public well-water supplies, even if extraction takes place many kilometers away from the original source of pollution. Hydrogeological information on groundwater movement gives the appropriate tools to plan development and extraction in a sustainable and safe way. However, not only urban areas contribute to potential pollution but also pesticides and fertilizers used in agriculture within Mexican irrigation districts can enter groundwater systems and become a potential expanding pollution source, so it can have an environmental impact.

Groundwater protection includes evaluation of the potential amount of water that can be extracted without risking permanent deterioration of its quality and quantity. This means extracting reserves more quickly than rainfall replenishes them or by acting to induce artificial percolation systems should not be permitted. It also includes the surrounding environment and the interaction of surface and underground waters. Over-extraction can damage directly the streams and wetlands systems. The interaction of possible sources of pollution with groundwater and the complete hydrologic system must be considered.

One way to protect groundwater is through public education, especially of those who use it. Groundwater is located within the aquifer, which is open along the entire basin. This means that every government in Mexico, municipal council, industry manager, rancher, or farmer who extracts groundwater in one or other way or makes decisions about land use and waste management has a role in water protection.

The hydrogeology of Mexico includes an evaluation of climatic conditions, the rainfall pattern, water chemistry, and physical rock features, such as permeability, porosity, fracturing, chemical composition, geology, and geotectonics. The geology and tectonic features of Mexico, as described by Moran Centeno (1984), include the description of the geological provinces, as well as the physiographic features. Therefore, the study of geological materials and the surrounding physiography determines the presence, distribution and water flow, groundwater relationship with the geological environment, conditions governing the movement of water, physicochemical features of groundwater and its development, exploration and exploitation, and a large list of related topics.

Groundwater contained in porous or fissured permeable rocks makes possible knowing water movement if there is a pressure excess. These rocks are called water-bearing, or aquifers, in contrast to impermeable rocks (aquicludes). Aquiclude is almost impermeable, or it allows water to move very slowly and divides or underlies rocks containing groundwater.

3.5 Hydrogeological Classification of Aquifers

According to the hydrogeological classification of the aquifers, which depends on permeability, hydrogeological collectors (aquifers) form, as well as impermeable rocks that act as hydrogeological isolators. The laws governing the location of groundwater location are defined by properties such as porosity and permeability.

Aquifers can be categorized into three categories according to the pore character: granular aquifers with pore-like voids; fissured aquifers with slit-like discontinuities such as fissured slates and shales, magmatic and metamorphic rocks; and aquifer containing mixed pores and discontinuities, for example, fissured sandstones and porous limestone, the volume of voids being defined by pores and paths of groundwater movement in the discontinuities.

The porosity degree defines the aquifer capacity. It is not just the porosity, but also the size of voids and the way they interconnect that determine the groundwater movement. Clay porosity reaches 50–60% and is practically impermeable to groundwater movement. However, sandstones and fissured rocks with a porosity of only 10–15% are good for aquifers. In fact, clay has subcapillary pores (diameters of less than 0.0002 mm), in which water cannot move, and most voids are isolated from one another.

Aquifers have capillary (0.0002–0.1 mm in diameter) and super-capillary (more than 0.1 mm) channels of interconnected pores. Groundwater moves mainly throughout the super-capillary pores. However, the hydrostatic pressure is capable of moving water through capillary pores. Therefore, the groundwater movement takes place under the influence of surface tension. In voids, the groundwater moves under gravity influence and by the difference in piezometric head.

Pores in rock may interconnect (open) or isolate (close). The total volume of open and closed pores is called the total porosity. It defines as the ratio of the total volume of pores to the total volume of rock. For hydrogeological purposes, the dynamic or effective porosity is the volume of open pores through which the movement of liquid takes place under the influence of pressure gradients. Effective porosity should not be confused with open porosity which characterizes all open pores. The liquid movements do not occur through the total volume of open pores. Part of the pore spaces occupied by capillaries weakly or strongly bound varieties of water. Open porosity is always greater than the effective porosity.

Permeability is the ability of rock to let liquids and gases move when a pressure gradient exists. Hydrogeology relates to hydraulic conductivity because this governs the water movement through aquifers. Quantitatively, the flow defines the hydraulic conductivity, which is a function of the porosity and viscosity of the concerned fluid.

The Darcy (d) is a unit of intrinsic permeability. This refers to a specimen of rock 1 cm in length and a cross-sectional area of 1 cm^2, through which flows 1 cm^3/s of liquid of viscosity 0.001 Pa-s, in which there is a pressure difference of one atmosphere. The hydraulic conductivity, which has the dimension of m/day or m/sec, is the flow velocity per unit hydraulic gradient. The following relationship exists between the coefficient of intrinsic permeability (Kp) and the hydraulic conductivity (K): $K = (Kp)(\gamma/\mu)$ in which γ is the specific weight in g/cm^3 and μ is the coefficient of dynamic viscosity in Pascal-seg.

Fresh groundwater at a temperature of 20 °C has an intrinsic permeability in darcys, which could correspond to the hydraulic conductivity of 0.85–0.9 m/day. Rock can be extremely permeable (with Kp from ten to several hundred thousand millidarcys (md)), semipermeable (10–0.1 md), and practically impermeable (less than 0.1 md).

Sandstones, well-sorted sands, and carbonate rocks with solution cavities and fissured crystalline rocks are classified as permeable rocks with pore spaces from 10% to 35%. Porosity distributes throughout the rock and is enabled by super-capillary or large capillary pores. Other pore dimensions have no significance.

Unsorted semipermeable rocks like sandy-clay sediments, sandstones with clay cement, and chalk-like limestones have pores of various sizes, partly of fine capillary and subcapillary dimensions. Rocks with subcapillary or closed pores, such as clays, argillites, sane shales/slates, and compact crystalline rocks, are practically impermeable. In nature, however, impermeable rocks do not exist because, with an excess of pressure, even thick layers of clay may become permeable to water. Moreover, dissolved substances can be squeezed together with water through such layers. The migration of dissolved substances through layers of clay is possible due to molecular diffusion under normal hydraulic gradients (Smirnov 1971). Taking geological time scales into account and certain hydrogeological conditions, clay formations are permeable. However, with the temperatures and pressures that rule in depths of 2–5 km, rocks with permeability less than 0.1–0.001 md may be regarded as aquicludes. Clay layers greater than 10-m thick must also be regarded as impermeable.

As the pressure increases at deeper levels, the porosity and the permeability decrease. There are exceptions because sandstones or limestones found at depths of 3–6 km have porosities in the range of 1–20% – like oil deposits. However, it is an order of magnitude less at those depths than at the surface.

Sedimentary rocks and sedimentary–volcanoclastic rocks differ sharply from crystalline rocks like magmatic and metamorphic rocks. The hydrogeological stratification of sedimentary rocks is the result of the alternation of aquifers consisting of layered, porous layered, fissured layered, or more complex types over a wide area. The available definition of the term aquifer, in the hydraulic sense, is an underground region with consistent and unified thicknesses or strata of permeable rock with pores, fissures, or voids, which fill with groundwater. An aquifer is composed of homogeneous or nearly homogeneous rock from the lithofacies point of view. It may consist of one or several layers of water-saturated rocks. In the first instance, it is a single rock bed, while, in the second, it is complex, double, or multilayered.

The aquifer has a hydraulic surface, which is free, and it is named the piezometric level. Aquifers are isolated from higher- or lower-lying aquifers by aquicludes. This does not exclude the possibility of hydraulic connections between separate aquifers, such as when wedging out of aquifer rocks occurs, or in the presence of hydrogeological structures like faults and tectonically weakened zones.

There are fundamental aquifer varieties according to bedding natures and hydrogeodynamic features, for example, groundwater horizons from ground level, the first layer of rock fed with infiltration water, lying above an aquiclude, or rocks above of a different degree of permeability. Groundwater horizons within stratified deposits are enclosed between aquicludes confined aquifers. They can be under no pressure (free aquifer) or under pressure (artesian water). Groundwater occurring in the various parts of the section is assigned to different aquifer horizons. However, aquifer alluvium and the groundwater zone of intensive fissured bedrock should recommend combining them into a single groundwater zone if they form a single hydrologic unit.

An aquifer complex is a larger unit of rock in which it is impossible to designate independent aquifer horizons. This is because of rapid changes in the composition and properties of the rocks with depth. It is a consequence of complex structural features of the geology, and it is difficult to study the formation. The aquifer complexes are differentiated according to lithofacies and hydrogeological features.

The water-bearing rock units are considered as a very important part of the stratigraphy through a cover of Mexican basins, especially within the physiographic province of basins and ranges. There is several water-bearing rock units constituted by: A. sandy-clayey friable rocks with pore-granular water; B. sedimentary-volcanoclastic cover or layers with fissure-granular and granular-fissure water;

C. indurated sandy-clayey rocks with fissure-granular water; D. carbonate formations with fissure-granular or karst-granular water; and E. evaporate deposits containing stratified aquifer horizons between evaporate beds.

Sometimes several similar aquifer layers alternate in a stratigraphic section and are divided regionally by confining beds. Such a situation is characteristic of a section through the basin-and-range province.

The aquifer capacity of igneous and crystalline rocks is characterized by a great variety of factors, such as the fissuring type and intensity degree. Magmatic and metamorphic aquifers can in the first place be assigned to the upper weathered horizons of these rocks. The zone of intensive fissuring or macro porosity is formed by the exogenetic processes, such as weathering and leaching, and it is characterized by water in fissures. It corresponds to the zone of regional rock fissuring. Other types are the system of differentiated pores, the zone of tectonic faults, of intrusive igneous bodies and the intense fissuring which accompanies them. Fissuring can be traced down from the surface into the depths. They contain water of the vein type.

The irregular fissuring of crystalline rocks changes the capacities of aquifers that usually distribute irregularly throughout the section. Even so, there are no hydraulic connections between them. The water infiltration into the crystalline rocks considered in the fissuring zone is the fracture aquifer or permeable zone. The permeable zone consists of magmatic or metamorphic rocks quite fissured to various degrees. Its shape varies, and its attitude ranges from horizontal to vertical.

Groundwater in the zone of intensive exogenetic fissuring assumes wide distribution. Individual accumulations to some extent connect with each other, and it is characterized by the absence of pressure. The permeable thickness varies enormously, from several meters up to several tens of meters, sometimes more than 100 m. Groundwater in zones of tectonic faults and intrusions is characterized by its localized and usually linear distribution, forming vein-type aquifers. The width of such veins or linear structures may reach 0.5–1 km. Fault zones in shallow deposits (up to 300 m) contain freshwater. However, deeper deposits contain mineral, thermal, carbonated, and salty waters, etc. Deep water is under high pressure and temperature when water wells from the boreholes penetrate the aquifer.

Aquifer fault zones occur in pyroclastic and crystalline rocks; however, they also are found in indurated sedimentary rocks. They widely distributed throughout the sedimentary-volcanoclastic cover of the basin and range province. Therefore, the laws governing the localization of groundwater in sedimentary-volcanoclastic rocks depend mainly on the lithofacies of the rocks. Meanwhile in crystalline rocks, the lithology is relatively insignificant; the fissuring is the major factor. There are other factors such as the location of the aquifer rocks.

3.6   Hydrogeological Rock Distribution in Mexico

The hydrogeology of Mexico begins by describing the rock distribution throughout the country, which defines the geological provinces. Rocks form structures, which result in topographic features. Each type of rock has hydrogeological characteristics

Fig. 3.3 Geology of Mexico (Ortega-Gutierrez et al. 1992)

that define their hydraulic conductivity as permeability and porosity (p & p). Igneous and metamorphic rocks use to have very low values of permeability and porosity parameters. However, near the surface, these rocks fracture easily and acquire very high values of fissuring that increases their super capillarity. The clastic sedimentary-volcanoclastic rocks have high values of super capillarity. Sometimes, near the surface, the high evaporation conditions precipitate salts, which seal voids between grains decreasing pores size and permeability drastically.

The distribution of rocks shown in the geologic map of Mexico (Fig. 3.3) indicates the prevalence of granular or fractured porosity of the rocks. Sedimentary-volcanoclastic rocks with yellow, orange, green, and brown colors have a predominately granular porosity, while pyroclastic, crystalline, and meta-morphic rocks of pink, red, and purple colors predominately have fractured porosity. Most fractured rocks have high values of permeability because of the intense fissuring. However, at depth, this is not necessarily true since the fissuring decreases drastically, as well as porosity and permeability. On the other hand, the clastic sedimentary-volcanoclastic rocks have granular porosities, which show a variated value of permeability, according to the grain-size diameters.

Igneous and metamorphic rocks have very high surficial fissuring; therefore, they exhibit super capillarity. Meanwhile, as depth decreases, the fissuring decreases and consequently porosity is depleted strongly, and, in consequence, permeability is practically nil. In addition, clastic sedimentary-volcanoclastic rocks have plenty of granular porosity. The basin-and-range physiographic province constitutes a thick alluvial deposit, which reaches thickness of more than 2000 m.

The classification of groundwater according to the way in which it is deposited (deposition of groundwater is not strictly speaking true because it is not only deposited but also moves) is widely accepted. This connects with the fact that the physical–geographical, structural, geological, and thermodynamic conditions determine, in the final analysis, the features of the formation, movement, and distribution of groundwater, and the character of the bedding most fully combines these natural conditions and features of the life of water underground. Groundwater is almost everywhere in natural conditions, varying in quantity and quality both temporally and spatially.

Starting from about 1860, the French term "artesian water" became quickly established in the literature regarding the widely practice of using spring groundwater. In contrast to the term artesian water, which described water under pressure, the term subsurface corresponding to the concept of nonpressured water arose. Subsurface water is that water that formed because of the absorbed atmospheric precipitation in the aquifer layer nearest the surface, which is located in the subsoil or in the deeper bedrock.

This also refers to the impermeable layer nearest the surface: water which remains free after the last rain as content in the rocks. Groundwater not only accumulates in unconsolidated deposits but also as water in dense and fissured rocks, considered as another variety. Subsurface water, soil water, or nonpressured water can be grouped as phreatic water. Nevertheless, some other terms survive besides subsurface water, such as percolating water, strata water, and fissure water. Percolating water is the infiltration water that is in the process of percolating and that has not yet attained the state of groundwater below the water table. Groundwater in the pores and voids of sedimentary rocks from the time of its formation is called strata water. Water that concentrates in the spaces of fissured rocks of secondary origin is classified as fissure water.

Groundwater is also categorized according to its origin into unsaturated or vadose, juvenile, fossil and mixed, and according to the hydrogeological and stratigraphic features, such as the upper (suspended water) and lower (phreatic water), and pressure water (subartesian and artesian). Within each group of groundwater, a distinction can be made between strata water, associated with sedimentary rocks, and underground water currents. In addition, groundwater describes water of the zone of aeration, subsurface water, and artesian water, groundwater in fissured and karstic rocks, deep-lying water, groundwater of recent volcanic activity, and groundwater under the floor of the oceans or seas.

The interstices of soils and rocks situated above the water table contain air, water vapor, hygroscopic water, and capillary moisture. Periodically, at the end of spring when the early rains fall or the normal rain period in summer, the soil moisture content increases to a point at which it is greater than the maximum molecular moisture content, and then subsurface water appears in the zone of aeration. This water percolates under the influence of gravity and may accumulate in the soil and locally above impermeable or semipermeable horizons or reach the water table. Therefore, groundwater may be divided into soil water, infiltrating water, and perched water.

3.7 Soil Water

In the soil, there are various forms of soil moisture. Soil moisture means soil water, and only free (gravitational) water enters into this category. Not all soil moisture is available to plants. Water vapor and hygroscopic (tightly bound) forms of moisture cannot utilized by plants. Pellicle water can be taken up partially by plants. By contrast, capillary water plays an extremely important role. In regions in which the water table is situated far below, capillary water is the main source of moisture for vegetation.

The presence of free (gravitational) water in the soil is caused by the percolation of significant quantities of atmospheric precipitation and the condensation of water vapor. The significant quantity of soil water is indicative of excess moisture, which leads to marshy conditions in the ground. A deficiency of soil water also decreases the fertility of the soil. Drainage also includes the regulation of the free-water content in soils. Soil water comes from both above (rain) and below (capillaries). By the transfer of free capillary rise water, especially in the case when the subsurface water lies near the surface of the ground, a considerable amount of organic material passes into this water from the soil, together with many microorganisms.

The soil–water regime is defined by the ratio of water inflow to outflow. *A circulating, a compensated,* or an *evaporating* regime exists according to which of these two factors predominate. The circulating regime is observed when the quantity of water arriving in the soil exceeds the sum of the amount taken up by the vegetation cover and that is lost by subsurface evaporation. A compensated regime exists when the inflow and outflow are equal. Finally, in the evaporation regime, the evapotranspiration rate exceeds that at which the water goes to the soil from the atmosphere.

If the evaporation regime persists, as occurs in northern Mexico, the mineral content of soil water and groundwater increases, forming a surficial caliche crust. Composition water passes from calcium bicarbonate to sodium bicarbonate, then to sodium sulfate and end up to sodium chloride.

3.8 Infiltration Water

Below the soil cover lies the unsaturated zone or aeration zone of considerable thickness, usually ranging from 5 to 50 m in the basin-and-range province of Mexico, it reaches larger thickness. It lies between the suspended and the capillary rise or capillary fringe water; it is the *intermediate layer* or dead horizon. Here, during rainy season, free or gravitational water reaches this intermediate layer, which moves downward or sideways.

Surface water that enters the zone of aeration uses up a considerable amount in moistening the soil and in the formation of physically bound and capillary water. After that, infiltration water occurs which moves downward under gravity. When the pores are completely saturated, downward movement proceeds because it is not absorbed in moistening the soil. In dry soils, infiltration reduces with increasing depth. Infiltration is how free or gravitational water moves within unsaturated soils.

Free percolation instead the normal infiltration occurs in northern Mexico. In free percolation, water movement is accompanied by the partial filling of the pores in the form of isolated streams. Meanwhile, normal infiltration takes place over a significant area with a continuous flow, in which all the interstitial spaces of the soil fill with water.

Groundwater feeds through the zone of aeration even in regions of desert dunes, which happens at about the same rate as the regions in which the moisture content fluctuates, or even when excess moisture exists. This is because in the absence of vegetation and the presence of a deep zone of aeration, no water is removed by plants, and evaporation within the rocks reduces sharply with increasing depth. A zone of aeration of greater than 5 m thick and the amount of underground evaporation do not exceed 0.5 mm/year.

Infiltration is maximum if the zone of aeration consists of soils with large interstices or super capillaries and fissures. On watersheds with poor retention levels, it reaches 20–80 mm/year, but decreases to <10 mm/year on slopes and in valleys where a layer of silty loam or mud and the soil and vegetation cover obstructs infiltration.

Perched water occurs when infiltration water meets a relatively impermeable horizon between water-saturated soils. The impermeable layer may be a clay lens of loamy soil in a layer of sand, parts of the weathered rock over bedrock or even alluvial soil horizons, when the surface water is soil water. The perched aquifer forms on such an impermeable layer. It is characterized in the first place by its seasonal nature because surface water arises only during rains and, secondly, by local geology. Consequently, it is not a continuous aquifer horizon. A perched aquifer sits above the water table. It usually has no hydraulic continuity with ground or river water, although in some areas, it can interact with them.

Northern Mexico has wide and thick alluvium basins between narrow and elongated sierras, where perched water bodies are common. They are often 0.4–1.0 m thick, rarely reaching 2–5 m and usually found in sandy-loamy soils. Homogeneous, very permeable beds, and poor water-retaining properties, for example, coarse-grained sand or fissured hard rock, are unfavorable to the formation of perched water.

The formation of perched water is strongly influenced by the relief: Slopes, especially steep slopes, where the surface runoff exceeds the infiltration, do not contain perched water. In an extreme case, it forms a thin aquifer layer for a short time. The best conditions for the formation of perched water arise on flat water divides and plains with local depressions or lakes into which rainwater runs. Perched water is often found in sections of river terraces.

In northern Mexico, in arid regions where perched water forms close to the surface, some of the water may evaporate. In this case, it is not only exhausted rapidly, but it is also easily polluted. Hence, near towns and industrial areas, where leakage of domestic and industrial waste occurs, the perched water becomes polluted. In natural conditions, its quality varies greatly. In regions of high rainfall like the southeastern Mexico, it is polluted to some extent with a relatively high content of organic material. In contrast, in northern regions of deficient rainfall, the subsurface evaporation increases the total dissolved solids to form saline water and even brines.

3.9 Subsurface Water in Mexico

In Mexico, subsurface water means free or gravitational water in the permanent aquifer layer situated nearest the surface. It changes over time, and it can be located in unconsolidated deposits or in the upper-fissured horizon of the bedrock. The surface of the subsurface water is mainly free or nonpressured water. When subsurface water is reached in boreholes or wells, its level is established at the same level at which it was found. Only in places where lenses of impermeable rock lie above the level of the subsurface water does it acquire low-scale local abnormal pressure. In Mexico, the surface of the subsurface water is also called the water mirror and occasionally a water table. A subsurface aquifer horizon or a part of it, in which the water movement takes place under the action of gravity in the direction of water table slope, is called a subsurface stream.

Shallow deposits of subsurface water, which has a connection with meteoric water, are characterized by the following:

1. It is nonpressured water, usually having a free surface. The pressure on it is equal to atmospheric pressure. It moves under gravity in the direction of the water table slope.
2. It is recharged mainly by infiltration of atmospheric precipitation and moisture condensation in the aeration zone. The recharge region coincides with the distribution region.
3. Discharge takes place at the base of the slope or into surface reservoirs, but also into water currents of such as streams, to which the subsurface water is hydraulically connected.
4. Large time variations because of the direct influence of surface factors, the level, water inventory, temperature, and other parameters are associated with subsurface water.
5. Shallow deposits and intensive underground runoff give rise mainly to fresh, nonsaline subsurface water. However, when there is a deficiency of moisture in an area, water evaporates and mineralizes.

The subsurface water recharges through the aeration zone by the infiltration of atmospheric precipitation (rain and floodwater) over its total area of distribution. In addition to the condensation of water in the aeration zone, in any region, there may be other recharge sources (river water, the entry of water from irrigation canals, inflows of artesian water from deep-lying aquifer horizons, etc.).

The level rise in the subsurface water and the increase of the summer flow that feeds from the subsurface water depend on the quantity of atmospheric precipitation and local conditions. Infiltration takes place over various periods after intensive rainfall. An example of a sharp increase in the summer flow is observed a month or more after the maximum rainfall occurs. This is characteristic of fissured and karstic rocks in which groundwater moves at greater velocities than in sandy aquifer strata. The greatest summer flow occurs several days or even hours after rains. In many cases, especially in deserts and semideserts, the subsurface water is feed by condensation. It is replenished because of the condensation of water vapor from the air that precipitates in those parts of the rock that have cooled. Nevertheless, the atmospheric precipitation in an extremely dry climate cannot provide any noticeable source of groundwater recharge. Such an extremely dry climate does not occur in Mexico.

As pointed out before, subsurface water is replenished by currents of artesian water from underlying strata. This form of recharge is possible in places where the impermeable cover of artesian aquifer horizons has cut through structural or stratigraphic windows. It occurs when the pressure head exceeds the level of the piezometric surface of the subsurface water. Discharge of groundwater takes place through dispersed and concentrated outlets, strata seepage, or marshy areas. Springs are usually in places where the aquifer horizons are exposed by erosion or by the wedging out of the aquifer beds.

If the discharge consists of aquifer strata made of fine- or medium-grained sands, then minor water outlets concentrate in small local hollows. Sometimes, strata seepage takes place near the outlets, usually on a slope, which is wet over the whole area of the aquifer strata outlet. Such places often continue along the slope as marshy strips where marsh vegetation grows, and water accumulates in slumps. In arid regions of Mexico, because of the evaporation of water, a layer of salt forms on the surface of such springs, sometimes further forming a caliche bulge.

Springs from alluvial deposits are often in the steep sides of terraces or river bends where a river undercuts its bank, where the cross section of the subsurface flow is reduced. Water outlets can also connect with the changes in the composition of the alluvial deposits. For example, within internal drainage basins of northern Mexico, where the composition of the alluvial slopes changes from sandy to sandy clay, the velocity of the subsurface water reduces, and it is forced to rise to the surface and produces springs within ground depressions.

The flow of all these springs is usually small and strongly time-dependent. The greatest flow is in fault-fissured and especially karstic rocks; they often amount to several hundreds of liters and even several cubic meters per second.

The way in which subsurface water forms depends on several factors, such as the recharge conditions, the permeability of the rocks, the configuration of the reservoir banks, the streams with which the subsurface water has hydraulic connections, and the position of the confining bed. The water table of the subsurface water is shown on a map of water-table contours (equipotential curves), which are lines joining points of equal water-table head. Water-table maps are constructed from data on the water-table heads taken simultaneously (because they vary with time) in boreholes, excavations, and wells. Apart from this, they use data on the position of springs, swampy hollows, and bogs that arise because of the groundwater. These emerge to the surface, up to the levels of the water in the rivers, lakes, and other bodies of water, and streams with which the subsurface water is hydraulically connected.

The depth of the water table often depends on the local relief. The top surfaces of the subsurface aquifer horizons are, for the most part, uneven, undulating, and often follow the curved shape of the relief. However, in certain places of Mexico, for various local reasons, such a relationship between the ground surface and the water table may be absent.

In river valleys, gorges, ravines, and other low points of the relief, groundwater is comparatively near the surface, but at water divides, the depth to groundwater may reach several tens of meters. Groundwater moves from higher to lower areas when gravity prevails. In the Sierra Madre, high plateaus, canyons, and high ground with broken relief regions, groundwater, as a rule, drains toward the river.

The coastal plains of Mexico are an example of where most of rivers (e.g., Panuco, Coatzacoalcos, Papaloapan, Grijalva, Balsas, Lerma-Santiago, Culiacan, Sinaloa, El Fuerte, Yaqui) drain the groundwater. On the other hand, in the basin-and-range province in northern Mexico (rivers like Carmen, Casas Grandes, Santa María, and others), especially where the climate is arid and the water table is deep, the rivers often feed the groundwater. Here, the water table sinks away from the line of the riverbank.

There can also be a more complex interrelationship between subsurface water and river water. For example, in north-central regions, subsurface water may flow from one slope of the river valley into the riverbed, and the other slope may absorb river water (Conchos, Nazas, Aguanaval, etc.). During periods of high water and floods, when the river water level is high, the water table near riverbanks rises. The curve of the head of water extends outward into the land between the rivers for several hundred meters and even for kilometers. When the water level falls, quite a sharp fall of the water table, close to the bank, results.

In large dams like the Aguamilpa, La Boquilla, Lazaro Cardenas, because of the considerable head of river water, the water table rises over a greater distance on the banks of the reservoir. The new position of the water table in the pressure zone established for large rivers remains for a period of several months.

The recharge of the subsurface water at any given place is determined by the infiltration of meteoric water, the condensation of water vapor, and subcurrents or side currents from neighboring areas. It discharges because of the outflows, evaporation from the subsurface water and vegetation transpiration. The quantitative determination of the elements of inflow and outflow in the subsurface water balance enables distinguishing the basic factors forming the hydrogeological regimes of the subsurface water. They define the seasonal and long-term variations of its level, volume, and direction of the formation process of the water's chemical composition.

Subsurface water is categorized by the way in which it forms, its distribution, recharge, and its regime into interfluvial and watershed areas, river valleys, alluvial fans, and talus of foothills and seashores. Characteristic of the interfluvial and water-divide areas is an infiltration-runoff or watershed regime, even if it forms in rocks that differ in composition. The water increment is defined by the infiltration of atmospheric precipitation and surface water and the decrease of outflow and local drainage.

The subsurface-water regime in valleys associated with streams is different. The recharge comes from atmospheric precipitation, for both river water and groundwater streams from watershed areas. The decrease part predominates the surface rivers and outflows. If the aquifer water table is less than 5 m in depth, evaporation is the dominant factor.

Large amounts of subsurface water accumulate in alluvial fans, cones, and lines of foothill, where the thickness of permeable layers reaches tens of meters. The water balance has significant increases along with infiltration, rain, condensation, and absorption of the surface runoff. Meanwhile in the decreasing water balance dominates the evaporation and underground runoff.

The subsurface water of coastal areas comes from atmospheric precipitation percolating from the surface, and the condensing moisture accumulates in the porous strata above the saline water, which is usually expelled into the sea. The thickness of the freshwater varies considerably. Mixing with saline water takes place extremely slowly but is accelerated by an irregular withdrawal regime, as happens on Sonora coast.

One should distinguish between dry plains, semideserts, and deserts, all of which are characterized by arid climates. One should also define the features of subsurface water as the dominant evaporation in the decreasing part of the water balance: the underground outflow in deep-lying deposits (more than 10 m), which combines with insufficient moisture. Because the depth of the deposit determines the amount of evaporation, the relief and the drainage of the area become the basic factors controlling the regime.

3.10   Zoning of Subsurface Water in Mexico

A broad zonation is characteristic of subsurface water. The latitudinal zones correspond to certain landscapes and succeed one another from northwest to southeast. The zonation of subsurface water is influenced mainly by climatic factors and the degree of land wetting. Important factors are the depth of the erosional down cutting, the collecting properties, and composition of the aquifer rocks.

Features such as relief and lithology are subordinate to the climatic factor, categorized as azonal or intrazonal water. Thus, it is the lithologic factor that is predominant in the formation of azonal water. Karst water, fissure water, and alluvial water belong to this category and are examples of azonal water. The saline water of saltmarshes and salt lakes and springs or lenses of freshwater, which are deposited above saline-aquifer horizons, are examples of intrazonal water.

Consequently, from the northernmost latitudes to the arid regions, not only subsurface water mineralizes but also its ion salt composition changes. The processes of formation of the subsurface water find expression in the variation of mineralization degree and composition. It distinguishes two major zones, which correspond to the genetic types of subsurface water: leaching and continental salination.

Hydrostatic pressure causes the water level to rise above the upper confining bed when horizons meet in boreholes and excavations, known as artesian water. Sometimes, it is called interstitial pressure water. Subsurface water in sedimentary strata belongs to this category, although, as been said before, it can also be found as fissure water in crystalline and igneous rocks.

In favorable structural-geological and hydrogeological conditions, boreholes produce artesian water. However, not everywhere does an artesian aquifer yield artesian water. This does not occur when the pressure head is below the surface of the ground. Furthermore, earlier artesian water is associated with ideal confined basin shape. Later, it was found in a much more varied and complex arrangement, such as monoclinal slopes and saturated fault zones located in aquifer horizons.

The artesian water found in Mesozoic rocks, although it is covered with, or underlain by, confining beds, is also found in Tertiary deposits. Characteristic features of artesian water are its interstitial pressure water and horizons isolated from above and below by confining beds. The artesian water and the region in which it is distributed do not coincide and are often separated from one another by great distances. When an artesian aquifer is drilled to make a well, the appearance of water in the borehole always is deeper than the established level. Often, the level becomes established at a higher level than the ground surface, and then water emerges from the borehole. The artesian water regime is more stable than the subsurface water because the surface waters have much less influence. Artesian water is fresh in the upper part of the cross section, but its mineralization degree increases with depth and it becomes saline water and even brine.

The natural reservoirs of artesian water are artesian basins of strata water and artesian slopes or monoclinal formations of strata water. An artesian basin holds pressure-water-bearing horizons or complexes, which form in folded structures where subsurface water is under hydrostatic pressure such as in the Sierra Madre Oriental and Sierra Madre de Chiapas. Complex water-bearing horizons of the artesian type possess relatively small recharge areas compared to their runoff areas. In Mexico, where there are artesian waters within high plateaus and coastal plains, the recharge region may be situated amid various favorable forms of relief of the interfluvial areas and Sierra Madre's uplands. These are characterized by the convergence of the pressure surface of the aquifer horizons and the fall of the head with depth.

Sometimes, the recharge of artesian water derives from an undercurrent of the low-lying horizon to the higher one. This occurs when pressures of alternating aquifers increase with depth. An undercurrent occurs along tectonically weakened zones or through relatively impermeable confining beds.

The intensity of the underground runoff is very varied. It depends on the hypsometric position of the recharge area and the discharge area and on the location within the artesian basin. Artesian basins are often found where the recharge area is at the same altitude, and there are no visible discharge areas. The subsurface water discharge takes place along tectonically weakened zones and over the whole area through the rocks of the upper relatively impermeable confining bed, because of considerable range of pressures.

Artesian water usually gravitates toward the marginal parts of foothill bends and intermontane valleys or toward the coast. They are also found within the slopes of synclines and depressions on platforms, especially when the layers of aquifer rocks are thick. In this sense, the coastal plain of the Gulf of Mexico in the Saltillo–Monterrey area is an artesian slope.

Although the regime of artesian water is more stable than that of other subsurface water, it is subject to the influence of natural factors, the first of which concerns the shallow aquifers. Surface factors have an influence on the deep-lying horizons. The more isolated are the artesian aquifers, the weaker is their connection with the surface. Artesian-type reservoirs contain huge quantities not only of freshwater but also mineralized subsurface water, which can be used for commercial purposes.

Subsurface water in fissured and karst rocks can be categorized by its distribution, movement, and accumulation features. It is difficult to establish the paths along such water move, to designate the aquifer zones, to define their thickness and the depth of the zones of intensive fissuring, and to choose parameters for hydrogeological calculations. This is related to the elementary cells in which the water moves. There are differences in void size (fissures, caverns, veins) of various origins and character, which form an anisotropic permeable medium.

3.11   Hydrogeological Evaluation of the Degree of Fissuring

How to estimate the degree of fissuring to solve hydrogeological problems has been studied by Davies and De Wiest (1967), Matthess (1970), and Castany (1963).

For igneous, metamorphic, and sedimentary rocks, the permeability and water-bearing capacity are governed by the presence of fissures, faults, and pores.

The hydrogeological significance of porosity in such rocks is extremely small. Matthess (1970) relates the permeability of dense rocks to their porosity, which includes the degree of fissuring. However, for some varieties of basalt, the permeability and the water-bearing capacity are determined by the presence of large pores, that is, macropores. The ability to transmit water correlates with the voids in karst rocks.

The hydrogeological study of fissuring requires an explanation of their origin, structural features, morphology and fissure orientation, and an evaluation of parameters to define the water-bearing capacity of rocks. These parameters are the orientation of the fissures, their aperture, and the void ratio of the rocks.

The orientation of the fissures in a block of rock depends on the nature of the tectonic forces, under the action of which the fissuring of the rocks took place. There are several major systems of subparallel fissures. These systems, crossing each other, form a water-conducting network. There are three main fault and fissure systems and three complementary fault and fissure systems in most of Mexico.

Rocks can be divided into homogeneous and heterogeneous groups depending on their reaction to mechanical forces. Anisotropic rock strength and structure are most apparent in foliated rocks such as slates and gneisses because of schistosity phenomenon. This is reflected in the lengths and widths of the predominant fissures compared with the minor ones. Igneous rocks are more isotropic, while the most anisotropy is found in high-grade metamorphic limestone. The differences in the permeability of a rock mass are defined by the orientation of the fissures and the extent of the main fissure systems.

The volume coefficient of fissuring, which is analogous to the coefficient of porosity, can be obtained from boreholes. Three basic groups of fissures can be differentiated: petrogenetic (syngenetic in sedimentary rocks), tectonic, and exogenetic, which reflect the temporal succession of fissure formation in rocks. Petrogenetic or syngenetic fissures, which originated during the rock formation, are the least permeable. Tectonic fissures define the general trituration degree of the rock, transverse fissures and faults develop beside intra-layer fissures, and they increase the rock's ability to conduct water. Finally, the exogenic factors related to weathering and stress-relief expansion lead either to a significant increase in the opening of tectonic fissures or conversely in rocks such as schists or even basalts, in which fissures in the near-surface layers become filled with silt.

The general absence of open fissures occurs with greater depth because of the larger rock pressures. Exogenic factors only operate as a general law of near-surface rock formations. However, the reduction in the degree of fissuring occurs at varying rates for different fissure systems; transverse fissures are preserved at significantly greater depths than are interlayer fissures and longitudinal fissures.

Analogous data have been obtained from yield measurements and experimental work on the infiltration in boreholes in the ore-bearing regions of the Sierra Madre Occidental and in the Sierra Madre del Sur (Davies and De Wiest 1967).

3.12   The Water-Bearing Capacity of Fissured Rocks

Fissure water occurs widely in hydrogeological massifs and is present to the least degree in the compressed parts of strata water basins. It may be nonpressured or pressured water. Horizons of nonpressured water (the water-bearing fissured zones) are usually connected to the upper parts of hydrogeological massifs, which correspond to exogenic processes where underground currents formed. They are directed from the watersheds to the slope bottoms. Because the permeability of the water-bearing rocks can vary over the area of a horizon, very permeable zones and practically impermeable zones can be found in close juxtaposition. The total thickness of the water-bearing zones is usually comparable to the thickness of the region of open fissuring and varies from 30 to 200 m.

In many regions in which the various types of igneous, metamorphic, and dense sedimentary rocks occur, it has been established that the reduction in permeability is associated with the fissure closing. The saturation of the rocks also decreases with depth. There is a direct relationship between the rock infiltration parameters and the fissuring degree with depth. With rare exceptions, fissured rocks are poorly saturated, as can be seen in the Sierra Madre Occidental. The loss of water through springs brings the fissure water out to the surface. The yield of boreholes is also small. The specific yield has the same magnitude as the outflow from the springs.

The amount of water that contains crystalline rocks does not depend on the mineralogical composition or structural features. The limits of the yield of boreholes in various rocks are very close, and their variations are defined by the degree of weathering and fissuring of the rocks.

Against the general background, there are certain varieties of extrusive rocks (andesite and basalt) in which currents of nonpressured and pressured water stand out. In local depressions, this water appears at the surface in the form of high-yielding freshwater springs. An example is the Neo-Volcanic Axis of México.

The water supply in México City drains from underground sources. Fissure water of hydrogeological andesite-basalt massifs is recharged from atmospheric precipitation, surface streams, and surface accumulations of water. In the basement zones of strata water areas, it may be replenished from aquifers, which overlie sedimentary rocks. Fissure water is usually nonpressured, and its movement obeys Darcy's linear infiltration law. Discharge from horizons of fissure water takes place into the depression river network. However, fissure water may be pressured water. Pressure arises from the presence of an impermeable or nearly impermeable upper-confining beds in the form of a weathered crust. There is no uniform closing of the various fissure systems with depth.

The crystalline rocks of the hydrogeological massifs usually interact only slightly with the low-temperature water, which infiltrates through them. The fissure water in most of cases is fresh.

3.13 Vein-Fissure and Vein Water

Vein and vein-fissure water, associated with the zones of tectonic disturbance and faulting, are an especially important category of groundwater in both hydrogeological massifs and strata basins. Tectonic disturbances at intersecting faults distribute most widely in the fold mountain regions, more rarely in crystalline shields, exposed basements, and glassy volcanic bodies. Such faults, having a tectonic and gravitational nature, are represented by channels, which are open or filled with unconsolidated debris material. They are accompanied by large zones of brecciated rocks with intensive developed parallel fissuring, which have a feather-like arrangement. Fault zones form a basic type of hydrogeological reservoir. Vein-fissure and vein water are more prevalent than elongated and relatively narrow stream forms.

In many folded structures, saturated fault zones often form huge deposits of freshwater, which are used for water supply. Such deposits are well known in the Basin and Range Physiographic Province and many other regions of Mexico. Large water inflows into mines come from vein and vein-fissure water, which reach hundreds and thousands of m^3/hour. Most of mines in Mexico suffer from flooding. At the same time, the amount of water contained in the fault zones of rock can vary significantly, even in a restricted area. The degree of mineralization and the composition of vein-fissure water vary widely. In shallow fault zones (100–200 m), it is usually fresh and differs little from fissure water.

When the faults penetrate more deeply into the rock formations, the water may be carbonated, thermal, saline, or even brine. Exotic vein-fissure waters form chemical–hydrogeological anomalies. Large deposits of mineralized water of various compositions are associated with fault zones. The tectonically active regions with mineral water of the vein and vein-fissure type may be widely distributed and defined by the specific character of the geological regions.

The leaching of carbonate rocks increases if the water contains carbonic acid. Both the internal voids and the surface-karst formations are usually associated with the degree of fissuring of the rocks, and karst formation proceeds most rapidly along the direction of the fissures. The forms of karst on the surface include pits, hollows, cavities, crevices, wells, shafts and dry valleys, or underground rivers. In mountainous regions in which karst is widespread, the water of small surface streams, which arise periodically during periods of heavy rainfall, is usually almost completely swallowed by fissures and other voids. Usually, karst cavities are distributed along the contacts of rock layers of that karst, also along lines of fracture destruction, where carbonate rocks are broken down. Some karst cavities can swallow large volumes of rain. Karst formation can reach several hundred meters.

The underground forms of karst, such as caverns, open fissures, and the various forms of channels, extend for many kilometers, forming a network of voids and cavities. They often are filled completely or partly with groundwater, and sometimes, they form real underground rivers.

The groundwater of the karst is called karst water or fissure-karst water. It may be pressured or nonpressured. The formation of karst needs, in addition to soluble rocks, water movement to dissolve it. Karst formation proceeds more aggressively where the water velocity is greater. In underground basins with poor water movement, the process of karst formation progresses slowly. The formation of karst proceeds more intensively with increasing differences in the levels separating the recharge and discharge regions. It increases rock permeability. The karst formation proceeds along fissures produced by weathering and tectonic activity. Water moves through them with larger velocity.

The ability of groundwater to dissolve decreases with depth because the carbonic acid is depleted on the rock leaching along the path of movement. Several mines in Mexico show that the degree of the rock karst process also decreases with depth, such as the Crystal Cave and others of Naica Mine in Chihuahua. The movement of water in karst, just as in fissured rocks, obeys the linear infiltration law. However, in areas of intense karst formation with huge caverns and channels, the water moves in a turbulent way.

Thus, the hydrogeological conditions of karst and fissure water are similar in many aspects. However, in karst conditions, especially in the upper zone, the groundwater movement advances more intensively. The stream yield in karst rocks is usually much higher; some large springs give rise to rivers such as in Cacahuamilpa Cave in Morelos, Mexico. The deposits in karst conditions are considered the most saturated in mineral deposits such as in Crystal Cave in the Naica mine, Chihuahua.

A karst-water regime is characterized by great variations in flow and level because of the unstable nature of the recharge. The flow of springs shows rapid variation and maximum and minimum values differing by factors of tens or even hundreds. Mineralized water content and temperature do not show large variations.

The largest karst area in Mexico is that of Yucatan Peninsula, located in southeast Mexico. It can be said that karst flow supplies the water requirements of the population of the Yucatan Peninsula. These territories with karst limestone features make impossible to have large rivers.

Therefore, the interrelationship of the surface water and the groundwater in regions with karst formation is not only close but complex too. If it is assumed that the total flow of all emergent springs in the region is the magnitude of the underground runoff, then it is possible to come to the wrong conclusions because some springs emerge more than once.

The chemical composition of karst water is varied. High-flow springs from karst limestones emerge as freshwater with a calcium bicarbonate composition. Calcium sulfate water is found in gypsum deposits. In some regions in which saline karst occurs, strongly mineralized water and sodium chloride brines are found in boreholes drilled into deep-lying strata.

Groundwater in karst regions has more pollution from surface than fissure water. In karst regions, after rainfalls, turbidity appears in the groundwater several hours later, without mentioning the increase in bacterial pollution.

3.14 Hot Springs and Hydrothermal Activity in Mexico

Mexico is a territory of active volcanoes or has been active in historical times, so it is considered a region of volcanic activity. Volcanoes are situated in the active Neo-Volcanic-Axis oriented east-west, in the central part of the country. Toward the southeast of Mexico, another volcanic region is along Veracruz-Tabasco-Chiapas, which is associated with the colliding edge between two tectonic plates and the deep Acapulco trench. The two tectonic structures, the transform fault defined by the Mexican Neo-Volcanic Axis and the colliding tectonic plates defined by the volcanic alignment Veracruz-Tabasco-Chiapas, include 98% of all active volcanoes in the region; only 2% of all volcanoes are associated with young faults of the ancient continental blocks such as the volcanoes in the Motagua transform fault. In such regions, the heat flow from the depths to the surface is often 50–100 times greater than the average heat loss from the Earth (White 1965). Here, the groundwater, being the active agent of heat and mass transfer, is involved in the water-bearing systems, which have specific hydrogeological conditions. It has become the regular practice of hydrogeologists when referring to such water, which is characterized by a wide range of high temperatures and a combination of various phase states, to use the term hot springs.

The highest concentration of geysers in Mexico is in the Neo-Volcanic Axis. The largest of them is Los Humeros, Puebla. It discharges a great stream of water to a height of 40 m and a column of steam to several tens of meters. However, the total number of geysers is small. A comprehensive summary of the main regions of geyser activity was compiled by Federal Commission of Electricity.

Hydrothermal phenomena occur in natural hot springs, not just in regions of active volcanoes but also in areas undergoing tensional stress, such as the northcentral part of the country. They have a wide temperature range. The fluctuating hot springs include all small geysers with spurts, boiling spring, and hot-water flows of 25 °C or greater (White 1967). The hot-spring flows can reach large magnitudes. In Mexico, boiling springs are known to have flows of 10–18 l/s.

Mexico has some hot springs related to the interbedded pyroclastic deposits and lava sheets of stratovolcanoes characterized by high, but uneven, permeability. Usually on the surface lava sheets, there is no surface runoff despite the abundant precipitation. Depending on whether unconsolidated deposits or lava predominates, a complex system of porous water and fissure water is formed.

3.15   Conclusion

The distribution of water in Mexico results from the distribution of rocks and geological structures aligned with the tectonic-plate limits. Mexico is divided into five geological provinces defined by rock type and age as follows: the Western Sierra Madre (Sierra Madre Occidental) comprised of a large volcanic sequence of 2000-m thick; the Eastern Sierra Madre (Sierra Madre Oriental) and the Sierra Madre of Chiapas, formed by a thick sequence of sedimentary rocks, which intercalates limestone and shale beds with a total thickness of more than 8000 m; and the Sierra Madre of Sur, which parallels the Neo-Volcanic Axis. Water distribution and its availability, in general, are restricted to the porous granular sedimentary rocks. However, there are large quantities of fissure water found in igneous and metamorphic rocks, which show up due to outcropping structures.

References

Castany G (1963) Traite Pratique Des Eaux Subterraines, ed. Dunod, 657 pp
CONAGUA (2016) Deputy director General's Office for Water Management
Davies SN, De Wiest RJM (1967) Hydrogeology, 2nd edn. Wiley, New York. 463 pp
EAM (2016) Estadisticas del Agua en Mexico, edición 2016; Comisión Nacional del Agua, Subdirección General de Planeación
Gleick PH (2002) The world's water 2002–2003: the biennial report on freshwater resources. Island Press, Washington, DC
Guzmán EJ, De Cserna Z (1963) Tectonic history of Mexico: Am Assoc Petrol Geol Mem 2:113–129

INEGI (2013a) Sistema de Clasificación Industrial de América del Norte 2013. Consulted on: https://www.inegi.org.mx/contenidos/app/scian/tablaxiv.pdf

INEGI (2013b) Estadísticas a propósito del día mundial de la Lucha contra la desertización y la sequía. Consultado en: http://www.inegi.org.mx/inegi/contenidos/espanol/prensIa/Contenidos/estadisticas/2013/sequia0.pdf (15/07/2015)

INEGI (2016) Anuario estadístico y geográfico de los Estados Unidos Mexicanos 2015. Consulted on: http://www3.inegi.org.mx/sistemas/biblioteca/ficha.aspx?upc=702825077280 (15/04/2018)

Klimentov PP, Bogdanov GY (1977) General hydrogeology. Moscow, Nedra, p 357

López-Ramos E (2018) Geología general y de México. 8a edición, Editorial Trillas S.A. de C.V., México, p 272

Matthess G (1970) Bezlehungen zwischen geologischem Bau und Grundwasserbewegung in Festgestelnen. Wiesbsden, 105 pp.

Morán–Centeno D (1984) Geología de la República Mexicana: México, Universidad Nacional Autónoma de México, Instituto Nacional de Estadística, Geografía e Informática (INEGI), 88 p

Ortega-Gutiérrez F, Mitre-Salazar LM, Alanís-Álvarez S, Roldán-Quintana J, Aranda-Gómez JJ, Nieto-Samaniego AF, Morán-Zenteno DJ (1991) Geological Provinces of Mexico—A new proposal and bases for their definition: Universidad Autónoma de México, 120 Instituto de Geología: Universidad Autónoma de Hidalgo, Instituto de Investigación en Ciencias de la Tierra; Sociedad Geológica de Mineralogía; y Secretaría de Educación Pública, Subsecretaría de Educación Superior e Investigación en Científica, Convención sobre la Evolución Geológica y Primer Congreso Mexicano de Mineralogía, Pachuca, Hidalgo, Memoria, pp 143–144

Ortega-Gutiérrez F, Mitre-Salazar LM, Alanís-Álvarez S, Roldán-Quintana J, Aranda-Gómez JJ, Nieto-Samaniego AF, Morán-Zenteno DJ (1992) Carta Geológica de la República Mexicana, 5a edición, Instituto de Geología de la Universidad Nacional Autónoma de México

Pinneker EV (1977) Problems of regional hydrogeology. Laws governing the occurrence and formation of groundwater. Nauka, Moscow. 196 pp

Raisz E (1959) Landforms of Mexico: Cambridge, MA, Mapa con texto, escala 1:3,000,000

Salas GP (1975). Metallogenetic chart of Mexico. Geological Society of America, Map and Chart Series MC13, scale 1:2,000,000

Smirnov SI (1971) On the relationship of molecular and filtration diffusion of minerals in groundwater. Trudy VIII Hydrogeol Geol collection 21:12–36

White DE (1965) Hot springs of volcanic origin. In: The geochemistry of recent post-volcanic processes. Mir, Moscow, pp 78–100

White DE (1967) Some principles of geyser activity, mainly from Steamboat Springs, Nevada. Am J Sci 265:641–684

Chapter 4
The Water–Energy–Food Nexus in Mexico

Carlos R. Fonseca-Ortiz, Carlos A. Mastachi-Loza, Carlos Díaz-Delgado,
and María V. Esteller-Alberich

Abstract The interdependence of water, energy, and food systems is widely recognized. Recently, this interdependence has generated significant concerns around
the world, as global projections indicate that the demand for water, energy, and food
will significantly increase over the next few decades. Mexico will confront several
challenges given the simultaneous pressures of population growth, urbanization,
climate variability, and climate change. Therefore, in this work, water, energy, and
food resources in Mexico are quantified in terms of their availability and demand.
Also, the disparities between the northern, southern, and central regions of the
country are highlighted. Specific challenges to water, energy, and food systems are
described, as well as the lack of an efficient approach for establishing the nexus
between these systems. Finally, one approach for effectively measuring and
establishing the water–energy–food nexus through emergy is proposed and exem-
plified. In this respect, emergy is an expression of all the energy used in the work
processes that generate a service (water, energy, or food) in units of one type of
energy (emjoules).

Keywords Water security · Energy security · Food security · Emergy

4.1 Introduction

Currently, the interdependence of water, energy, and food systems has increasingly
generated concerns around the world because of the correlations among these
sectors. Negative impacts on one system may affect one or all the other systems.
The present focus aims to integrate the concepts of water security (Ws), energy
security (Es), and food security (Fs) based on their connections and to therefore

C. R. Fonseca-Ortiz · C. A. Mastachi-Loza · C. Díaz-Delgado (✉) · M. V. Esteller-Alberich
Instituto Interamericano de Tecnología y Ciencias del Agua, Universidad Autónoma del Estado
de México, Carretera Toluca - Ixtlahuaca km 14.5, Toluca 50120, State of Mexico, Mexico
e-mail: crfonsecao@uaemex.mx; camastachil@uaemex.mx; cdiazd@uaemex.mx;
mvestellera@uaemex.mx

© Springer Nature Switzerland AG 2020   65
J. A. Raynal-Villasenor (ed.), *Water Resources of Mexico*, World Water Resources 6,
https://doi.org/10.1007/978-3-030-40686-8_4

characterize the water–energy–food nexus (*WEFn*). The concepts of *Ws* and *Es* are still being developed and debated in the literature, and differing definitions have been proposed (Ang et al. 2015; Cook and Bakker 2012). However, the concept of *Fs* is widely accepted (Pinstrup-Andersen 2009). In addition, several problems impede the development of a *WEF-n*, for example, the inflexibility of research institutions, the lack of greater interdisciplinary collaborations, the complexity of the topic, current political economy, and the ambitious objectives of this approach. The identification of connections among these three overarching systems requires distinct disciplines and scales to be united (Leck et al. 2015). In the following, the most accepted definitions of *Ws*, *Es*, and *Fs* within the *WEFn* focus are presented:

Water security is defined as "the capacity of a population to safeguard sustainable access to adequate quantities of water of acceptable quality" (UNU-INWH 2013). In addition, water security is a key aspect of sustaining livelihoods and human well-being, which includes basic resources for a good life, health, happiness, freedom of choice and action, good social relations, and safety. Also, secure access to water can promote socioeconomic development and can protect individuals from water-related disasters and water-borne pollution. Finally, water security enables ecosystem preservation and promotes political stability and a climate of peace.

Energy security is defined as "the uninterrupted availability of energy sources at an affordable price" (IEA 2014). The four main characteristics of secure energy are availability, affordability, accessibility, and acceptability (Cherp and Jewell 2014).

Food security exists (i) when people have constant access to food, independently of economic factors, political instability, or adverse weather conditions; and (ii) when food is constantly available, which is related to the level of food production, stock levels, and net trade. In addition, food security is characterized by (iii) the physical and economic access to food, which is determined by available income, expenditure, markets, and prices; and iv) the sufficient availability of safe and nutritious foods that meet the dietary needs and preferences of a population, thereby enabling individuals to sustain active and healthy lives (FAO 2008).

According to the definitions of *Ws*, *Es*, and *Fs*, the *WEFn* focus can be defined as the establishment of the connections between water, energy, and food systems with the objectives of recognizing their interdependency and ensuring the future availability and supply of WEF. In addition, the access to WEF of an adequate quantity, quality, and price is important for sustaining livelihoods and protecting against disasters that could affect the equilibrium among the various systems (Fig. 4.1).

Despite the challenges presented by an integrative *WEFn* focus, worldwide population growth will increase the demand for energy, water, and food. For this reason, it is important to identify the connections among these sectors and to find an effective means to correlate them within Mexico and around the world, which could be one means of improving their management and efficiency in meeting the needs of distinct populations from an integrative perspective.

Fig. 4.1 Conceptual model of the water–energy–food nexus

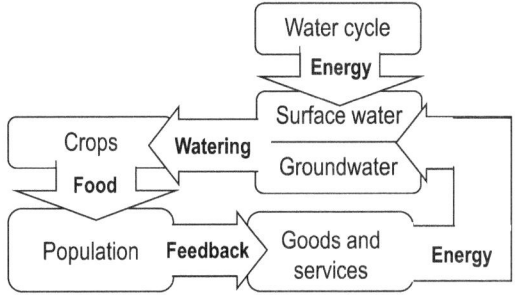

4.2   Water, Energy, and Food Systems in Mexico

Currently, large amounts of data relevant to water, energy, and food systems have been integrated in databases at the national and international levels. Related indicators, including those developed by distinct institutions, are also widely accessible. The FAO previously developed a system called the *WEFn* Rapid Appraisal based on the Nexus Assessment (Giampietro 2013), which provides a quick way to assess specific interventions to promote development goals, such as food, water, and energy security. An online application was developed to perform this assessment (http://www.fao.org/energy/water-food-energy-nexus/, 2017), although Mexico is not on the list of the included countries.

To analyze indicators related to water, energy, and food systems, and their evolution in Mexico, data from the World Bank (2017), FAO (2017a, b) and Enerdata (2017) can be used. In Table 4.1, some indicators for Mexico are shown for 2015. In Fig. 4.2, the behavior of these indicators over time is shown.

One of the most important variables in Mexico that influences water, energy, and food systems is population because population growth exerts a significant pressure on resources. In 2015, Mexico had more than 127 million inhabitants and an annual growth rate of 1.3%. By the year 2030, a population of 137.5 million inhabitants is expected (CNA 2016). Nearly all variables related to consumption in the energy sector have also increased. Even so, energy consumption does not exceed production; therefore, food and water systems, which are dependent on energy, are not yet put at risk because of this factor. Notably, regarding the water system, access to sanitation facilities has markedly increased over time. In the year 2015, 95% of the population had access to sanitation facilities.

However, the population indicators, water resources, and GDP of Mexico at the national level hide the large variability of distinct indicators across various regions of the country (Cervantes-Jiménez et al. 2017). From an overall perspective, it might appear that the water, energy, and food systems of Mexico are not at risk. However, the reality at the regional level is different, and the wide variability in indicators is

Table 4.1 Indicators of the *WEFn* in Mexico for 2015

Water	
Long-term average annual precipitation (mm/year)	758
Water resources (m^3/person/year)	4000
Proportion of water devoted to agriculture (% of total)	77
Access to improved water sources (% of population)	96
Energy	
Energy consumed by power irrigation (million kWh)	985
Crop area irrigated (% area equipped for irrigation)	86.1
Subsidies for electricity used to pump irrigation water (%)	60
Food	
Agriculture, value added (% GDP)	3
Cultivated land (ha) per capita	0.20
Food exports (million US$)	16,230
Food imports (million US$)	21,503
Employment in agriculture (%)	13.4
Employment in agriculture, female (%)	3.6
Prevalence of undernourishment (%)	<5.0
Underweight, children under 5 (%)	2.8
Dietary intake of cereals/roots/tubers (%)	44
Cereal-import dependency ratio	30.7

Enerdata (2017), FAO (2017a, b) and World Bank (2017)

visible in Fig. 4.3. For example, two-thirds of Mexico is considered arid or semiarid (central-northern region) and has an annual precipitation of less than 500 mm. This region also contains 80% of the population (Becerril-Piña et al. 2016). Meanwhile, the southeastern region has an annual precipitation that exceeds 2000 mm per year and contains two-thirds of the renewable water resources of Mexico, too. The availability of water is seven times greater in this region than in the central-northern region. However, this region only contains one-fifth of the population. Also, in this contrasting panorama, the southeastern region presents the highest score on the Nutritional Risk Index despite having the greatest water availability. Meanwhile, the highest consumption of water for agriculture occurs in the central-northern region, despite this region having lower water availability and experiencing aquifer overexploitation. The use of water in this region has low efficiency because of the existence of several subsidies for energy (indirect) and water (direct) for both production and consumption. For example, users who utilize electricity to pump irrigation water are offered a 60% subsidy on the cost, regardless of their scale of production or income. Furthermore, the adoption of irrigation technologies in areas traditionally cultivated with rain-fed crops has not increased over the last 40 years, and the existing infrastructure has simultaneously deteriorated, generating usage inefficiencies (FAO 2016).

Also, in terms of food supply, Mexico imports a large quantity of food (US $21,506 million), especially cereals. As 40% of the Mexican diet is composed of

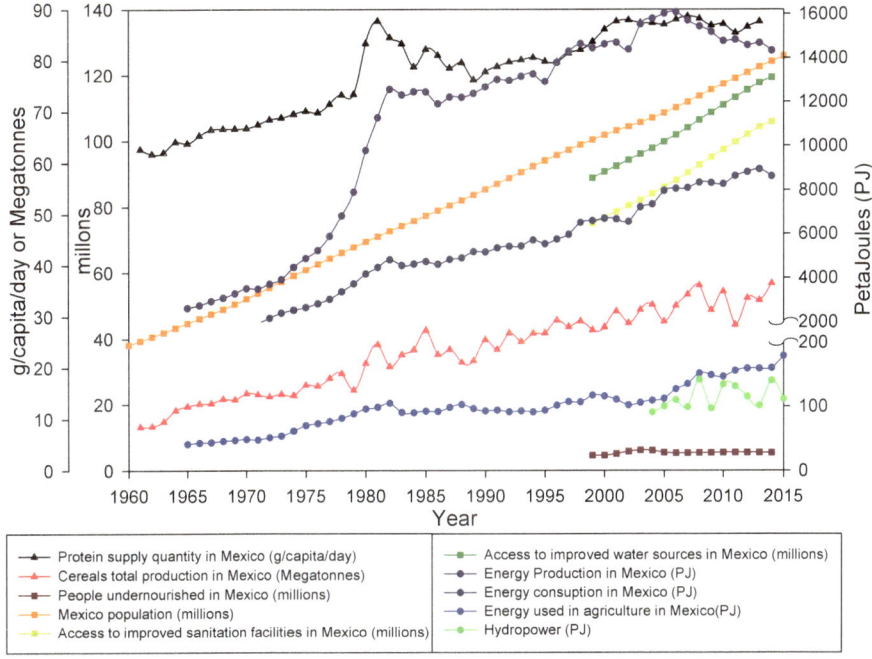

Fig. 4.2 Historical evolution of several indicators of the *WEFn* in Mexico. (Enerdata 2017; FAO 2017a, b; World Bank 2017)

cereals, roots, and tubers, this constitutes a large vulnerability in the food system. In addition, 46% of the population is living in poverty, and a large portion of the population is suffering from obesity, diabetes, and/or undernourishment.

The population distribution of Mexico is also uneven, and the population density varies widely from 14 to 5967 inhabitants/km^2. In the highly populated urban regions, water consumption levels exceed 400 hm^3. In addition, the areas devoted to livestock and agriculture are largely located in regions with restricted aquifers that are overexploited. For this reason, these activities are increasingly insecure because of their dependence on decreasing water sources. The interrelationships among these indicators highlight the importance of using a *WEFn* focus at regional and local scales, in addition to considering regional and local socioeconomic conditions.

Furthermore, certain natural phenomena present challenges for future water, energy, and food systems. For example, extreme changes in the climate variability (e.g., droughts, floods, hailstorms, and frost), climate change, and desertification can compromise the connections and equilibrium between these systems. Changes in temperature, precipitation, and, consequently, evapotranspiration can exert greater pressures on agriculture, which compromise *Fs*. In particular, *Fs* is directly related to environmental conditions because food production, distribution, storage, and markets are affected by climate variability, mainly that of water availability (Mastachi-Loza et al. 2016). In one analysis of climate change scenarios in the irrigation

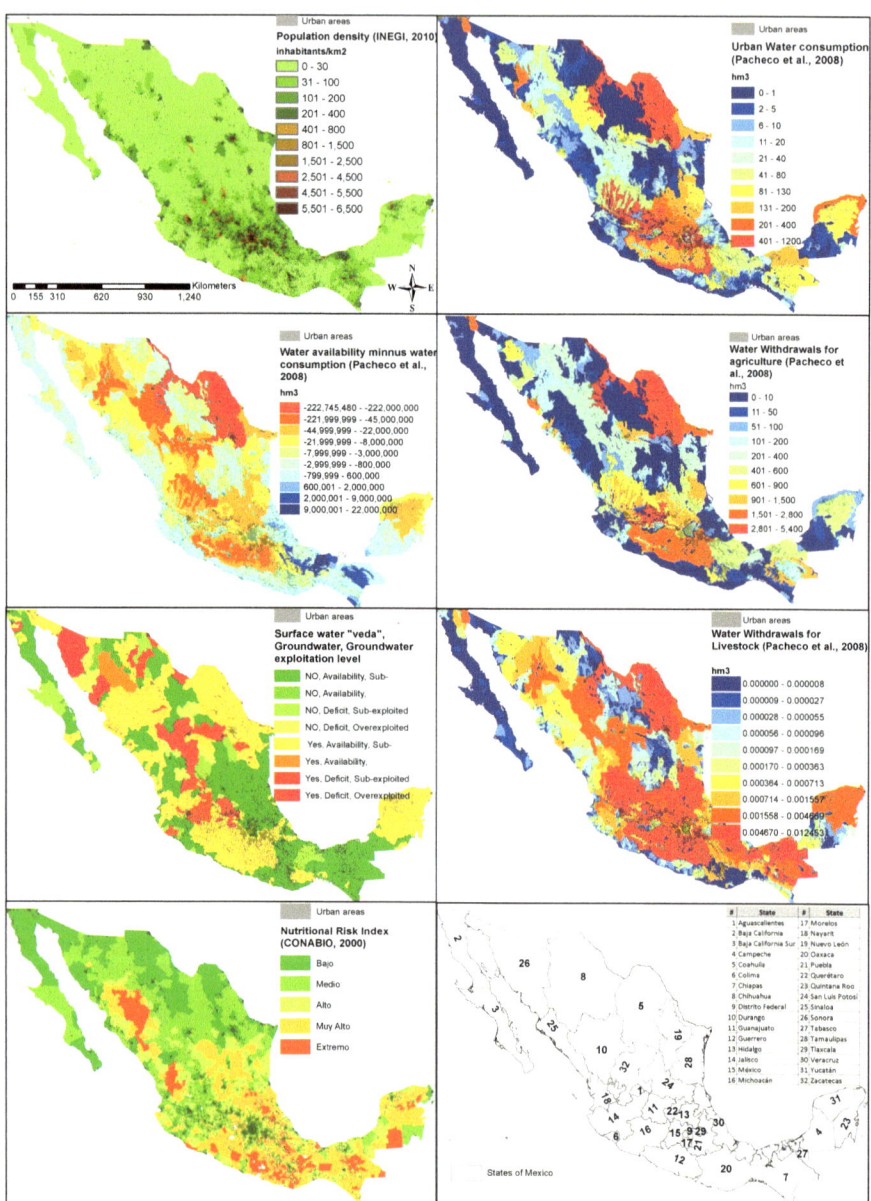

Fig. 4.3 Spatial distribution of several indicators of the *WEFn* in Mexico

districts of Northern Mexico, according to the representative concentration pathways (*RCP*) 4.5 and 8.5, crop-water requirements were expected to increase by 8.5% and 21%, respectively, by the year 2100 (Paredes-Tavares et al. 2018). In addition, desertification or degradation of land as a result of several factors, such as anthropic activities and changes in climate, leads to a reduction or loss of economic productivity (UNCCD 1994). Becerril-Piña et al. (2015) analyzed the risk of desertification in central Mexico by examining the probabilities of various scenarios. In the last 20 years, a high rate of change of land use (34 km^2 per year) has been observed. In addition, urban zones increased by 40% and agricultural areas by 30%. Overall, the region presented a high risk of desertification, mainly along the agroindustry corridor of the central region of the country.

4.3   Challenges to the Water–Energy–Food Nexus in Mexico

One difficulty in establishing relationships between water, energy, and food systems involves evaluating these systems at a common scale. Also, traditional and, according on Kapp (1975), questionable economic indicators of development and production have taken precedence at the policy level to the disregard of social and environmental indicators.

In economic terms, the cost of supplying water for agricultural uses is determined by the investment in works and infrastructure to deliver water, as well as the operating costs required for water to arrive at its destination (Balairón Pérez 2002). However, the price of water is not completely reflective of all of the costs and benefits associated with water services (UNESCO 2006). For instance, the scarcity of water for human consumption has been increasingly recognized worldwide (increasing the supply and demand for water), but water is also largely considered as a renewable resource. In addition, the value of water cannot be solely measured in monetary terms because water also holds unique ecological and cultural values in distinct regions of the world.

The ecological value of water is associated with the water requirement of ecosystems and the services provided by ecosystems (physical environments, climate regulation, and biodiversity) that maintain the status and quality of water. Meanwhile, the cultural value of water is related to the inherent or "sacred" value of water resources, which is a reflection of certain societal beliefs and values (De Groot et al. 2002).

In response to the presented dilemma, several theories have created models for water management based on nature and natural water dynamics, rather than the economic aspects of water resources. For example, "opportunity cost," defined as the loss of potential gains that could have been obtained from other alternatives when one alternative is chosen (Stevenson and Lindberg 2009), has been used to evaluate the various elements and scenarios of a system, including their effects on water supply and management.

One relevant theory is the water footprint. The method to calculate the water footprint, as well as the considered parameters, is distinct from that used to calculate the economic footprint. The water footprint is determined in terms of "virtual water," which is the quantity of water resources required to produce particular goods or services (Allan 1993; Hoekstra 2011). In this method, comparisons are made between products or services based on the quantity of water required to create and/or transform them. Thus, distinct commercial transactions can be compared in terms of units of water volume. For example, according to Mekonnen and Hoekstra (2011), Mexico exports 42.5% of its national water footprint (1978 m^3/year per capita), while the United States exports only 20.2% of its water footprint (2842 m^3/year per capita).

The water footprint has been promoted by worldwide organizations, such as the Water Footprint Network, yet presents one great disadvantage: It does not consider other natural resources that are equally necessary for society. For example, mineral products, such as carbon (e.g., a source of energy), iron, and copper (e.g., used as construction materials and to create tools), among others, are disregarded.

In this context, some models represent the behavior of physical systems through modeling the input/output of energy to a certain system based on the principles of thermodynamics (Bastianoni et al. 2007), especially according to the implications of the second law of thermodynamics: "Processes occur in a certain direction, and energy has quality as well as quantity" (Cengel and Boles 2002). In general, these models are based on the premise that classic parameters (the monetary cost of production and amount of capital) are not sufficient or significant indicators of the optimality of a production line or design (Sciubba and Ulgiati 2005). However, the best method to evaluate energy inputs/outputs has been widely debated, for example, the use of emergy versus exergy (Bastianoni et al. 2006; Herendeen 2004). Exergy analyses were mainly developed to study the processes involved in systems of energy conversion, while emergy analyses are commonly applied to the processes involved in generating products or services at a large scale (Lazzaretto 2009). Although some large-scale studies on water resources are based on exergy, such as the study of Chen et al. (2009), emergy analyses are more frequent at regional scales (Lv and Wu 2009).

4.3.1 Alternative Approach for the Assessment of the Water–Energy–Food Nexus (Emergy Accounting)

As Odum (1996) refers "Emergy is the available energy required, directly or indirectly, to generate a service or product." In emergy accounting, various forms of energy may also be represented by the solar emjoules (seJ) or equivalent solar emergy. In this sense, emergy is the amount of solar energy required per product or energy unit, expressed in seJ/J or other units. In addition, mass (seJ/g) can also be expressed in unit emergy values (Brown et al. 2010; Pulselli et al. 2011).

Systems evaluated in terms of emergy are usually represented by energy-flow diagrams. Figure 4.4 shows a water-supply system for urban users with urban demands but without restrictions for other uses, such as agricultural uses. In the socioeconomic subsystem, which is related to developed lands, water is transported from surface water and groundwater bodies toward through supply processes fed by goods and services.

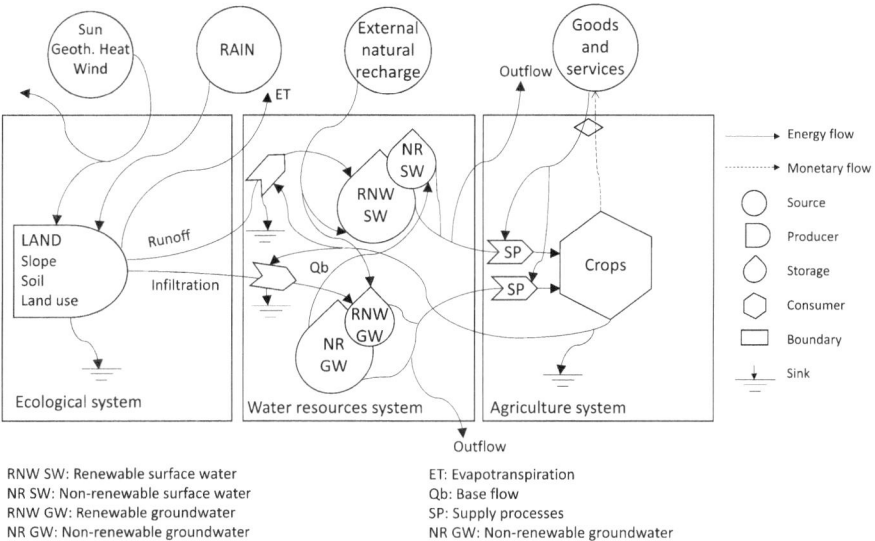

Fig. 4.4 The water supply system using a diagram of Emergy. (Díaz-Delgado et al. 2014)

Fonseca et al. (2017) propose three indicators to evaluate these systems: (a) the water deficit, (b) the environmental sustainability index (*ESI*), and (c) the economic impact. The water deficit is the difference between the supply and demand of water. The ESI (Eq. 4.1) is the relationship between the output of emergy and the environmental load (Almeida et al. 2007; Brown and Ulgiati 1997; Lv and Wu 2009). It is a function of the emergy flows associated with renewable resources (*R*), nonrenewable resources (*N*), and feedback (*F*) in the form of goods and services. The four thresholds of this index are as follows: (a) $ESI = 0$ (null sustainability), (b) $0 < ESI < 1$ (environmental load greater than the output), (c) $1 < ESI < 10$ (persistent influence of nonrenewable resources and socioeconomic feedback), and (d) $ESI > 10$ (greater influence of renewable resources) (Fonseca et al. 2017).

$$\text{ESI} = \left(\frac{R + N + F}{F}\right)\left(\frac{R}{N + F}\right) \tag{4.1}$$

With respect to the discussion on whether water resources should be considered as renewable, the current proposal favors the arguments provided in Díaz-Delgado et al. (2014), who classified water under an integrated management framework. In particular, various types of nonrenewable and renewable water resources were distinguished. Nonrenewable water resources were geographically and temporally delimited and were mainly associated with groundwater flows directly originated from precipitation. In addition, these researchers proposed the use of the filling index (Pernia et al. 2005; Vrba et al. 2007) to identify the exploitation of groundwater in areas where groundwater reserves are declining.

The economic impact, the third indicator proposed to evaluate water-supply systems, is "the sum of the variable energetic requirements in relation to the annual volumetric flow of water and the monetary production cost of electrical energy" (Fonseca et al. 2017).

The estimation of the unit emergy value (*UEV*) of water resources has been defined by some authors (Buenfil 2001; Díaz-Delgado et al. 2014) as the proportion in relation to the precipitation volume of both the flow and potential chemical energy (given by the Gibbs free energy) associated with the recharge of water bodies (via run-off or infiltration). For more detail, see Díaz-Delgado et al. (2014).

Meanwhile, the emergy flow associated with supply processes, such as water extraction and treatment, can be estimated as a polynomial function of the water supplied flow. This emergy flow tends to represent the emergy of installation maintenance and type and work requirements and the piezometric head to overcome (regarding even friction losses). For more detail, see Fonseca et al. (2017).

4.3.2 Emergy-Accounting Case Study

In the energy–water–food nexus framework, the emergy accounting of an agricultural district in the Upper Course of the Lerma River (*UCLR*) basin, Mexico, is presented as a case study. The *UCLR* basin has an average altitude of 2600 m.a.s.l. The valley has a mean annual rainfall of 900 mm and a subhumid temperate climate, while the mountainous region has a mean annual rainfall of 1200 mm with a semicold or cold climate (Esteller and Diaz-Delgado 2002). The aquifer of the Valley of Toluca (*VTA*) is freely developed and reaches depths of over 500 m in the valley (Esteller et al. 2012). According to the DOF (2003), the *VTA* (with a discharge of 53.6 hm^3/year, recharge of 336.8 hm^3/year, and extracted volume of 422.4 hm^3/year) has a deficit of 152.4 hm^3/year, which is obtained from groundwater reserves.

The corresponding irrigation district (Fig. 4.5) has a crop area (maize) of 50.32 km^2 and a surface water concession of 10,500 m^3/year. According to Díaz-Delgado et al. (2014), the approximate supply from deep wells is 151,330 m^3/year, and the water demand is 4.4 hm^3/year.

Under the conditions of the current scenario, an emergy analysis of crops was carried out for one calendar month (July). All water subsidies (water or energy) were omitted from the analysis. The monthly water demand was 100,000 m^3, and the water deficit was 61% (Table 4.2). The *UEV* associated with the 875 m^3 of supplied surface water was 1.00E+12 seJ/m^3 (ID = 1903). With respect to groundwater resources, the piezometric level of the study area indicates that the aquifer is being recharged. In this respect, 238.12 m^3 of water was estimated to be renewable from 37,832.75 m^3 of groundwater that was supplied from two deep wells (IDs = 26 and 451) (Díaz-Delgado et al. 2014; Fonseca et al. 2017). The *UEV* associated with the extraction of groundwater was estimated using the method proposed by Fonseca et al. (2017) considering the following operation characteristics of the wells:

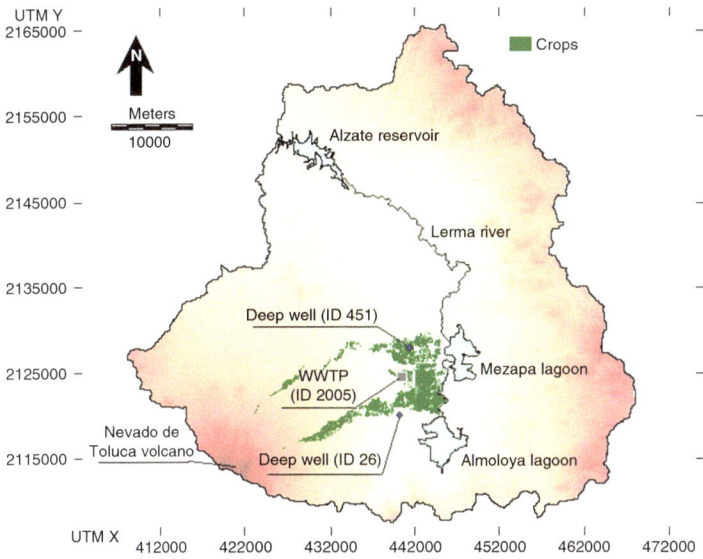

Fig. 4.5 Upper Course of the Lerma River (UCLR) basin, Mexico (case study)

(a) extraction depths of 11 and 35 m, (b) 0.30 m in diameter, (c) storage coefficient of 0.15, and (d) transmissivity of 0.005 m²/day.

The resulting ESI score for the *UCLR* basin under these conditions was 2.62. According to Fonseca et al. (2017), this score resulted from a large emergy flow because of the economic feedback and the use of nonrenewable versus renewable resources. Energy consumption (1.36E+10 J) in this scenario had an economic cost of US$408 resulting from the production cost of electrical energy (Ávila et al. 2005; Fonseca-Ortiz et al. 2013).

The economic impact, just of energy consumption because of water extraction, does not appear to be significant. However, this consumption is not representative considering that the water supply deficit is greater than 50%. If the entire water demand can be satisfied by deep wells under current conditions, the approximate economic cost to extract groundwater would be US$1075 monthly.

One alternative is to use renewable water resources recycled at a wastewater treatment plant to satisfy the water demand. In the study area, the nearest wastewater treatment plant treats water using stabilization lagoons and has a capacity of 37 L/s.

Table 4.3 shows the emergy count considering the use of renewable water resources. In comparison to current conditions, the emergy flow is 7.5 times greater, but the ESI increases 7.8 times. The percentage of renewable resources rises to 74%, and the economic impact is USD $1,591 monthly. Summing up, current practices for agricultural watering depict lower economic impacts. Nevertheless, the observed *WEFn* from an emergy-accounting perspective (Fig. 4.6) highlights environmental benefits associated with the increased use of renewable resources.

Table 4.2 Emergy accounting of the case study under the current conditions

Source node	Source description	Input	Flow (m³)	C_1	C_2	C_3	C_4	Emergy (seJ)	Type	C_{2e}	C_{3e}	C_{4e}	Energy (J)
Demand (m³)	100,000												
26	Deep well	Water	21245.85	0	1.00E+12	0	0	2.12E+16	N	0	0	0	0
		Infrastructure	21245.85	2.11E+14	0	0	0	0	F	0	0	0	0
		Energy	21245.85	0	1.49E+10	1.58E+05	8.05E-03	3.87E+14	F	1.49E+05	1.58E+00	8.05E-08	3.87E+09
26	Deep well	Water	120.57	0	1.00E+12	0	0	1.21E+14	R	0	0	0	0
		Infrastructure	120.57	2.11E+14	0	0	0	0	F	0	0	0	0
		Energy	120.57	0	1.49E+10	1.58E+05	8.05E-03	1.80E+12	F	1.49E+05	1.58E+00	8.05E-08	1.80E+07
451	Deep well	Water	117.55	0	1.00E+12	0	0	1.18E+14	R	0	0	0	0
		Infrastructure	117.55	2.13E+14	0	0	0	2.13E+14	F	0	0	0	0
		Energy	117.55	0	3.79E+10	1.61E+05	1.97E-02	4.45E+12	F	5.61E+05	1.83E+00	1.12E-07	6.60E+07
451	Deep well	Water	16348.78	0	1.00E+12	0	0	1.63E+16	N	0	0	0	0
		Infrastructure	16348.78	2.13E+14	0	0	0	2.13E+14	F	0	0	0	0
		Energy	16348.78	0	3.79E+10	1.61E+05	1.97E-02	6.62E+14	F	5.61E+05	1.83E+00	1.12E-07	9.67E+09

1903	Surface water	Water	875.00	0	3.80E+12	0	0	3.33E+15	R	0	0	0	0
		Consumable	875.00	0	2.82E+09	0	0	2.47E+12	F	0	0	0	0
Total	38,707.75		4.26E+16										1.36E+10
Deficit	−61,292.25												
	61%												
ESI	2.62												

Table 4.3 Emergy accounting of the case study under the alternative scenario

Source node	Source description	Input	Flow (m³)	C_1	C_2	C_3	C_4	Emergy (seJ)	Type	C_{2e}	C_{3e}	C_{4e}	Energy (J)
	Demand (m³)	100,000											
26	Deep well	Water	1.21E+02	0	1.00E+12	0	0	1.21E+14	R	0	0	0	0
		Infrastructure	1.21E+02	2.11E+14	0	0	0	2.11E+14	F	0	0	0	0
		Energy	1.21E+02	0	1.49E+10	1.58E+05	8.05E-03	1.80E+12	F	1.49E+05	1.58E+00	8.05E-08	1.80E+07
26	Deep well	Water	2.12E+04	0	1.00E+12	0	0	2.12E+16	N	0	0	0	0
		Infrastructure	2.12E+04	2.11E+14	0	0	0	0	F	0	0	0	0
		Energy	2.12E+04	0	1.49E+10	1.58E+05	8.05E-03	3.87E+14	F	1.49E+05	1.58E+00	8.05E-08	3.87E+09
451	Deep well	Water	1.18E+02	0	1.00E+12	0	0	1.18E+14	R	0	0	0	0
		Infrastructure	1.18E+02	2.13E+14	0	0	0	2.13E+14	F	0	0	0	0
		Energy	1.18E+02	0	3.79E+10	1.61E+05	1.97E-02	4.45E+12	F	5.61E+05	1.83E+00	1.12E-07	6.60E+07
451	Deep well	Water	1.63E+04	0	1.00E+12	0	0	1.63E+16	N	0	0	0	0
		Infrastructure	1.63E+04	2.13E+14	0	0	0	0	F	0	0	0	0
		Energy	1.63E+04	0	3.79E+10	1.61E+05	1.97E-02	6.62E+14	F	5.61E+05	1.83E+00	1.12E-07	9.67E+09

1903	Surface water	Water	8.75E+02	0	3.80E+12	0	3.33E+15	R	0	0	0
		Consumable	8.75E+02	0	2.82E+09	0	2.47E+12	F	0	0	0
2005	Wastewater treatment plant	Water	6.13E+04	0	3.80E+12	0	2.33E+17	R	0	0	0
		Infrastructure	6.13E+04	4.01E+16	0	0	4.01E+16	F	0	0	0
		Energy and consumable	6.13E+04	0	4.83E+10	0	2.96E+15	F	6.43E+05	0	3.94E+10
Total	100,000		3.19E+17								5.3E+10
Deficit	0	0%									
ESI	20.58996										

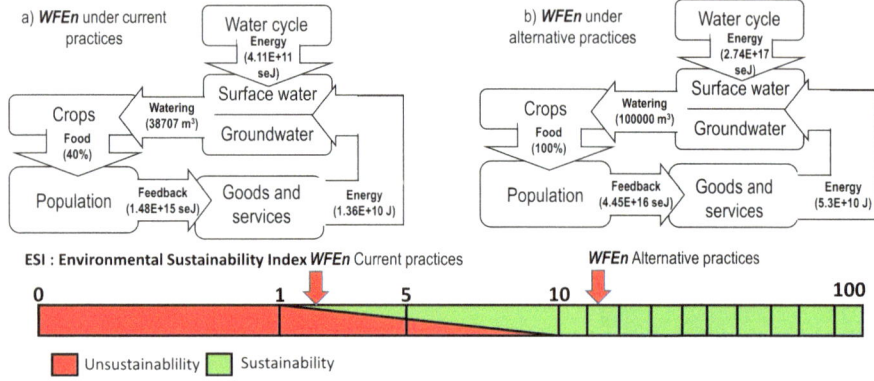

Fig. 4.6 *WEFn* evaluation through the environmental sustainability index (*ESI*) under current and proposed alternative practices for the case study

Despite the greater economic impact of this alternative supply scenario, the following factors should be considered: (a) the relationship between energy consumption, extracted water volume, and extraction depth is not linear. For this reason, the extrapolated economic impact of deep wells may be greater than determined in this first attempt to calculate the associated economic impact; (b) the extraction of greater volumes of water from deep wells results in the depletion of aquifers over time. This impact should also be contemplated in the analysis at a rate of US\$0.03/m^3.

Acknowledgments This study was carried out with financial support provided by CONACYT (248498 and 248327) and UAEM-SIEA-Quebec-Canada 4212/2016E.

References

Allan JA (1993) Fortunately there are substitutes for water otherwise our hydro-political futures would be impossible. Priorities Water Resour Alloc Manag 13:26

Almeida C, Barrella FA, Giannetti BF (2007) Emergetic ternary diagrams: five examples for application in environmental accounting for decision-making. J Clean Prod 15:63–74

Ang BW, Choong W, Ng T (2015) Energy security: definitions, dimensions and indexes. Renew Sust Energ Rev 42:1077–1093

Ávila S, Muñoz C, Jaramillo L, Martínez A (2005) Un análisis del subsidio a la tarifa 09. Gaceta Ecológica, México

Balairón Pérez L (2002) Gestión de recursos hídricos. Edicions UPC, Barcelona

Bastianoni S, Pulselli FM, Rustici M (2006) Exergy versus emergy flow in ecosystems: is there an order in maximizations? Ecol Indic 6:58–62

Bastianoni S, Facchini A, Susani L, Tiezzi E (2007) Emergy as a function of exergy. Energy 32:1158–1162

Becerril-Piña R, Mastachi-Loza CA, González-Sosa E, Díaz-Delgado C, Bâ KM (2015) Assessing desertification risk in the semi-arid highlands of central Mexico. J Arid Environ 120:4–13

Becerril-Piña R, Díaz-Delgado C, Mastachi-Loza CA, González-Sosa E (2016) Integration of remote sensing techniques for monitoring desertification in Mexico. Hum Ecol Risk Assess Int J 22:1323–1340

Brown MT, Ulgiati S (1997) Emergy-based indices and ratios to evaluate sustainability: monitoring economies and technology toward environmentally sound innovation. Ecol Eng 9:51–69

Brown MT, Martínez A, Uche J (2010) Emergy analysis applied to the estimation of the recovery of costs for water services under the European Water Framework Directive. Ecol Model 221:2123–2132

Buenfil AA (2001) Emergy evaluation of water. University of Florida Gainesville, Gainesville

Cengel YA, Boles MA (2002) Thermodynamics: an engineering approach. Sea 1000:8862

Cervantes-Jiménez M, Mastachi-Loza CA, Díaz-Delgado C, Gómez-Albores MÁ, González-Sosa E (2017) Socio-ecological regionalization of the urban sub-basins in Mexico. Water 9:14

Chen B, Chen GQ, Hao FH, Yang ZF (2009) Exergy-based water resource allocation of the mainstream Yellow River. Commun Nonlinear Sci Numer Simul 14:1721–1728

Cherp A, Jewell J (2014) The concept of energy security: beyond the four As. Energy Policy 75:415–421

CONAGUA (2016) Estadísticas del agua en México. CONAGUA

Cook C, Bakker K (2012) Water security: debating an emerging paradigm. Glob Environ Change 22:94–102

De Groot RS, Wilson MA, Boumans RM (2002) A typology for the classification, description and valuation of ecosystem functions, goods and services. Ecol Econ 41:393–408

Díaz-Delgado C, Fonseca CR, Esteller MV, Guerra-Cobián VH, Fall C (2014) The establishment of integrated water resources management based on emergy accounting. Ecol Eng 63:72–87

DOF, (Diario Oficial de la Federación) (2003) Acuerdo por el que se dan a conocer los límites de 188 acuíferos de los Estados Unidos Mexicanos, los resultados de los estudios realizados para determinar su disponibilidad media anual de agua y sus planos de localización. Secr Medio Ambiente Recur Nat SEMARNAT Viernes 31:90–91

Enerdata Y (2017) Global energy statistical yearbook 2017. https://www.enerdata.net/. Accessed 20 Nov 2017

Esteller MV, Diaz-Delgado C (2002) Environmental effects of aquifer overexploitation: a case study in the Highlands of Mexico. Environ Manag 29:266–278

Esteller MV, Rodríguez R, Cardona A, Padilla-Sanchez L (2012) Evaluation of hydrochemical changes due to intensive aquifer exploitation: case studies from Mexico. Environ Monit Assess 184:5725–5741

FAO (2008) An introduction to the basic concepts of food security. FAO, Rome

FAO (2016) Mexico, Country fact sheet on food and agriculture policy trends. Food Agric. Organ. U. N. Rome Italy. http://www.fao.org/3/a-i6006e.pdf. Accessed 16 Nov 2017

FAO (2017a) FAOSTAT database. Food Agric. Organ. U. N. Rome Italy. http://www.fao.org/faostat/en/. Accessed 16 Nov 2017

FAO (2017b) Aquastat: FAO's information system on water and agriculture. http://www.fao.org/nr/water/aquastat/main/index.stm. Accessed 16 Nov 2017

Fonseca CR, Díaz-Delgado C, Esteller MV, García-Pulido D (2017) Geoinformatics tool with an emergy accounting approach for evaluating the sustainability of water systems: case study of the Lerma river, Mexico. Ecol Eng 99:436–453

Fonseca-Ortiz CR, Díaz-Delgado C, Hernández-Téllez M, Esteller-Alberich MV (2013) Demanda hídrica urbana en Mexico: modelado espacial con base en sistemas de información geográfica. Interciencia 38:17

Giampietro M (2013) innovative accounting framework for the food-energy-water nexus. Food and Agriculture Association of the United Nations, c2013, Rome

Herendeen RA (2004) Energy analysis and EMERGY analysis—a comparison. Ecol Model 178:227–237

Hoekstra AY (2011) The water footprint assessment manual: setting the global standard. Earthscan, London/Washington, DC

IEA (2014) Energy security. IEA, Paris

Kapp KW (1975) Socio-economic effects of low and high employment. Ann Am Acad Pol Soc Sci 418:60–71

Lazzaretto A (2009) A critical comparison between thermoeconomic and emergy analyses algebra. Energy 34:2196–2205

Leck H, Conway D, Bradshaw M, Rees J (2015) Tracing the water–energy–food nexus: description, theory and practice. Geogr Compass 9:445–460

Lv C, Wu Z (2009) Emergy analysis of regional water ecological–economic system. Ecol Eng 35:703–710

Mastachi-Loza CA, Becerril-Piña R, Gómez-Albores MA, Díaz-Delgado C, Romero-Contreras AT, Garcia-Aragon JA, Vizcarra-Bordi I (2016) Regional analysis of climate variability at three time scales and its effect on rainfed maize production in the Upper Lerma River Basin, Mexico. Agric Ecosyst Environ 225:1–11

Mekonnen MM, Hoekstra AY (2011) National water footprint accounts: the green, blue and grey water footprint of production and consumption. UNESCO-IHE, Delft

Odum HT (1996) Environmental accounting: emergy and environmental decision making. Wiley, Chichester

Paredes-Tavares J, Gómez-Albores MA, Mastachi-Loza CA, Díaz-Delgado C, Becerril-Piña R, Martínez-Valdés H, Bâ KM (2018) Impacts of climate change on the irrigation districts of the Rio Bravo Basin. Water 10:258

Pernia J, Lambán L, Molinero A (2005) Indicadores e índices sobre el estado cuantitativo de las aguas subterráneas en función del nivel piezométrico. Aplicación al acuífero de la Sierra de Estepa (Indicators and indices on the quantitative status of groundwater based on the groundwater level. Application to the aquifer of Sierra de Estepa). In: Presented at the VI Simposio del Agua en Andalucía. IGME, pp 843–853

Pinstrup-Andersen P (2009) Food security: definition and measurement. Food Secur 1:5–7

Pulselli FM, Patrizi N, Focardi S (2011) Calculation of the unit emergy value of water in an Italian watershed. Ecol Model 222:2929–2938

Sciubba E, Ulgiati S (2005) Emergy and exergy analyses: complementary methods or irreducible ideological options? Energy 30:1953–1988

Stevenson A, Lindberg C (2009) New Oxford American dictionary. Digit Version 2, 80.4. Oxford University Press, Oxford

UNCCD (1994) United Nations convention to combat drought and desertification in countries experiencing serious droughts and/or desertification, particularly in Africa. UNCCD, Paris/New York

UNESCO (2006) Water: a shared responsibility, World Water Assessment Programme. Berghahn Books, Paris, New York and Oxford

UNU-INWH (2013) Water security & the global water agenda. United Nations University, Hamilton

Vrba J, Hirata R, Girman J, Haie N, Lipponen A, Shah T, Wallin B (2007) Groundwater resources sustainability indicators. UNESCO, Paris

World Bank (2017) World Bank data base. https://data.worldbank.org/. Accessed 23 Nov 2017

Chapter 5
Data Models for River-Basin Management in Mexico

Carlos Patino-Gomez and Paul Hernandez-Romero

Abstract One of the main problems about water-related data in Mexican hydrological regions is related to the different kinds of structure that are used to handle it. In addition, in many cases, the geographic and temporal information is disjointed, inaccurate, or incomplete, and it is difficult to establish a topological relationship among the elements that represent the basin in an integral way. For this reason, the development of relational data models (geodatabases) for each of the hydrological regions of Mexico is fundamental. This kind of structure would enable integrating georeferenced and temporal information, establishing several connections among geographic and historical data. One of the main advantages of these relational datasets resides in the specific structure to manage millions of records, establishing a relationship with gauging stations, such as hydrometric or meteorological stations, from which the simulation models could read and input all the needed data to generate various alternatives for making recommendations for water allocation in the Mexican hydrological basins. Another technique proposed is watershed regionalization, that is, dividing large hydrological regions to achieve the goal of watershed parameterization.

Keywords Hydrological regions · Data model · Water management · Relational databases · Watershed parameterization

5.1 Introduction

During the past two decades, water resources in Mexico have been under pressure due to extreme meteorological events like droughts, particularly in the northern part of Mexico. This situation has created several kind of problems among the users in the basins, particularly between the urban and agricultural water uses. Many national

C. Patino-Gomez (✉) · P. Hernandez-Romero
Universidad de las Americas Puebla, Cholula, Mexico
e-mail: carlos.patino@udlap.mx; paul.hernandezro@udlap.mx

© Springer Nature Switzerland AG 2020
J. A. Raynal-Villasenor (ed.), *Water Resources of Mexico*, World Water Resources 6,
https://doi.org/10.1007/978-3-030-40686-8_5

and international institutions and researchers have tried to develop strategies and methodologies to generate various scenarios to face this water-scarcity problem, but one of the main obstacles is related to the lack of reliable information from the irrigation districts or institutions who are in charge of measuring or generating the information needed to achieve a good analysis and to identify the kinds of alternatives. In hydrological study and analysis of watersheds, it is imperative to have data of various types, such as climatic data, land use, soil type and consumptive uses in quantity and quality, in order to carry out simulations that generate results for proposing adaptative measures to face extreme events, such as droughts and floods, as well as compliance with water demands for various purposes, generating public policies that guarantee the availability of water in the required quantity and quality. This production of basic information should be incorporated within a standard structure that enables its management in an efficient manner.

For many years, in the Mexican hydrological basins, a particular situation of having access to information related to water uses has been identified, in particular the updated water balance that enables knowing water availability for improving the content of the Mexican databases. Based on these updated data models, researchers could have access to all information needed to create water management models to generate scenarios for a better water allocation. Due to an increase in irrigation areas and population in big cities, the pressure on water resources is increasing to satisfy the water demand of the all users in the Mexican basins.

Binational watersheds, such as the binational basins of the Colorado River and the Rio Grande that are shared by Mexico and the US, represent a very special challenge due to their complexity. It is extremely important to establish protocols and joint mechanisms for the generation of hydrological, meteorological, and hydraulic information, among others, that allow the generation of data models in a joint manner that considers a standard structure to facilitate the exchange of information as established in various documents and international conventions for basins shared by more than two countries, such as the Helsinki Rules (ILA 1966) and the United Nations Economic Commission for Europe during the Convention in 1995 (UNECE 1995). Having the same data model for all the countries that share a hydrological basin will allow carrying out comparative analyses of various results produced by their respective simulation and management models, with the goal to propose measures that improve the distribution of water resources in these hydrological basins. These types of datasets are fundamental to generate water-management models for improving water allocation in the transboundary basins of Mexico, both in the north and in the south of the country.

As in many watersheds around the world, the information related to the physiographic characteristics in the Mexican hydrological basins are produced using traditional techniques and old or discontinued maps, field surveys and delineations of the watershed division created by hand using photogrammetry. Fortunately, computational techniques supported within geographic information systems have been developed in the past decades, which help hydrologists to make the parameterization of watersheds in a faster, systematic, and efficient way, reducing time in hydrological analysis and avoiding errors (Spence et al. 1995).

From the development of information in raster format, such as digital elevation models, it is possible to generate hydrographic networks or physiographic parameters of watersheds in places where the river system is not available, or where the quality of the hydrographic network is deficient (Jenson and Domingue 1988). The matrix arrangement in the structure of the raster format offers a great advantage for hydrological analysis and generates hydrological information that can be used in simulation models in basins of various sizes (Lacroix et al. 2002). In some hydrological regions of Mexico, hydrological parameters are being generated from digital elevation models, such as the calculation of the drainage area, transit time, and hydrographic network, among others, as established by Tribe in 1992.

Another significant change in the generation of hydrological parameters is the appearance of automated processes using remote sensing, which enables obtaining precipitation data automatically and transmitting them to a distributed hydrological simulation model, to estimate the corresponding runoff and thereby know the amount of water available.

Taking into account the need to have a standard structure for the management of water data, the Center for Research in Water Resources (CRWR) of the University of Texas at Austin developed a particular data model structure called ArcHydro. This will be used as a basic structure for managing all data related to water management, incorporating the spatial and temporal information contained in a relational geodatabase. This data model has the ability to assign attributes and establish relationships and connections through the creation of geometric networks between hydrological information, incorporating all elements in a geographic information system (Maidment 2002).

For this reason, it is imperative to attend to the problem related to the availability of water data, both quantity and quality, to be used by the researchers to generate public policies for improving water management in the Mexican hydrological basins. All the water data need to be incorporated into a standard structure to enable easy access to the simulation models for various kinds of analysis across many sciences, such as the hydrological, environmental, economic, and social, and then to generate recommendations for the decision makers considering various scenarios.

5.2 Current Conditions of the Hydrological Basins in Mexico

For many years, the Mexican federal government has invested a lot of money to improve the hydraulic infrastructure in the hydrological watersheds, seeking to improve water management, without the desired results. This situation is the result of sporadic and isolated actions because the decision makers did not take into account other factors such as the opinion of society and local conditions and did not include the experience of researchers who have developed water management plans taking into account all these variables. These water management models have included data models and their relationship with the simulation models to produce results that could be used to establish public policies regarding Mexican water law. An increase in the population is expected in most of the Mexican hydrological regions, which will result in problems in the supply of drinking water and handling of sewerage. Currently, there are already serious problems in this type of infrastructure, so the three levels of government in Mexico must work together to find integral solutions that make it possible to deliver the required service. For this, the integral management of water resources in the Mexican basins is considered, which also enables satisfying the demand for the diverse uses of water, not only the water supply for urban areas. The development and implementation of data models that enable the provision of reliable information in the appropriate formats will be fundamental for the development of these water-resource management models. Due to the climatic variability in the Mexican hydrological basins, it is fundamental to analyze in detail the behavior of precipitation, for example, because in the northern part of the country it could be around 160 mm per year and in the southern more than 4000 mm per year, for example, in zones of Chiapas state. Another important topic to be considered is the migration of people from rural to urban areas, demanding more water for urban use in the megalopolis such as Monterrey, Guadalajara, Mexico City, and Puebla. Also, the main part of the gross domestic product is generated in zones of Mexico where there is not enough water to satisfy the demand for consumptive uses, generating big risks for the local economies. For these reasons, it is imperative to take into account many factors to establish public policies, such as who is creating more pressure in hydrological basins by demanding more and more water, which sectors are participating in the water distribution and how they could improve their processes to save or reuse water to satisfy the demands. Also, it is important to review the hydrological basins that have restrictive water-use rules and reconsider these restrictions using the information contained in the relational models and the results of the simulation models. Another important scenario to be reviewed is the participation of private companies in water distribution in the urban areas because this participation could make water management more efficient in these zones, improving the water quantity and quality for the population. There are some good examples in Mexico, such as the water company of Leon, Guanajuato or Monterrey, Nuevo Leon.

Unfortunately, in many of the Mexican basins, the spatial and temporal information related to hydrology are disconnected. For example, there are serious errors defining the physical boundaries, and in other cases, the time series related with flow rates, climatic information or water quality are not complete within their registration periods, leaving many gaps in the structure.

For the reason just mentioned, hydrological, meteorological, hydraulic, and all information related to water administration need to be public and supported on standard platforms by the Federal Government of Mexico, throughout the National Water Commission (CONAGUA) and the state agencies of Mexico. This information should include states covering the watersheds, the hydrographic network, lakes, dams, hydrometric and meteorological stations, and the polygon describing the watershed's boundary of the Mexican hydrological basins, provided by the

Table 5.1 Summary of the main information in the Mexican hydrological regions

Description of data
Political boundaries: It includes states covering the Mexican hydrological regions.
Basin delineation: This element describes the hydrological divisions of Mexican basins and sub-basins.
Hydrography: It describes the stream network of the basins, at various scales.
Waterbodies and dam locations: This feature describes lakes, lagoons, and the main dams in the hydrological regions of Mexico.
Gauging stations: This element describes the location of climatic, water quality and hydrometric stations, among other points of interest.
Historical information: All information related to time series of runoff, temperature, evaporation, precipitation, water quality, etc., that should be considered.
Digital Elevation Models: It describes the elevation in the basins for the watershed parameterization.
Control Points: This feature includes water rights, return flow points, diversions, etc.

CONAGUA, research institutes as the Mexican Institute of Water Technology (IMTA), universities, international agencies such as the Mexican International Commission of Water and Boundaries (CILA), and the National Institute of Geography and Information (INEGI), among others.

As a first approach, Table 5.1 shows the main information that should be included to standardize the water information in Mexico.

According to the author's experience, errors exist in some geographic information related to the hydrography data in the Mexican watersheds, which includes inaccurate positions of the gauging stations; bad editing processes to generate the hydrographic network producing disconections in the river system; and errors during the projection process that generate a wrong position of many of the geographic information, such as lakes and dams. Figure 5.1 shows some of the described errors so that users can see the wrong direction of some rivers in the Rio Conchos watershed.

The Mexican agencies use the Geographic Coordinate System and Lambert geographic projection to create their geographic information. This is the INEGI, the agency in charge of producing this information in collaboration with other institutions.

5.3 Methodology

5.3.1 Developing the Scheme of the Hydrological Data Models

As the first step to create a data model, known as a geodatabase in GIS jargon, it is necessary to create a framework, perhaps using the Visio software, to serve as a standard framework that incorporates specialized structure such as a UML file. This

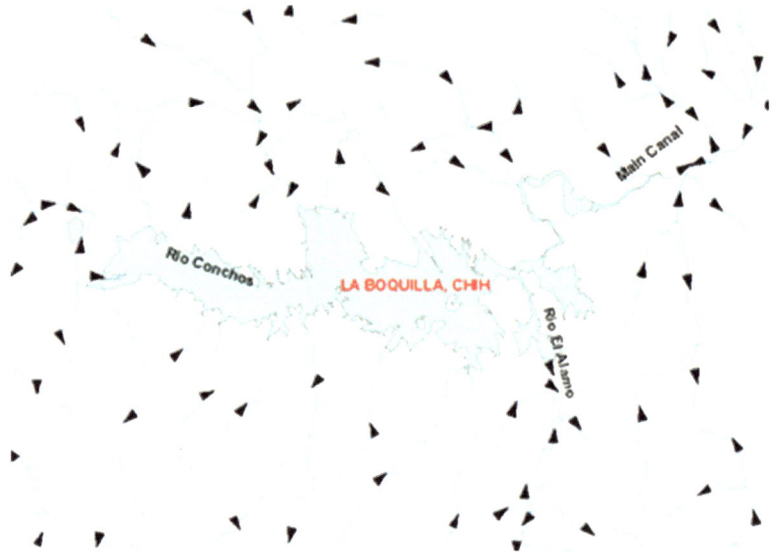

Fig. 5.1 Common errors in the Mexican hydrography

framework follows a specific and predefined data-model philosophy, incorporating changes in the structure of the attribute tables to add as many fields as required by the simulation models, or to satisfy the specific information required by the decision makers, primarily in the Mexican federal government.

To generate a personal geodatabase for the Mexican watersheds following the predefined structure of the UML file, an Access version of the UML can be applied to an empty geodatabase to recreate the feature datasets, feature classes, tables, and all the relationships among them (Fig. 5.2).

5.3.2   Mexican Hydrological Basins: How to Improve Their Temporal and Geographic Information Format

It is fundamental and very important to create hydrological and water management models in the Mexican watersheds that would provide the required information to avoid flooding problems, to mitigate the negative effects of droughts or to improve the water quality in the hydrological basins (Patino-Gomez 2005). However, the data models must incorporate all the elements related to the water resource, such as sources of availability, climatological variables, orographic characteristics of the hydrological basins, type and use of land, and economic activities, among others. All this interrelated information must establish the necessary arrangements and structures to generate a first simple model of water management.

Fig. 5.2 Standard data model structure proposed for the Mexican basins

In Mexico, various efforts have been made to create or improve, in a more efficient way, databases related to water resources, but these datasets incorporate water quantity and quality information separately. Another problem is the quality of the available information because, while there is sufficient information for some Mexican hydrological basins, sometimes the quality of this information is not correct, presenting gaps in the historical series and geographic positions, or including nonsense data that affect the result of simulation models. Unfortunately, the hydrological regions of Mexico do not have a standard structure that could store water quality and quantity information to establish a relationship among the various elements and thus would enable identifying actions to improve or make more efficient the water allocation in the entire hydrological watershed. Also, it is necessary to incorporate climate-change information in these databases since it is a phenomenon that will significantly affect the water availability and the water quality in the Mexican basins, putting at risk meeting the demands allocated for various uses.

A communication structure and protocols should also be proposed for sending and receiving information between the data models and the simulation models in the hydrological basins. This sequence in the simulation will enable an integrated simulation system to make decisions in the presence of an extreme event, such as a hurricane, for example.

The ArcHydro data model seems to be one option because this is a standard structure that allows very easy access to water information required by simulation models in a very efficient way. Indeed, the most recent version developed at the Universidad de las Americas Puebla can establish the required communication with

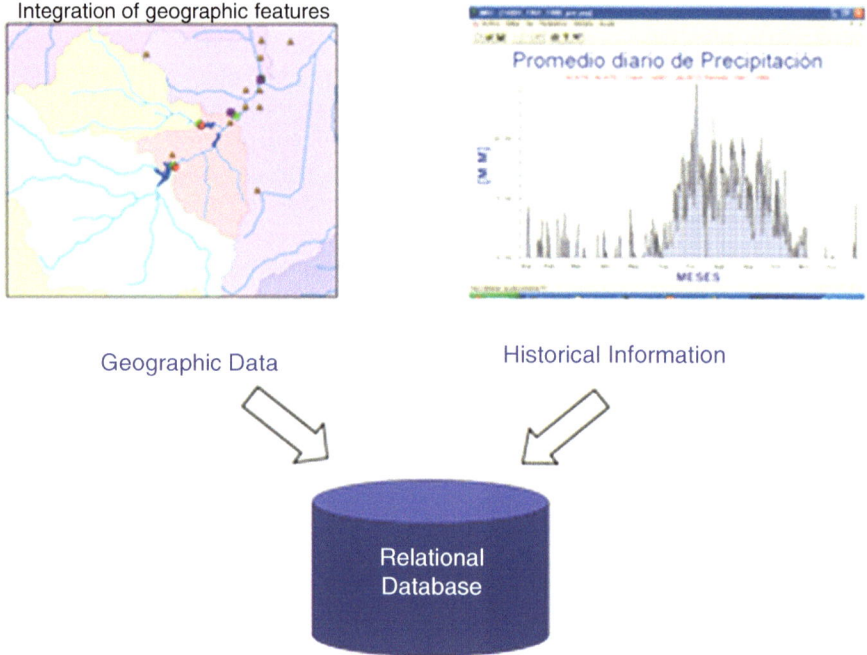

Fig. 5.3 Integration of water related data for the Mexican basins

climate change scenarios to analyze precipitation and temperature anomalies. This special structure is based on the geodatabase concept contained in the GIS platform, which incorporates hydrological, geomorphological, hydraulic, and all the spatial and temporal information related to the water resources of the Mexican hydrological watersheds, as shown in Fig. 5.3.

As the first part of the process to harmonize the water data in the Mexican hydrological watersheds, it is fundamental to create the standard structure of a data model or the ArcHydro data model for each of the 13 administrative hydrological regions following the division used by CONAGUA, using Geographic Information System technology that includes spatial and temporal information from various sources. These geodatabases should preserve the names used by CONAGUA to maintain the institutional identifiers.

Each of these geodatabases just mentioned would be a particular data structure to include georeferenced geographic information, as well as historical information from the hydrometric and meteorological stations located in the watersheds. One advantage of the use of relational databases is related to the creation of geometric networks. This network establishes the connection between the rivers and the gauging stations, for example, and enables finding any errors or disconections in the hydrographic network. The connectivity in the network makes possible tracing the contaminants in the rivers and provides the required information to calculate the concentration of contaminants in the water. Summarizing, the Mexican data models would contain all the information related to water resources in a systematic and standard structure, which makes it possible to modify the framework for specific conditions, depending on the zone of the country where the watershed is located.

5.3.3   Data Collection

As was mentioned, information created by different agencies or institutions in Mexico does not follow the same criteria or structures, so it is usually produced and published in different formats. This is a big problem for the users because, before they can incorporate this water data into simulation models, a preprocessing step to correct errors or to prepare the information in an efficient way is required. It is important to mention that all the official water information in Mexico is produced and disseminated usually by the INEGI and CONAGUA. In some cases, users could find information from other water projects developed by companies, universities, or research institutions, but it is neccesary to review and analyze it to check if it is appropriate and does not contain the original errors. Part of the original geographic information for one of the Mexican hydrological regions is shown in Fig. 5.4.

Temporal information in the hydrological basins of Mexico comes from various institutions, who report it in different structures and in many cases with several gaps

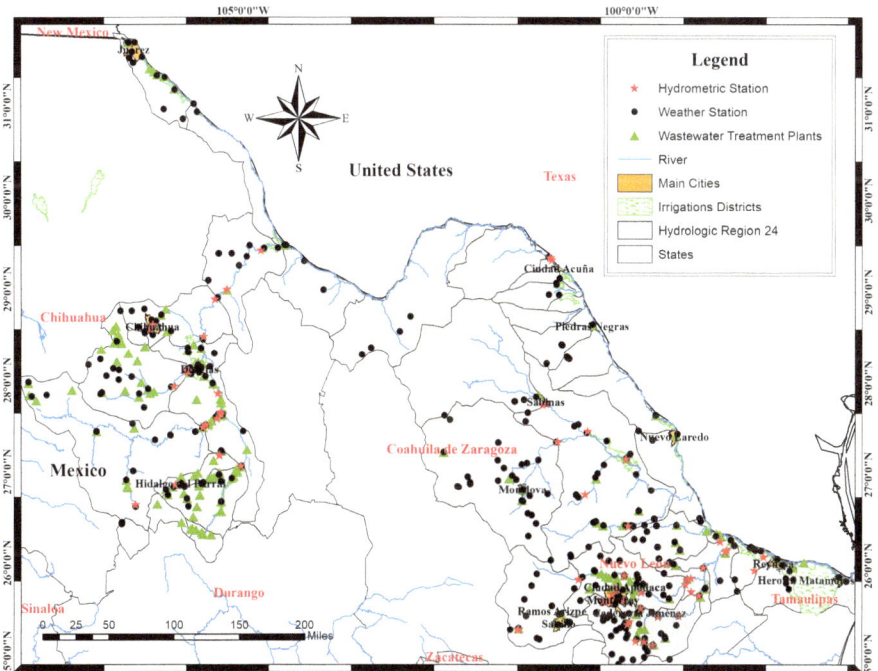

Fig. 5.4 Original water data in the Mexican hydrological regions

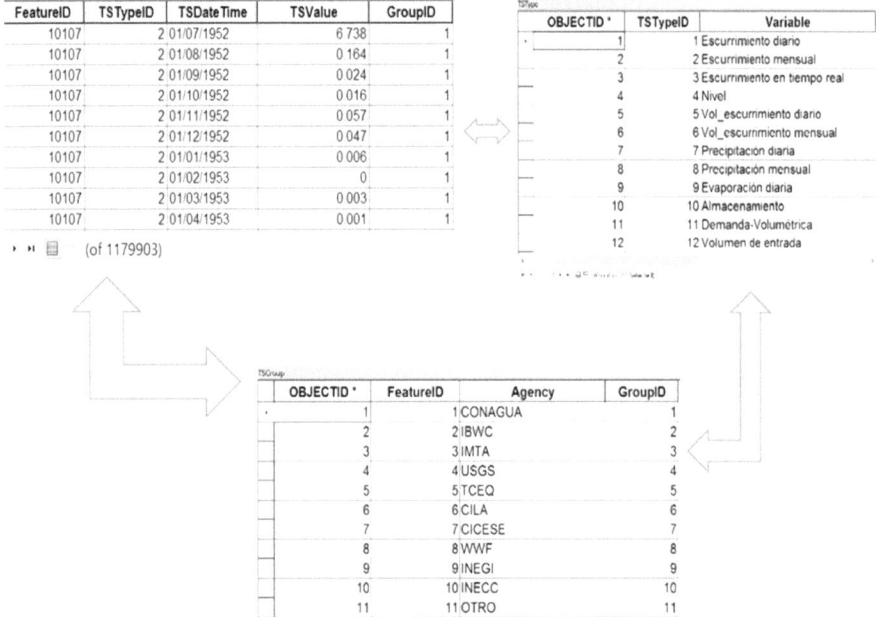

FeatureID	TSTypeID	TSDateTime	TSValue	GroupID
10107	2	01/07/1952	6 738	1
10107	2	01/08/1952	0 164	1
10107	2	01/09/1952	0.024	1
10107	2	01/10/1952	0.016	1
10107	2	01/11/1952	0.057	1
10107	2	01/12/1952	0.047	1
10107	2	01/01/1953	0.006	1
10107	2	01/02/1953	0	1
10107	2	01/03/1953	0.003	1
10107	2	01/04/1953	0.001	1

(of 1179903)

OBJECTID *	TSTypeID	Variable
1	1	Escurrimiento diario
2	2	Escurrimiento mensual
3	3	Escurrimiento en tiempo real
4	4	Nivel
5	5	Vol_escurrimiento diario
6	6	Vol_escurrimiento mensual
7	7	Precipitación diaria
8	8	Precipitación mensual
9	9	Evaporación diaria
10	10	Almacenamiento
11	11	Demanda-Volumétrica
12	12	Volumen de entrada

OBJECTID *	FeatureID	Agency	GroupID
1	1	CONAGUA	1
2	2	IBWC	2
3	3	IMTA	3
4	4	USGS	4
5	5	TCEQ	5
6	6	CILA	6
7	7	CICESE	7
8	8	WWF	8
9	9	INEGI	9
10	10	INECC	10
11	11	OTRO	11

Fig. 5.5 Proposed structure for the historical data in the Mexican basins

in it. It would be more useful to have all the historical data related to gauging stations in one place, following a predefined structure that makes possible its topological connection with the spatial georeferenced information. This is another important advantage of using relational database to have all water-related data interacting among their elements in a dynamic way and within a standard structure. Figure 5.5 shows a specific structure for handling the historical information related to the gauging stations located in the Mexican hydrological basins. It would enable knowing from where that information is coming and what kind of variable has been incorporated into the relational database.

The FeatureID field describes the public identifier assigned for the official agency, and the TSTypeID includes the kind of variable. Another important field is the GroupID that describes who is publishing the information. The TSDateTime and the TSValue field correspond to the collection date and its corresponding value.

5.3.4   Raster-Network Regionalization for the Mexican Basins

In some projects in small watersheds around the world, the use of a Digital Elevation Model (DEM) with high resolution to determine hydrological parameters is very common. Recently, the studies on bigger watersheds have generated a particular interest in new models and methodologies (Kite 1995). In particular, some Mexican

Fig. 5.6 DEM the
Papalopan basin in Mexico

hydrological basins are too big to be processed as one huge raster, which could be a
DEM or a climate-change raster. The raster-network regionalization technique
proposed by Patino-Gomez (2005) could be used to analyze these kinds of huge
basins. This procedure consists basically in dividing a very big hydrological watershed into smaller watersheds to avoid technical or computing problems during the
geoprocessing of a DEM. The raster information about the Mexican hydrological
basins is not common, and INEGI has published the whole DEM for the country, but
there is no methodology reported to generate one DEM or another kind of raster for a
specific hydrological region. In this direction, the authors have been proposing to
implement this kind of methodology in various projects to facilitate access to this
information. Figure 5.6 shows the DEM for the whole country and the result for a
specific Mexican hydrological region, including also a climate-change scenario that
could be very useful for vulnerability analysis in the climate-change context.

5.3.5   Development of the Mexican Geospatial Database

As was mentioned and as is common in many watersheds around the world,
knowledge and hydrological information available in some Mexican basins is not

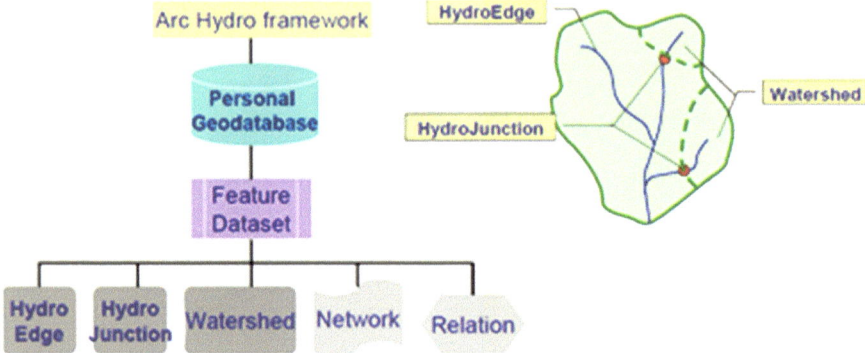

Fig. 5.7 Integration of water related data for the Mexican basins

enough or contains some errors and is of poor quality. For this reason, it is important to have a data model to compile, analyze, and make corrections, the geospatial database format being an excellent option. This kind of structure can handle the information in an efficient way and put all the water data in the appropriate structure required by the simulation models. Creation of Mexican geodatabases for the whole country is imperative. These geodatabases would manage all the water data in the corresponding watershed, incorporating the hydrographic networks, states covering the watersheds, lakes, dams, and the historical information from the gauging stations. Also, this kind of structure will make possible establishing relationships among georeferenced information and the time-series data (Fig. 5.7). Another important advantage of using this relational structure is the feasibility to link the geographic information with the time series of hydrometric or water-quality stations, so it would be possible to establish a topological relationship in the same data model for the 13 Mexican hydrological regions.

All the geographic information from the Mexican basins usually is created as a shapefile format, but this format does not allow establishing the rules and topology among the various elements that describe the hydrology in the basins, so these shapefiles would be incorporated into a feature dataset assigning and standardizing automatically the official projection parameters. Also, the personal geodatabase could have as many feature datasets as required by users or decision makers, and all information would be in just one access file.

The principal features of the proposed relational geodatabases are mentioned below:

- Monitoring points: Monitoring points are gauging stations, such as hydrometric and climatological stations, located in the appropriate position where it is important to measure the water or climatic variables.
- Waterbodies: These features include all dams, lakes, and lagoons, among others. Also, very wide rivers could be considered as a waterbody for the hydrological analysis of the watersheds.

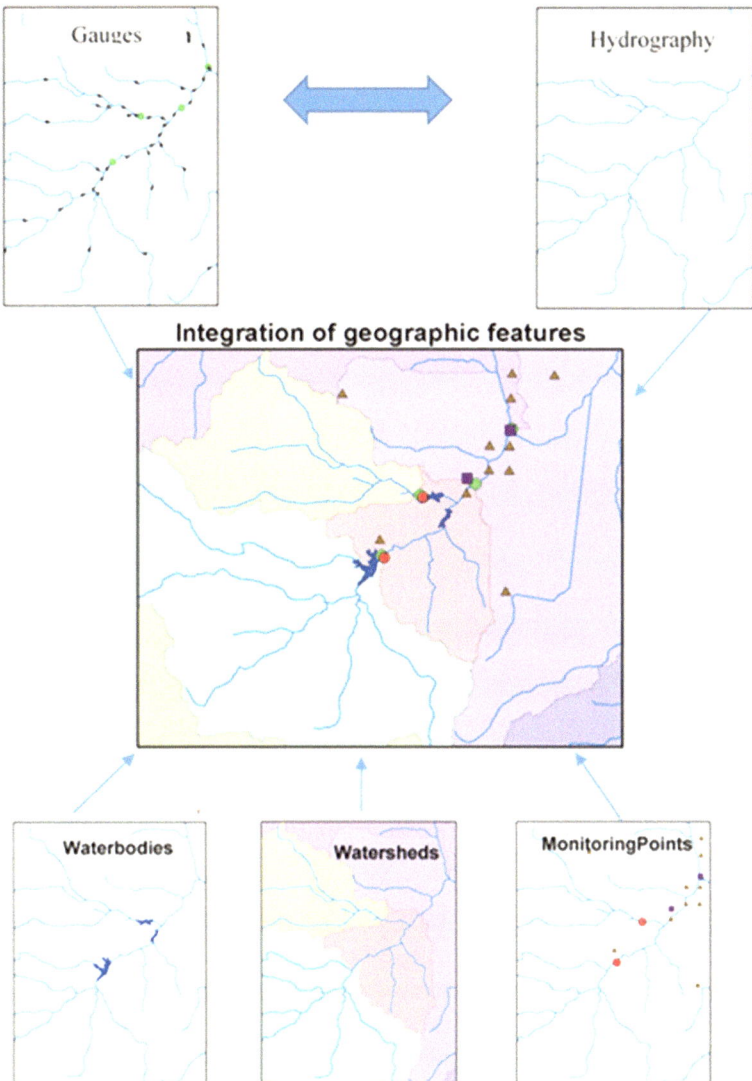

Fig. 5.8 Relational data model components

- Watersheds: This component corresponds to the drainage area flowing into the river system as part of the surface streamflow. If it is necessary to divide Mexican watersheds for specific purposes, these new drainage areas could be generated from the DEM.

To have the most robust data model for the Mexican basins, the authors propose establishing several relationships among the components just described and to link the time-series data to these elements (Fig. 5.8).

Fig. 5.9 Schematic
network proposed for a
Mexican sub-basin

To complete the data models for the Mexican hydrological regions, it is necessary to incorporate all the shapefiles related to the water data into the feature datasets, which are one of the most important components of the relational database. These feature datasets should have projection parameters and extent properties. One of the main recommendations is to add the largest geographic feature class into the feature dataset first, when a geodatabase is created; otherwise, some errors could be reported when new larger georeferenced information is added to this feature dataset.

Usually, the Mexican water data has a public identifier, but it is imperative to have an internal identifier within the data model for all the geographic features. These internal identifiers would be used to establish the relationships and interconnections among all the geographic feature classes contained in the geodatabase and enable communication with the simulation models in order to generate useful information, such as floodplains or alert systems, for decision makers.

Another important advantage of using relational database for the hydrological regions in Mexico is the creation of a schematic network that would enable connectivity among the hydrography and the gauging stations, as is shown is Figs. 5.9, 5.10, and 5.11, and from which hydrologic or water management models can read the water-related data.

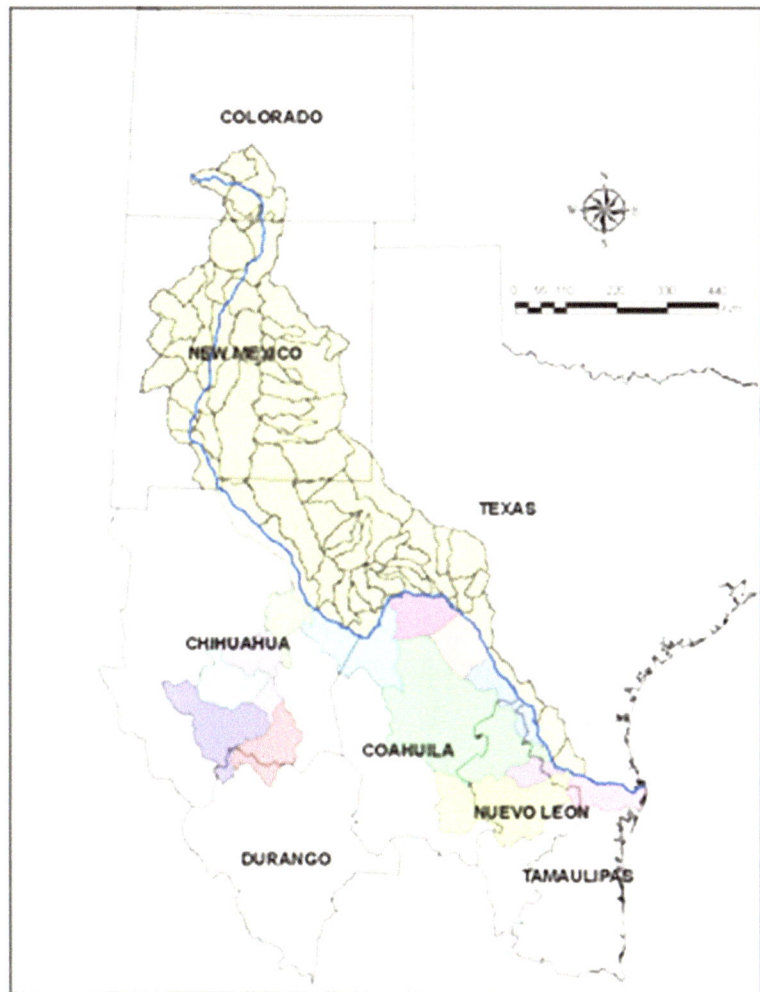

Fig. 5.10   Political division of the rio Bravo/Grande basin

5.4   A Case Study: The Rio Bravo/Grande Basin

The río Grande/Bravo basin, one of the most important binational basins, covers parts of five states in Mexico (Chihuahua, Coahuila, Durango, Tamaulipas, and Nuevo Leon) and three states in the US (Colorado, New Mexico, and Texas).

In general, the hydrological basins located in northern Mexico are characterized by having precipitation below the national average. This is a determining factor given that the main irrigation districts are located in these regions. In addition, the droughts that occur are more intense and recurrent, affecting the local economy of this basin because it affects the agricultural and livestock sector significantly. The

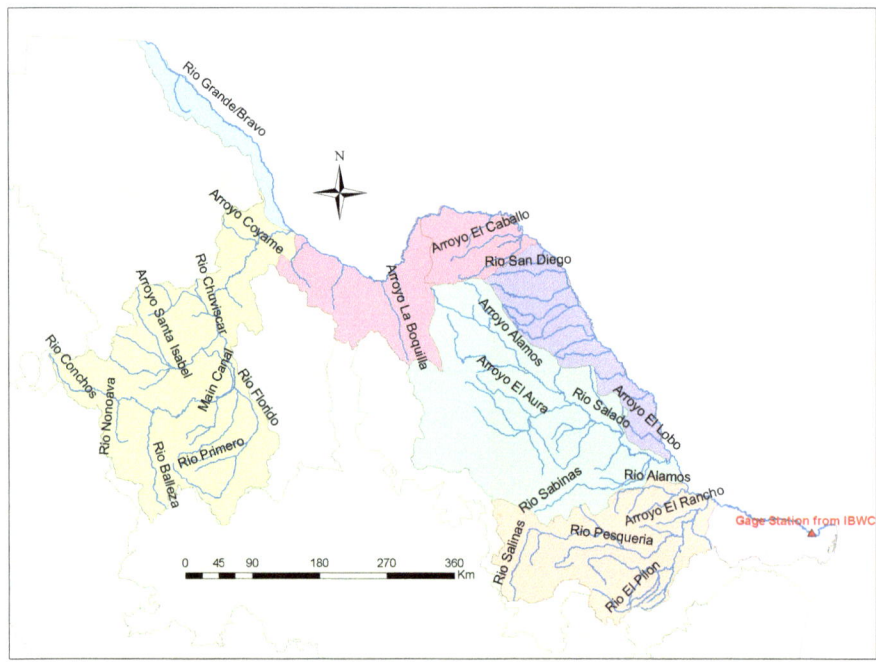

Fig. 5.11 Primary tributaries of the rio Bravo basin

other problem is the migration to urban areas, which generate a greater water demand to satisfy the needs of the population. Given the exposed conditions and association with the phenomenon of climate change, decision makers need to rely on integral studies that incorporate all the variables through the use of technology and automated systems, to improve water management in Mexico, and thus to guarantee the conservation of the ecosystems and sustainability in the watersheds.

The federal government of Mexico is aware of the importance of having water-management plans to allocate water resources in an efficient way under various scenarios of water scarcity or under a specific hydrometeorological event. This is so, in particular, due to the importance of the economic, social, political, and environmental aspects, in the transboundary basins. As part of the process to improve water management in the hydrological basins in Mexico, it is fundamental to generate or to improve the data models containing all the water resources data and climate-change scenarios in a standard structure, from which users can get the data for the simulation models in an appropriate format. This work is one of the main goals of a project financed by the Mexican National Science and Technology Council (CONACYT) that considers the development of a huge relational data model that could be the example for the rest of the Mexican hydrological regions.

During this project, it was possible to explore the possibility to incorporate hydrological information not just from the Mexican agencies, but also from the US agencies who use different formats and structures for handling this kind of information. All information coming from both countries were incorporated into a relational data model using the ArcHydro structure. This geodatabase could assist

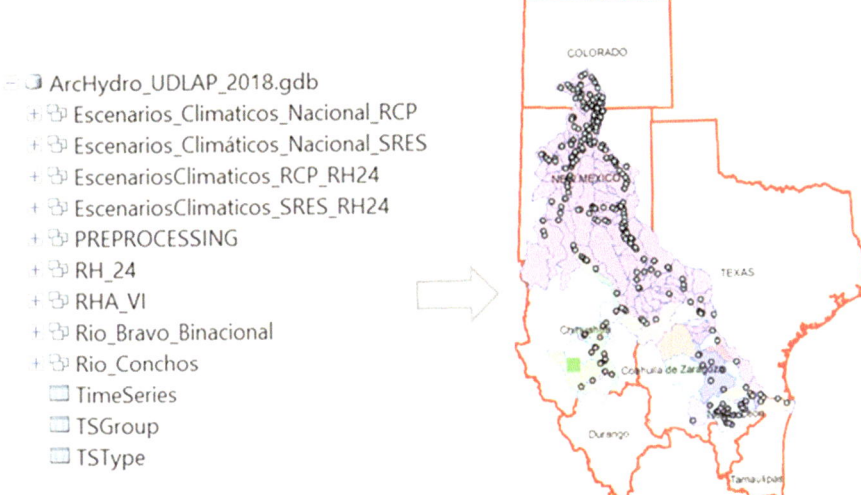

Fig. 5.12   Final relational data model for the rio Bravo hydrological region

reinforcing the negotiation process between Mexico and the US for the water allocation in this important binational watershed, providing the information required by the decision makers.

As occurs in most of the Mexican hydrological regions, the original information from the Rio Bravo basin was found to be in various different scales and structures. One of the most common structures to manage the geographic information by the CONAGUA is the shapefile. However, this particular structure has many limitations, so it was converted as a feature class and incorporated into a relational database created for the whole basin. Regarding historical data, the collection of temporal information began identifying the climatic, hydrometric, and water-quality stations located in this basin. The information was collected from governmental agencies and other research projects in progress that are available for the case study area. Following the ArcHydro structure, it was possible to have and relate georeferenced and time-series data in just one structure. The result is shown in Fig. 5.12.

Another advantage of using this relational data model is that regionalized climate-change scenarios can be projected and incorporated within the same geodatabase, thus providing the opportunity for decision makers to analyze, in space and time, anomalies of temperature and precipitation supported by a GIS platform (Fig. 5.13).

Finally, the connection of this kind of data model with simulation platforms is feasible and very useful. The authors propose an integration of hydrological processes as part of the water management plans in the Rio Bravo basin, distinguished by its approach to managing supply and demand, working on the water-balance principle. Based on the relational data contained in the geodatabase, it is possible to develop scenarios to improve water management under various hydrological conditions, taking advantage of the use of the hydrographic network created in this basin as is shown in Fig. 5.14. This network was created among 1.5 million reaches in the countries.

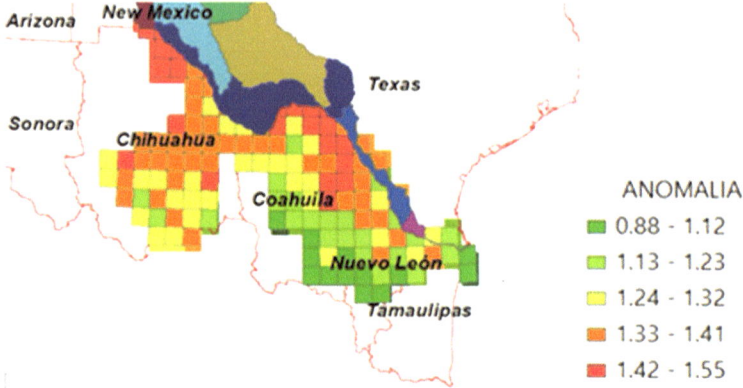

Fig. 5.13 Temperature anomaly projected to 2030 in the rio Bravo basin

Fig. 5.14 Hydrographic network in a binational hydrological region

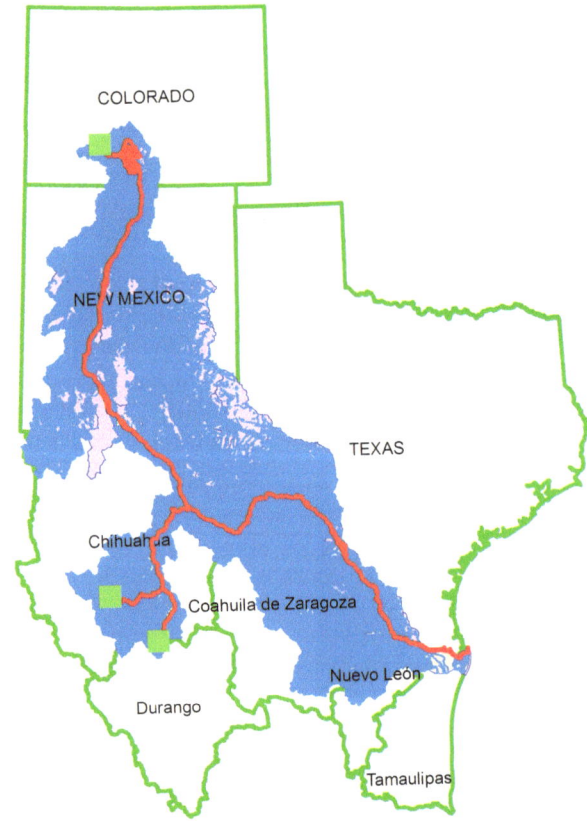

5.5   Conclusions and Recommendations

The Mexican hydrological basins suffer from a deficit of hydrological information and an adequate structure to carry out the studies of water availability necessary to determine the volumes of water for the various uses, as well as to face extreme events such as drought. It is essential to have the appropriate information and modeling systems to generate scenarios that make possible generating alternatives concerning the distribution of water. These proposed alternatives are fundamental mainly in the zones where water resources are over-allocated.

Climate change or extreme hydrometeorological events in Mexico, such as droughts, have caused conflicts among the various users due to the demand for water. Aware of this situation, the Mexican government has recognized the importance and urgency of the development and use of new technologies and simulation models to enable establishing public policies for improving water management in the Mexican watersheds. Unfortunately, water-related data from the Mexican hydrological regions have been produced and published in different structures, including errors and poor quality, so this original information needs to be analyzed to identify the errors and then fixed.

The development of structured data models containing georeferenced and historical data, such as hydrological, meteorological, and water-related data for the hydrologic administrative regions of Mexico is proposed by the authors to manage the water data in a more efficient manner in Mexico. These integral relational geodatabases, one for each Mexican hydrological region, should compile, integrate, and relate all the geographic information and its corresponding time-series data. They would enable finding in one place all the water resources data, which will provide the required information in the appropriate structure for the simulation models, generating various scenarios. This is particularly relevant for negotiating among the various levels of governments in Mexico for sharing water resources within the country. Another important advantage of using this kind of relational database is the application of an appropriate and useful methodology for the automatic hydrological analysis of very large to huge Mexican watersheds, making possible the use of computational tools to create systematic processes as part of the hydrological modeling in the watersheds.

References

International Law Association (ILA) (1966) The Helsinki rules on the uses of the waters of international rivers. In: Adopted by the International Law Association at the fifty-second conference, held at Helsinki in August 1966
Jenson SK, Domingue JO (1988) Extracting topographic structure from digital elevation data for geographical information system analysis. Photogramm Eng Remote Sens 54:1593–1600
Kite GW (1995) Scaling of input data for macroscale hydrologic modeling. Water Resour Res 31:2769–2781

Lacroix MP, Martz LW, Kite GW, Garbrecht J (2002) Using digital terrain analysis modeling techniques for the parameterization of a hydrologic model. Environ Model Softw 17:127–136

Maidment D (2002) Arc hydro: GIS for water resources. ESRI Press, Redlands, 203 pp

Patino-Gomez C (2005) GIS for large-scale watershed observational data model. PhD. Dissertation. The University of Texas at Austin

Spence C, Dalton A, Kite GW (1995) GIS supports hydrological modeling. GIS World 8(1):62–65

Tribe A (1992) Automated recognition of valley lines, drainage networks from grid digital elevation models: a review, a new method. J Hydrol 139:263–293

United Nations Economic Commission for Europe (UNECE) (1995) Helsinki convention: transboundary watercourses and international lakes. In: Adopted by the UNECE at the Helsinki Convention in July 1995

Chapter 6
Water Use and Consumption: Industrial and Domestic

Maria E. Raynal-Gutierrez

Abstract Mexico's annual average natural volume of water amounts to 446 km^3 (about 0.2% of the total volume of water in the world, which is 200,000 km^3). Mexico receives about 1489 thousand million m^3 a year as rain—67% of this volume falls between June and September. The precipitation distribution is unequal across the territory. The southern states of Chiapas, Oaxaca, Campeche, Quintana Roo, Yucatán, Veracruz, and Tabasco get 49.6% of the total volume of precipitation. Of the total precipitation, it is estimated that 73% is lost to evapotranspiration, 22% becomes runoff to surface-water bodies, and 6% infiltrates groundwater aquifers (CONAGUA, Atlas del Agua en México 2016. Comision Nacional del Agua, Ciudad de Mexico, 2016a).

In the Human Development Report of 2006 published by UNDP, Mexico's estimated per capita daily use of water was 388 liters, placing Mexico in the fifth place for water consumption (UNDP. J Gov Inf 28: 72–91, 2006). In 2015, 92.5% of households were connected to the local water-distribution system (95.7% in cities, 81.6% in rural communities). The water stress in Mexico is considered strong if it is greater than 40%. The average national in 2015 was 19.2% (maximum 138.7%, minimum 1.7%), which is classified as moderate. In Mexico in 2016, reported water-use distribution was as follows: 76.3% agriculture, 14.6% drinking water distribution, 4.8% thermoelectric, and 4.3% for industrial use. Of the water used in Mexico, 61.1% comes from surface-water sources and 38.9% from aquifers (CONAGUA 2016a).

Of the 76.3% of water used for agriculture, 61.1% of the water comes from surface-water sources. The annual surface used for crops production varied during 2008–2012 between 21.8 and 22.1 million hectares (11.25% from a total surface of 196.4 million hectares). The irrigated crop area was 6.5 million hectares (29.41% of total area of crops). On the other hand, of the 18.9% of water used for domestic and industrial use, 58.6% comes from groundwater. The National Water Commission in

M. E. Raynal-Gutierrez (✉)
Department of Civil and Environmental Engineering, Universidad de las Americas Puebla, Cholula, Puebla, Mexico
e-mail: maria.raynal@udlap.mx

© Springer Nature Switzerland AG 2020

J. A. Raynal-Villasenor (ed.), *Water Resources of Mexico*, World Water Resources 6, https://doi.org/10.1007/978-3-030-40686-8_6

103

Mexico reports that the percentage of homes with running water is almost 89% (CONAGUA 2016a).

Keywords Agriculture water supply · Industrial water supply · Surface water · Groundwater · Lakes and dams · Public water supply

6.1 Water Resources

Mexico receives an annual average natural volume of water of 446 km^3 (about 0.2% of the total volume of water in the world, which is 200,000 km^3), which becomes surface or groundwater. Mexico gets about 1489 thousand million m^3 a year as rain, 67% of this volume falling between June and September. The volume of precipitation falls in an uneven pattern across the Mexican territory due to its diversity of climates and its topography. The southern states of Chiapas, Oaxaca, Campeche, Quintana Roo, Yucatán, Veracruz, and Tabasco receive 49.6% of the total volume of precipitation. The main three parts of the water cycle in Mexico comprise the following percentages: 73% lost to evapotranspiration, 22% becomes surface runoff, and 6% groundwater that has infiltrated aquifer (CONAGUA 2016a).

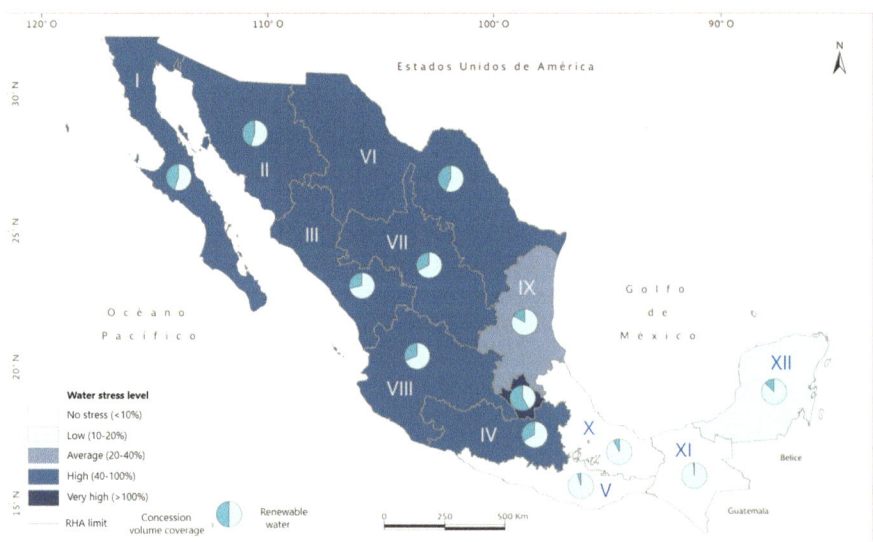

Fig. 6.1 Water stress in Mexico. (CONAGUA 2017)

The various levels of water stress in Mexico are shown in Fig. 6.1. Northern and central Mexico are under high water stress (≥ 40) are indicated as darker areas, whereas southern Mexico has low water stress (≤ 20) as indicated by lighter areas, (CONAGUA 2017).

6.2 Geography

Mexico is in the North Hemisphere and is considered a part of North America (between the west meridians $118°22'00''$ and $86°42'36''$, and between the north latitudes of $14°32'27''$ and $32°43'06''$). Mexico is bound on the north by the US (3152 km), the west and south by the Pacific Ocean (7828 km), to the east by the Gulf of Mexico and the Caribbean Sea (3294 kms—with a total costal line of 11,122 km), and to the southeast by Guatemala (956 km) and Belize (193 km). The total land surface of Mexico is 1,964,375 km^2. The continental territory has an extension of 1,959,248 km^2, while the insular territory has an extension of 5127 km^2 (CONAGUA 2016a).

Because of its size, global location, and topography, Mexico has an array of climatic conditions. The northwest and central part of the country (two-third of the total surface of the country) has a semiarid or arid climate, with an average annual precipitation of less than 500 mm. The southeast of the country lies south of the Tropic of Cancer and has a humid climate, with an average annual precipitation equal to or higher than 2000 mm. Seasonal temperature variations within the tropics are small, often only about 5 °C between the warmest and coolest months. In those areas, winter is defined as the rainy season, rather than the cold season (CONAGUA 2016b).

Mexico's topography is hilly; thus, elevation has a major impact on climate. As can be seen in Figs. 6.2 and 6.3, the elevation in Mexico can vary from 0 to 3000 m above sea level. From sea level to just over 900 m are areas with uniformly high temperatures. For example, Veracruz, located on the Gulf of Mexico, has an average daily temperature of approximately 25 °C. In the range 900–1800 m, the average daily temperature is about 19 °C. In the range 1800–3350 m, the average annual temperature is about 15 °C. Above this elevation, are the alpine pastures and the permanent snow line, which is found at 4000–4270 m in central Mexico. According to the 2010 national census, more than 50% of Mexico's population inhabited regions with an altitude higher than 1500 m above sea level (CONAGUA 2016b). The geographical location of Mexico and its topography have a direct influence on water availability throughout the country (CONAGUA 2016a).

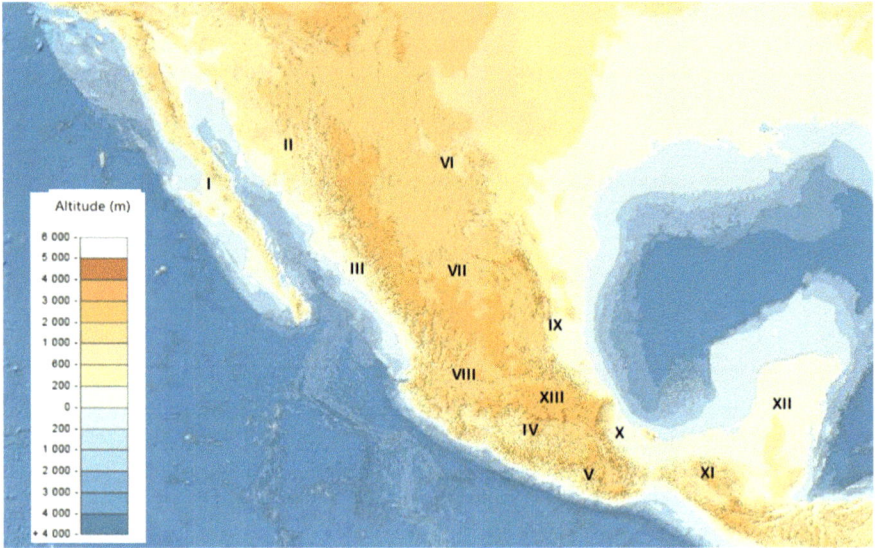

Fig. 6.2 Elevation profile map of Mexico. (INEGI 2016)

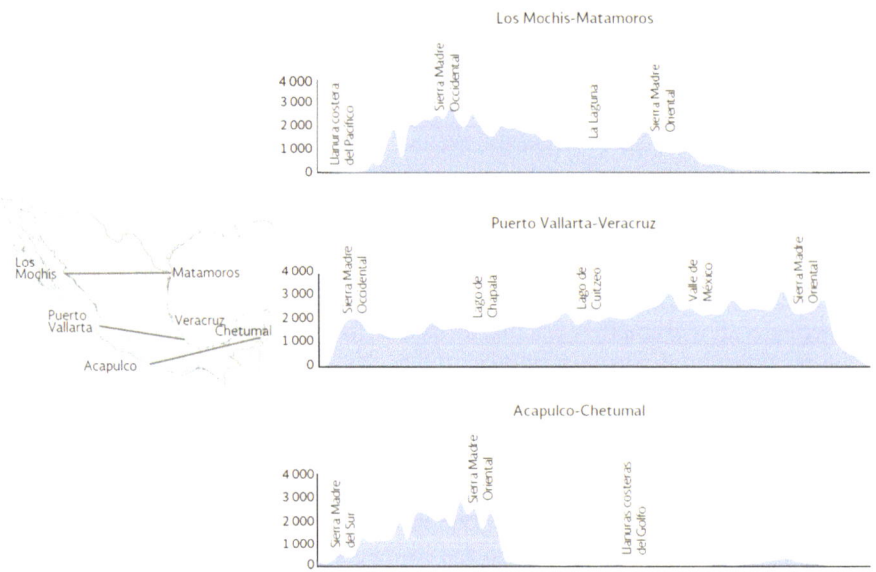

Fig. 6.3 Elevation profiles along different sections in Mexico. (CONAGUA 2016b)

6.3   Surface Water

Mexico has 51 main rivers, with a total length of 633,000 km, which capture 87% of the surface runoff and have watersheds covering 65% of the national territory. The location of the main rivers in Mexico is shown in Fig. 6.4. Rivers Grijalva-Usumacinta, Papaloapan, Coatzacoalcos, Balsas, Panuco, Santiago, and Tonala carry 66% of the total runoff, and their watersheds cover 22% of the area of Mexico (CONAGUA 2016c).

The Lerma River originates west of Mexico City and flows westward to form Lake Chapala, the country's largest natural lake. The Santiago River then flows out from the lake to the northwest, until it reaches the Pacific Ocean. The eastward-flowing waters of the Panuco River originate in the eastern part of the central area of the country and end its path at the Gulf of Mexico. The Balsas River, a major source of hydroelectric power, begins in the southern portion of Central Mexico. Farther southeast, on the Guatemala frontier, the Grijalva-Usumacinta river system drains most of the area of the Chiapas State. Along with the Papaloapan River, which enters the Gulf of Mexico, the Grijalva-Usumacinta carries two-fifths of the total volume of surface water contained in Mexico's rivers (see Fig. 6.4).

There are very few perennial streams and rivers in the north part of Mexico. The main river found along the border with the US is the Bravo River. Most of its tributaries discharge inland, instead of into the sea (see Fig. 6.4).

Fig. 6.4 Main rivers and lakes of Mexico. (INEGI 2016)

6.4 Lakes and Dams

There are seven main lakes in Mexico, all of which are in the central part of the territory. Table 6.1 and Fig. 6.5 present key information on the main natural lakes in Mexico. The total volume of water stored in lakes and lagoons is 14,000 million m^3, about 34% of the annual surface runoff of Mexico (CONAGUA 2016b).

The total volume of water stored in dams in Mexico is estimated to be 125,000 million m^3. There are more than 5000 dams, but only 180 are considered

Table 6.1 Names, location, watershed surface, and storage capacity of the main natural lakes in Mexico

Number	Name	State name	Watershed area (km^2)	Storage capacity (Mm^3)
1	Chapala	Jalisco/ Michoacán	1116	8126
2	Cuitzeo	Michoacán	306	920
3	Pátzcuaro	Michoacán	97	550
4	Yuriria	Guanajuato	80	188
5	Catemaco	Veracruz	75	454
6	Nabor Carrillo	México	10	12
7	Tequesquitengo	Morelos	8	160

CONAGUA (2016b)

Fig. 6.5 Map of main lakes of Mexico. (CONAGUA 2016b)

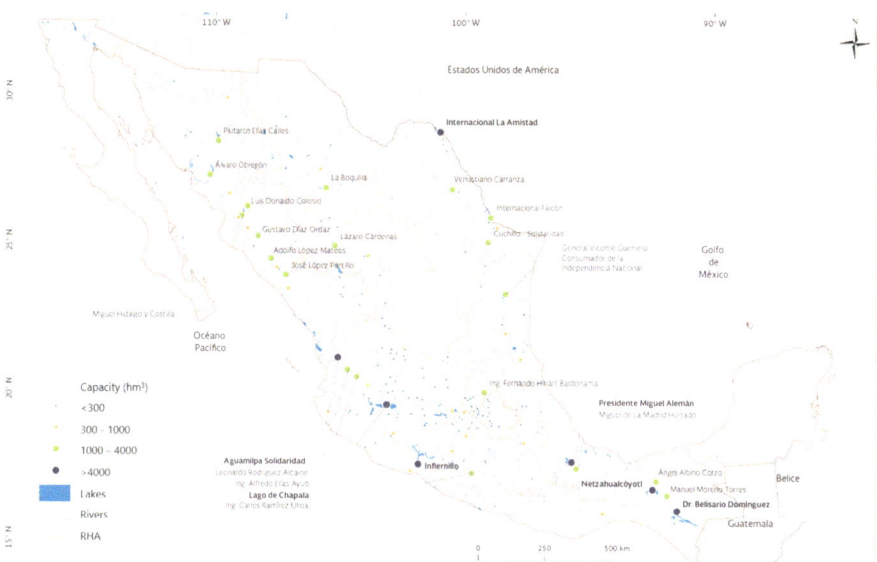

Fig. 6.6 Map of Mexico showing the main lakes and dams. (CONAGUA 2016a)

as main dams, and 80% of the total volume of water is stored in them. Figure 6.6 is a map showing the location of lakes and dams.

6.5   Groundwater

The water stored in the aquifers is an important source of water in many regions of Mexico. In northern Mexico, where the climate is arid, the main water source comes from groundwater. Two-thirds of the total groundwater used in Mexico is pumped form the aquifers in northern Mexico, while the rest is pumped from the aquifers in southern Mexico.

In 2015, 38.9% (33, 331 million m^3) of the water used in Mexico came from groundwater (CONAGUA 2016a). The natural recharge for groundwater in 2015 was estimated to be 91,788 million m^3. Of the 653 aquifers in Mexico, the number of overexploited aquifers in 2015 was determined to be 105 and 205 aquifers had no availability of water at all. In Fig. 6.7, the locations of the 653 aquifers are mapped. The aquifers that are overexploited are marked in red (CONAGUA 2016a, b). Of the 653 aquifers, there are 18 aquifers with saline intrusion and 32 with salinization of soil or saline groundwater (see Fig. 6.8). The latter are located mainly in Baja California and the central northern part of Mexico, where conditions of low precipitation, high solar radiation, and the presence of evaporite rocks converge. Just one aquifer has presented these three characteristics. It is located south of the Baja California peninsula (CONAGUA 2016b).

Fig. 6.7 Map of Mexico with the extension and location of the aquifers. (CONAGUA 2016b)

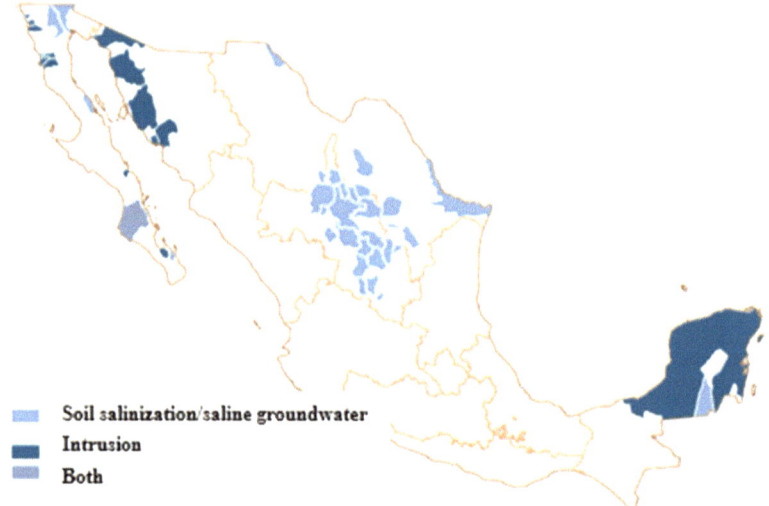

Fig. 6.8 Map of Mexico showing the aquifers that have either or both marine intrusion or soil salinization/saline groundwater. (CONAGUA 2016a)

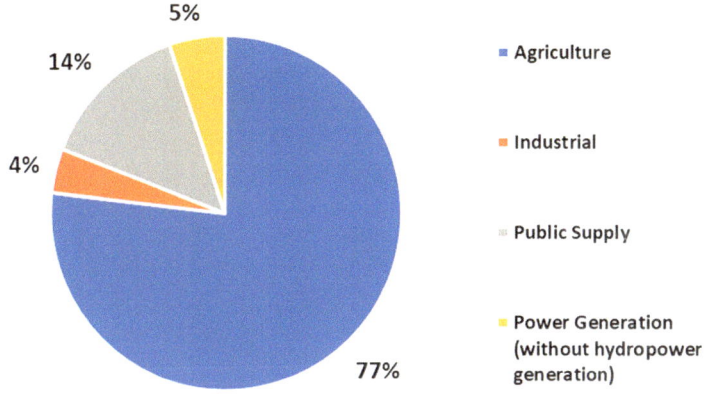

Fig. 6.9 Uses of water in Mexico. (CONAGUA 2017)

6.6   Water Use

In 2006, the Human Development Report published reported that the per capita daily use of water in Mexico was 388 liters, placing Mexico in fifth place for countries consuming the most water (UNDP 2006). It is estimated that 92.5% of the population has access to drinking water supplied by local governments (95.7% in cities, 81.6% rural communities). The water use in Mexico is grouped into two categories. The first category includes all economical activities where water is consumed: agriculture, the public water supply, industry, and power plants based on burning fossil fuels. In Fig. 6.9, the percentages of the uses of water in Mexico are shown (CONAGUA 2017).

In the year 2015, 61.1% of the water used for these economic activities came from surface-water sources and 38.9% from groundwater sources. Of the total volume of water used in the same year, 76.3% was used by agriculture, 14.6% was used for the public water supply, 4.3% by industry, and 4.8% for electricity generation by power plants. The second category, which is called "non-consumptive" uses, includes only the hydropower plants in the country. The water-stress level in Mexico is considered strong if it is greater than 40%. The average national level in 2015 was 19.2% (maximum 138.7%, minimum 1.7%), which is classified as moderate (CONAGUA 2016a).

6.7   Agriculture

Agricultural activities in Mexico include about 30 million hectares of agriculture land, of which 6.5 million hectares are irrigated and 23.5 million hectares use precipitation as the water source. Every year, about 22 million hectares are harvested. Of the total volume of water in Mexico, 76% is used for this economic

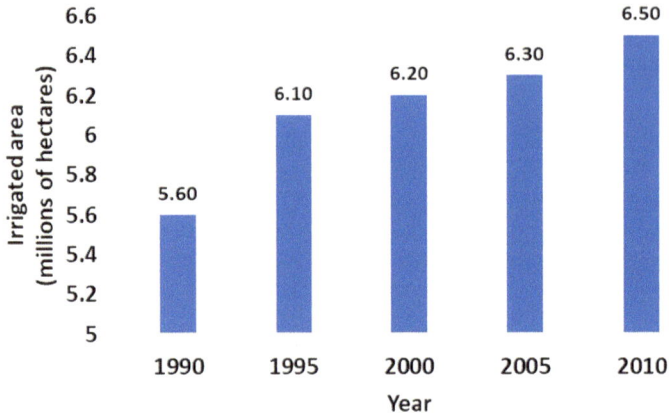

Fig. 6.10 Evolution of irrigated land in Mexico. (CONAGUA 2017; Escudero 1999)

activity. Of this volume, 64.5% comes from surface-water bodies and 35.5% comes from groundwater. About 33% of the water stored in dams is used for irrigation. Mexico ranks in seventh place worldwide in irrigated land surface (CONAGUA 2016b). In 2017, 13.26% of the population was engaged in agriculture, an economic activity that contributes 3.6% to Mexico's GDP (INEGI 2017). The recent evolution of irrigated land in Mexico is shown in Fig. 6.10.

6.8 Public Water Supply

Public water supply is defined in Mexico as the water used for domestic purposes, some industrial activities, and the service sector. The main water source for the public water supply in Mexico is groundwater (58.6%). The 2015 percentage of people with access to clean water connected to the local water distribution system was 92.5% (95.7% in cities, 81.6% rural communities). The recent evolution of water supply coverage in Mexico is shown in Fig. 6.11 (CONAGUA 2017).

There are 874 drinking water treatment plants in Mexico, each with an installed capacity equal to or greater than 1 m^3/s. They have a total installed capacity of 140.74 m^3/s, which treated a total flow rate of 97.90 m^3/s in 2015 nationwide. The recent evolution of drinking water treatment in Mexico is shown in Fig. 6.12 (CONAGUA 2017).

In Mexico, 91.4% of the population is connected either to a sewer collection system or to a septic tank (96.6% in cities, 74.2% rural communities). In 2015, there were 2447 municipal wastewater-treatment plants, which treated 120.9 m^3/s (57% of the total volume of municipal wastewater produced). The BOD_5 produced per year in 2015 was equal to 1.95 million metric tons, of which 0.84 million metric tons per year (56.9%) are removed through treatment. The treated effluent wastewater is then discharged into surface-water bodies (CONAGUA 2016a). The recent evolution of

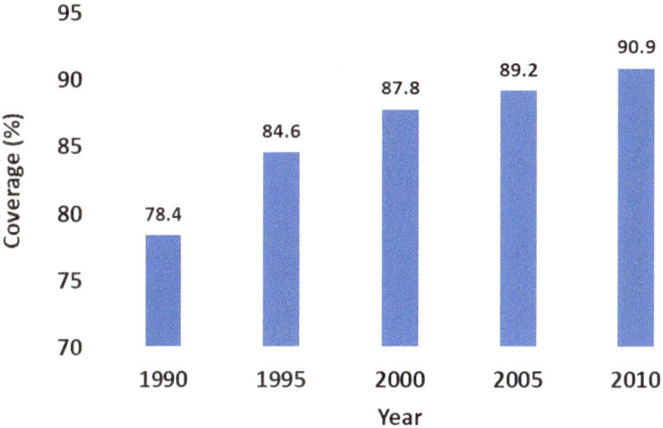

Fig. 6.11 Evolution of public water supply coverage in Mexico. (CONAGUA 2017)

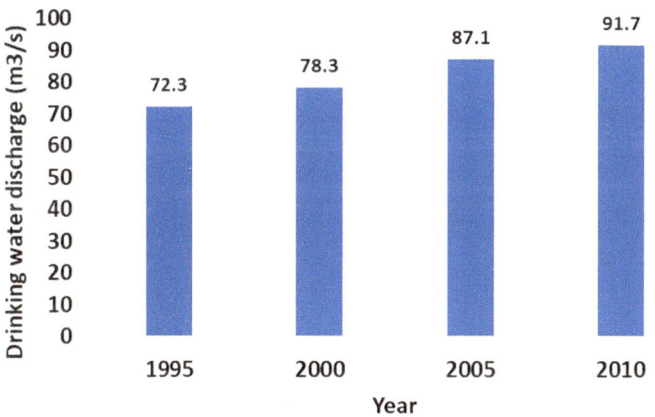

Fig. 6.12 Evolution of drinking water treatment in Mexico. (CONAGUA 2017)

public sewerage coverage in Mexico is shown in Fig. 6.13, and the recent evolution of wastewater treatment is shown in Fig. 6.14 (CONAGUA 2017).

6.9   Industrial Water Supply

The Mexican industrial sector is composed by economic activities such as mining, the generation of electricity, water supply, fuel production and supply, agro-industry, manufacturing industry, construction, commerce, and service sectors. In 2015, the industrial sector used 3676 million m^3 of water (4.3% of the total water consumed). Industry in Mexico should invest in its own water-supply infrastructure. The water sources for industrial activities in 2015 in Mexico were 43.7% surface

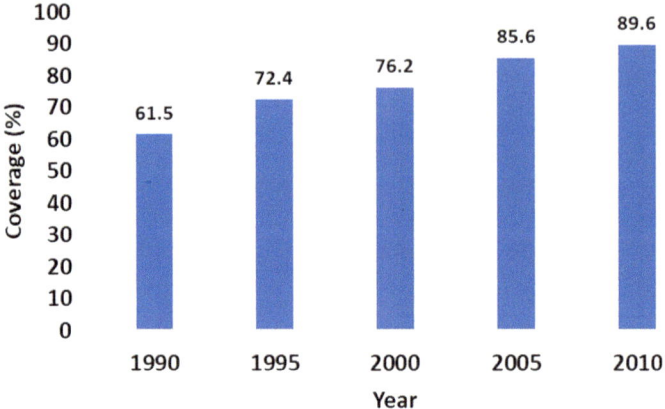

Fig. 6.13 Evolution of public sewerage coverage in Mexico. (CONAGUA 2017)

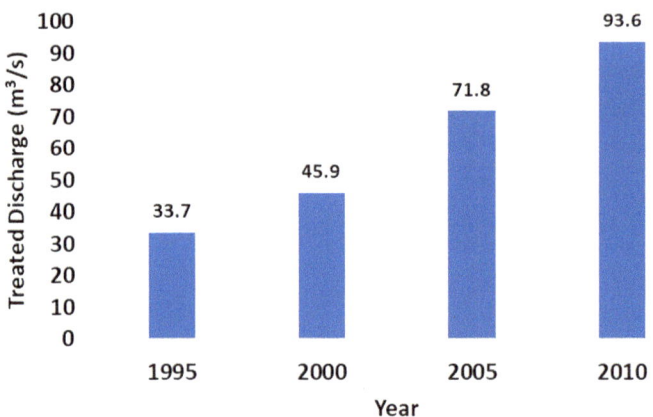

Fig. 6.14 Evolution of wastewater treatment in Mexico. (CONAGUA 2017)

water and 56.3% from groundwater (CONAGUA 2016b). In 2015, the industry sector generated 214.6 m^3/s of wastewater, of which 70.5 m^3 per second were processed in treatment plants (32.85%). The industrial wastewater produced had a BOD_5 load of 10.15 million tons per year, and the treatment facilities removed 1.49 million tons of BOD_5 per year (14.68%) (CONAGUA 2016b). In 2017, 60.5% from the total population is engaged in the industry sector of Mexico. The industrial economic activity contributes 32.4% to Mexico's GDP (INEGI 2017).

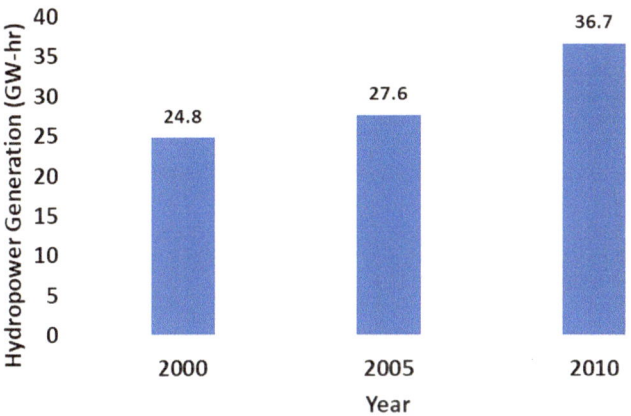

Fig. 6.15 Evolution of hydropower generation in Mexico. (SENER 2013)

6.10   Electricity Production

Of the total amount of electricity generated in 2015 in Mexico, 88.5% (231 tera-watts-hours) were produced by fossil-fuel power plants and 11.5% (30.1 terawatts-hours) by hydroelectric power plants. Of the total volume of water available in Mexico, 5% was used for electricity production in power plants. Of this portion, 80% of the water resources used came from aquifers and 20% from surface water bodies. On the other hand, 34 dams in Mexico are used to produce electricity; thus, of the total volume stored in dams, 37% (138, 662 million m^3) is used for electricity production. The largest hydroelectric power plants in Mexico are in the states of Guerrero and Chiapas (CONAGUA 2016a, b). In 2017, about 0.79% of the total population was employed by this economic sector (INEGI 2017). In Fig. 6.15, the recent evolution of hydropower generation is shown.

References

CONAGUA (2016a) Atlas del Agua en México 2016. Comision Nacional del Agua, Ciudad de Mexico
CONAGUA (2016b) Estadisticas del Agua en Mexico, edición 2016. Comision Nacional del Agua, Ciudad de Mexico
CONAGUA (2016c) Situacion de los recursos hidricos. Comision Nacional del Agua, Ciudad de Mexico
CONAGUA (2017) Atlas del Agua. http://www.conagua.gob.mx/atlas/
Escudero, G. (1999). Balance del estado general y la evolucion de la agricultura y el Medio Rural de America: retos y oportunidades en el Siglo XXI. San José: IICA- Costa Rica, Inter-American Institute for Cooperation on Agriculture.

INEGI (2016) Mapa digital de Mexico. Retrieved from http://gaia.inegi.org.mx/mdm6/?
       v=bGF0OjIzLjMyMDA4LGxvbjotMTAxLjUwMDAwLHo6MSxsOmMxMTFzZXJ2aWNpb
       3N8dGMxMTFzZXJ2aWNpb3M=
INEGI (2017) Población ocupada según sector de actividad económica, nacional trimestral. http://
       www.inegi.org.mx/sistemas/BIE/CuadrosEstadisticos/GeneraCuadro.aspx?s=est&nc=597&
       c=25586. Accessed 23 Nov 2017
SENER (2013) Prospectiva del Sector Electrico (2013–2027). https://www.gob.mx/cms/uploads/
       attachment/file/62949/Prospectiva_del_Sector_El_ctrico_2013-2027.pdf. Accessed 23 Nov
       2017
UNDP (2006) Human development report 2006 – beyond scarcity: power, poverty and the global
       water crisis. Journal of Government Information 28:31–66

Chapter 7
Water Resources in Mexico: Some Proposals for the Future

Humberto Marengo Mogollón

Abstract Water management in Mexico presents a problem that has significantly worsened over the years because population and economic growth have more severely impacted water reserves. The result is that in some regions of the country today the total volume demanded is greater than that supplied, which forces decisions on who should receive or not receive this vital resource. Since this causes serious distributive problems without raising substantive solutions, it is therefore considered to be a priority to discuss the problems and make proposals that could resolve the root cause of the problem in Mexico. In this chapter, the analysis is based on three fundamental axes: (1) the fundamental right of the population to have access to water, (2) the need to have fair and equitable access to both drinking water and sanitation services, and (3) the resilience/sustainability and management of hydro-meteorological risks when they affect the population. The aim of this chapter is to comment on the problem that has been identified from the perspective of this author and note some of the technical aspects, while also including some political and social aspects. A general proposal is then made regarding some action lines that can be followed in order to reach a solution within the near future.

Keywords Water management · Mexico · Scarcity · Pollution · Ecological ordering · Climate change · Desalination · Virtual water · Aquifer recharge

7.1 Introduction

Mexico has a territorial coverage of 1964 million km^2 with a population numbering 123 million (INEGI 2015), which places it 11th in the world in terms of population—with a growth rate of 1.4%. Of this population, 75% lives in urban areas, and the population projection up to 2030 does not indicate a significant change in the current trend.

H. M. Mogollón (✉)
Instituto de Ingeniería, Universidad Nacional Autónoma de Mexico, Ciudad de Mexico, Mexico

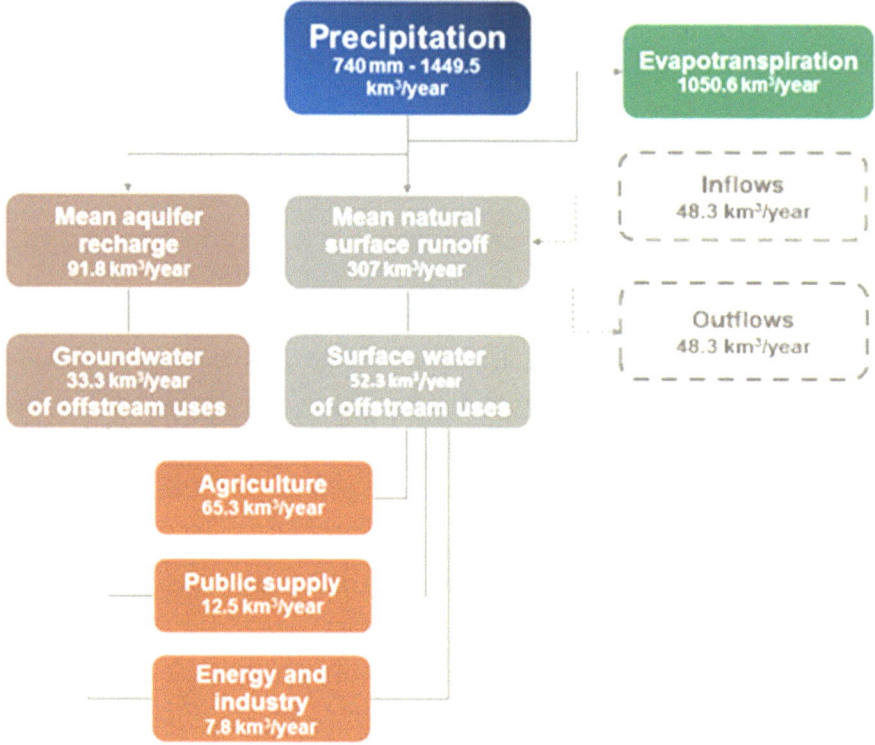

Fig. 7.1 Water cycle at the national level. Proprietary development, with data from CONAGUA (2017), Arreguin-Cortes and Cervantes-Jaimes (2018)

Within the national territory, the average rainfall is 775 mm (Fig. 7.1), which is equivalent to 1513 km^3. Of this amount, 1084 km^3 undergoes evapotranspiration (ET), the average surface runoff is 400 km^3, and 47 km^3 is actually utilized (which is just a little over 10%) (Arreguin et al., 2012). It is noteworthy that Mexico receives 50 km^3 of water from the US and Guatemala, which flows into the country via rivers with headwaters in those neighboring countries, and exports 0.44 km^3 of water to the US according the 1944 US–Mexico Water Treaty. On the other hand, aquifers receive a recharge of 78 km^3, and extraction equals 28 km^3.

Water storage in dams is on the order of 150 km^3, of which 113 km^3 is used in hydropower energy, constituting an exploitation rate of approximately 40%. However, 77% of this resource occurs in the southeast of the country, while poorly distributed and, in many cases, poorly managed resources exist in the rest of the territory.

Concentrations of population and economic activity in Mexico have created areas of great water scarcity, not only in regions with low rainfall but also in areas where this aspect was not perceived as a problem when urban growth starts or at the establishment of agriculture using irrigation. The competition for this resource is already causing conflicts of various levels of intensity and scale, and is present not only among users in the same community but between various communities, municipalities and states, and even in cross-border areas.

In this context, there are at least three reasons why knowledge and analysis of conflicts related to water can be critical factors in decision making with respect to environmental policy in Mexico:

1. In an attempt regulate water usage and to avoid conflicts, the institutional framework has been changed, but the changes have not brought substantive solutions and the framework has only been distorted by a complex and unclear process. This has been evolving without achieving a reform that is in keeping with the level of the problem. The design of prevention mechanisms and, where appropriate, mediation and conflict resolution would require a more in-depth knowledge of the way in which those conflicts arise and develop.
2. Some conflicts emerge as movements that reject any public decisions.
3. General conflicts are associated with a set of causes that vary according to geographical region or sector, and in some cases, the causes are repeated while in other cases there is simply a refusal to accept the causes.

In some areas, the precursor of a movement may be poor administrative management combined with the mobilization of organized social groups, while in other areas the recurrent drought is the focus of interest. Thus, it is useful to systematize conflicts by using a conflict typology that classifies relevant variables for each sector (urban or rural) and region (divided into states or basins). Systematization of existing conflicts and variables associated with each case constitute a preliminary public-policy agenda and an initial diagnosis of the conflict that must be promptly faced.

7.2 Access to Water in Mexico

Recently, in a forum convened at Universidad Nacional Autonoma de Mexico (UNAM), it was established that 71% of the available water bodies in the country have some degree of contamination, and 19% of the aquifers are overexploited because of human activities (González 2018).

Of the total amount of water with concessions for consumptive uses, 76% of the regulated water volume occupies first place among water concessions and is destined for irrigation of crops. Agricultural production units cover more than 100 million hectares, and, of this amount, 28 million hectares correspond to agricultural land (20% are irrigated and 80% are seasonal), producing about 280 million tons of food per year, part of which is exported to Canada and the US. However, the use of water in this activity is extremely inefficient. This is due to evaporation, infiltration, and improper practices, and between 40 and 50% of the water being delivered is lost. Likewise, around 500,000 hectares are deforested annually for agricultural production, which has severe effects on the hydrological cycle such as the reduction of recharging of aquifers and modifying runoff.

As a first reflection point, it is noteworthy that irrigation technology must be improved and then it is necessary to redistribute water in Mexico, which implies that there is a great problem to be solved. The next greatest consumptive use, in terms of concession volume, is the public supply, which is about 15% of the total water. Of this, approximately 40% of the water in the net lost due to leaks, while only 40% of the wastewater receives treatment. In fact, more than half of the wastewater spills directly into rivers and streams with no treatment.

In 2013, hydroelectric power plants used a volume of 112.8 km^3 of water to generate 10% of Mexico's electrical energy. Construction of dams for energy supply and other purposes is a controversial issue from environmental, social, and economic perspectives.

Secondly, and notwithstanding, dams offer storage of water bodies that enable water availability in the short term and will reduce the effects of droughts; they will also help in fighting the effects of climate change. This situation acquires great importance in the analysis and research related to public policies that lead to the topic of water security throughout the country in the next years.

Moreover, industry generates greater pressure on water resources due to the impacts caused by wastewater discharges and their polluting potential than solely by the quantity of water used in production. The main pollutants in industrial wastewater are heavy metals and organic matter, which are often discharged untreated into streams. On the other hand, it is common to find that garbage is dumped into the rivers and water bodies in both rural and urban areas; at beginning of the rainy season, this can exacerbate the pollution of rivers and streams throughout the country. This can further lead to blocking culverts and rivers and subsequent contamination.

The water-assignation system (in my opinion) must undergo radical change to make it flexible and to accord with water availability. A priority should be given to the human right to water over industrial and mining use—without forgetting what is required to satisfy their needs—and another priority should be given to enforcing the laws on pollutant discharges. Irrigation of agriculture must also become a more efficient use of the water it consumes.

Another fundamental point (by its own importance) is the study of the basin as a planning unit. In this regard, it is necessary to understand the functioning of ecosystems, to maintain hydraulic balances, to implement territorial ordering, and to establish realistic models of water use for drought years when runoff becomes scarce.

Social participation within localities should be encouraged be providing reliable, continuous information and accessible data. It is also necessary to modify regulations so that the human right to water is incorporated, but services offered should be remunerated to sustain systems that are efficiently managed and controlled. Water should be considered for what it really is – a valuable resource that functions as part of nature itself with mechanisms that promote environmental health. However, we must never forget the cost this incurs.

Academic research into the hydraulic sector is required to generate debate, form research groups, and train human resources, as well as to propose the selection and linking of projects to be promoted (Ruiz 2009).

It is urgent to define improvements in the water use asignation of existing mechanism, and that of change in the soil use, with the scope of aviod political advantages. From this point of view, the required studies and analysis most be established for recognnizing: (1) territorial ordering, (2) the need to have storage water (reservoirs), and then establish and promote the purpose of public-utility service that the water provide in cases when phenomena such drougths or floods ocurr, and (3) demonstrate the benefit that they can produce regarding the suitable exploitation of the requested water resource.

In summary, it is urgent maintain the structure and functioning of the basins to continue dependence on services that are provided by the hydrological cycle. On the other hand, is necessary to have both; surface and underground water storage to meet societal demands in such a way that the suply of water is guaranted wtih enough quantity and quality, making the assignation with a timely basis for its use such that the economic resources must be allocated in a timely basis.

7.3   Equal Access to Drinking Water and Sanitation Services

Mexico has made significant progress in terms of its extent of the areas benefiting from drinking water and sanitation services. According to the National Water Commission (CONAGUA 2014), at the end of 2013, coverage of the territory had been achieved for 92.4% for potable water and 91.0% for sewage. However, these global figures do not show the disparity that exists between various regions and states, between urban and rural areas, or between high-income areas and areas with higher rates of marginalization. These disparities are related not only to coverage, but also to the levels of quality, sustainability, and efficiency in provision of drinking water services, sewerage, and treatment and disposal of wastewater. The latter are activities which, according to the 1983 Reform of Article 115 of the Political Constitution of the United Mexican States, are assigned to the municipalities (González 2018).

There are more than 192,000 localities in Mexico which manage water using a variety of models (Arreguín et al., 2012) ranging from private participation schemes to decentralized public bodies, entities attached to state or municipal governments, and various forms of community organization. While each one of these modalities experiences particular challenges, however, the great majority are characterized by (1) their organizational inefficiency, technical and commercial management; (2) lack of qualified technical personnel in operation and administration; (3) structures and tariff levels that do not reflect the costs of the services; (4) politicization of decision-making; (5) poor transparency and accountability; and (6) low levels of citizen participation in the formulation and execution of hydraulic projects (González 2018). The final decisions can fall on individuals who often do not have the minimum required knowledge of hydraulic phenomena and their needs, but who are politically very conflicted.

The challenges of the drinking water and sanitation subsector are evident in that 86% of Mexicans have no drinking water service on a 24 hour-per-day basis and 48% do not receive water daily (González 2018). Because of these deficiencies, more than 75% of the population must purchase bottled water at a cost of around 150 pesos (just under US$7.5 per month (at December 2017 prices) This represents more than 40% of the individual fee for providing the service, and delivers an amount that is approximately 200 times less than the volume of water supplied by the operating agency.

In addition, the water is supplied by tanker supply or reticulation systems that are sometimes linked to interest groups that act against the improvement of those systems. Discussion and analysis of these problems must be contemplated and implemented with the goal of defining and achieving structural and nonstructural actions that must be undertaken to ensure an effective human right to water and sanitation.

It is necessary to discuss and analyze how the provision system of potable water can be transformed, given that it is now characterized by inequality and high consumption of bottled water and water from the pipes (tanker) system. Finally, it is necessary to define the technological and financing options that exist in order improve the drinking water subsector and turn wastewater into a resource that contributes to the development of the drinking water subsector.

7.4 Resilience and Hydro-meteorological Risk Management

Disasters are a determining factor in the persistence of poverty and effects of hydro-meteorological phenomena, especially droughts and floods, which represent an important challenge for Mexican society. In this context, climate change poses a greater intensity and frequency of extreme hydro-meteorological events, in addition to the rise in sea level, the impact on public health by vectors and the impact on agricultural production. In this regard, it is necessary to develop techniques and analysis criteria that incorporate the possible variations beyond the limits that are commonly considered. The following actions have been identified as essential: (1) ensure that disaster-risk reduction is a national and local priority, with a solid institutional base for its implementation; (2) identify, evaluate, and closely monitor disaster risk and improve warning systems; (3) use knowledge, innovation, and education to create a culture of security and resilience at all levels; (4) reduce the fundamental risk factors; and (5) strengthen disaster preparedness to achieve an effective response.

The Central National Disaster Prevention Center in Mexico reported that, during the last 13 years, the impact of disasters in Mexico has had an average annual cost of US$2 billion (CENAPRED 2017a). Beyond recognizing human vulnerability to the danger of hydro-meteorological phenomena, reference is made to resilience as the capacity of a society or a system to recover from adverse situations. In this sense, there is a tendency to recognize the need to go beyond the reactive vision of addressing risks and prevention as a basic principle of risk management. This vision seeks to establish a new national strategy in Mexico regarding water-risk management that includes reviewing public policies and the allocation of resources (CONAGUA 2012b). It has long been recognized that risk management "requires an integral treatment of possible solutions to prevent disasters and to protect the population, based on an analysis of the factors that determine the decision to settle in vulnerable areas" (CENAPRED 2017b). Instruments such as territorial ordering and law compliance have not been implemented due to the lack of governability.

7.5   Reflection upon Resilience

Among the issues of water in Mexico, the exposure of the population to droughts and extreme events must be considered from the perspective of water security, which was defined by Grey and Sadoff (2007a) as "the availability of an acceptable quantity and quality of water for health, livelihoods, ecosystems and production, coupled with an acceptable level of water-related risks to people, environments and economies."

According to Martínez-Austria (2007), it is helpful to complement this definition with criteria that includes the concepts of equity and sustainability and introduce the effects of climate change. So, Martinez-Austria (2007) states that water security can be dfined as a condition in which "Ensures sustainable water supply for all uses, including conditions of equality and at affordable prices, promoting health, economic development, production of food, and energy and promoting the environment conservation." This will provide protection, with an accepatble risk to the population and their productive systems against effects of hydrometeorological events mitigating their effects, and providing measures of adaptation to climate change.

In this sense, in many countries water security has not been achieved, but rather it has become increasingly threatened (World Economic Forum 2011a, b). Evidently, Mexico is no exception. Population growth, economic development, urbanization, climatic variability resulting from global climate change and environmental degradation itself continue to increase the pressure on water resources in such a way that permanent or recurrent conditions of scarcity are already registered in some regions of Mexico. Furthermore, inadequate water management can often exacerbate this problem.

At the Fourth World Water Forum of 2006, it was enunciated that "*There is a general consensus that water development is one of the bases of economic and social growth and development*" (Martínez-Austria and van Hofwegen 2006). The purpose of water management is precisely to ensure the production of well-being and the protection of people and their goods from extreme manifestations of climate, such as droughts and floods, while preserving the environment. Water management plays a fundamental role in human health and well-being in the fight against poverty and economic development. Therefore, at the present time, no nation can minimize the importance of achieving good water management. Achieving and maintaining water security has been one of the historic goals of each civilization in various phases (WRG 2009). However, currently, new challenges are being added to the historic challenges of water security.

Pressure on water resources is rapidly increasing in various parts of the world due to (1) population growth; (2) increased demand for energy and food; (3) changes in human diet related to economic development; and (4) spread of rapid urbanization, together with pollution of water sources. Moreover, climate change poses significant challenges for water management both now and in the future. These challenges are of such magnitude that in a document that was presented at the World Economic Forum (2010–2011) it was stated that *"We simply cannot manage water in the future as we have done so far, or the economic network will collapse."*

For many countries, water security has become a matter of national security, and there is also concern at the international level due to the numerous transboundary basins that exist and cover 43% of the Earth's land surface, an area which is inhabited by 40% of the world population.

To guarantee water security is one of the main challenges in many regions of the world, especially in countries located in arid and semiarid zones and which are confronted by recurrent or permanent scarcity of water resources, as well as in some regions that are subject to tropical meteorological phenomena and are prone to flooding. Grey and Sadoff (2007a, b) distinguish regions with a *"legacy of difficult hydrology"* as those in which there is an absolute scarcity (e.g., arid zones) and, at the other extreme, those located in low-lying lands with a continuous risk of flooding. The most difficult combination is where there are large hydro resources with inter-annual variations over the course of each year.

Unfortunately, in most of the Mexican territory, there is a *"legacy of difficult hydrology."* On the one hand, 70% of the territory is arid or semiarid with large seasonal and interannual variations of precipitation, and, on the other hand, the remaining 30% of the territory experiences the risk of recurrent floods caused by both tropical meteorological systems and cold fronts.

7.6   Main Challenges of Water Security

The achievement of water security faces its main challenges manifested in the scarcity of water, the contamination of water bodies, the adverse effects of extreme hydro-meteorological phenomena (i.e., floods and droughts), and the growing conflicts over water and the environmental deterioration of watersheds and aquifers. The main factors that induce or increase these risks for water security are the demographic processes, the growing demand for food due to both demographic growth and changes in diet, the demand for water for energy production, and the effects of climate change and poor water management (Montero et al., 2010).

7.7 Population Growth and Urbanization

By 2050, the world population will reach 9300–10,600 million people (UNFPA 2011). The United Nations Population Division envisages that the urbanization process, which has stabilized in developed countries, will continue in developing countries. Thus, the total urban population will increase by 72% from 3.6 billion people in 2011 to 6.3 billion people in 2050. The entire new urban population will be practically concentrated in the cities of the least developed countries. The rural population, on the other hand, will decrease between 2011 and 2050. These urbanization processes will pose enormous regional challenges for water management because, with very few exceptions, nature itself does not provide water resources that are needed to supply water to such human concentrations at the local level. This statement of the situation fact does not include the difficulties of treating and disposing of resulting wastewater, as well as the by-products of treatment, such as residual sludge.

In the 1950s in Mexico, the urbanization process began at an accelerated pace, which will continue up to the year 2050. According to these forecasts, by 2030, the country will reach an urban population of about 112 million inhabitants (82.6% of the total number of inhabitants) and in 2050 it will reach almost 124 million (86% of total number of inhabitants) (Rivas Acosta et al., 2010). The urban population of Mexico will grow by 35.7 million inhabitants in 2050, as compared to 2010, which represents a population greater than the sum of the current metropolitan areas of Mexico City, Guadalajara, Monterrey, and Puebla, which are the four largest urban areas in the country. The associated challenges of water supply and sanitation will be enormous and will require a highly efficient management of urban water, together with measures for environmental conservation that are necessary to preserve the sources of supply. Likewise, substantial investments in infrastructure, institutional development, and human-resource training will be required, especially considering that, according to 2010 data population census in Mexico, the drinking water coverage and sewage system among the population were 90.9% and 89.6%, respectively (CONAGUA 2012a).

Urban concentrations in megacities pose special problems. According to estimations by the United Nations Organization, cities with a current population of more than 10 million will reach more than a billion people, in contrast to the 148 million inhabitants who resided in those cities in 1970.

In Mexico, by the year 2030, about 50% of the population will live in just 31 cities of more than 500,000 inhabitants, with high concentrations in the Mexican mega-cities of Mexico City, Guadalajara, Monterrey, and Puebla (Villagómez and Bistrain 2008). Many of these cities, as in other parts of the world, are located in regions where almost all water resources are already fully used or even overexploited. The metropolitan areas of central and north-central zones, for example, depend on aquifers for their supply, and these are already being overexploited.

Without considering the effects of global climate change, by 2030, some of the main basins in Mexico will register conditions of big water stress. The condition of water scarcity is determined by means of per capita availability. In this way, when availability is less than 1700 m^3/inhabitant/year, it is considered that there is a shortage; when availability is less than 1000 m^3/inhabitant/year, it is considered that there is extreme scarcity; and when availability is less than 500 m^3/inhabitant/year, it is considered that there is absolute scarcity.

According to this classification, around 2030, in the Valley of Mexico, the condition of absolute scarcity that the area already suffers will be aggravated. The regions of the Rio Grande (northern Mexico) and the Baja California peninsula (northwest of Mexico) will suffer conditions of extreme scarcity, and the Lerma-Chapala (in central Mexico) will suffer conditions of scarcity (less than 1700 m^3/inhabitant/year). The Hydrological-Administrative Regions of the North Central Basins and Catchment will be close to conditions of scarcity, which would likely be reached as result of climate change.

7.8   Conflicts over Water in Mexico

Because analysis of water-related conflict has been the subject of a multiple disciplinary approach, numerous methods, definitions, and approaches are already available. Given the complexity, comprehensive knowledge requires multidisciplinary analysis involving the fields of psychology, philosophy, sociology, political science, anthropology, and law (Deutsch and Coleman 2000).

However, it is noteworthy that a model can simplify the reality so that a specific behavior or phenomenon can be explained by a reduced set of variables. Therefore, although data have been collected on economics, institutional politics and biophysical variables, there are studies from various disciplines yet to be used in the choice of which of these variables should be evaluated. So, the data processing is undertaken using statistical procedures that will determine which of the original variables explain hypotheses or not. If variables show significance, an explanation is required that would exceed the limits of the model, and it would be necessary for further intervention using a multidisciplinary analysis.

Another process that affects conflict and which is vital for determining the nature of it is the presence of violence (Oberg 1996). Here, it is important to ask which factors lead to violence and what kind of interventions could reduce its probability. The factors that determine the presence of violence can be of various types:

(a) After all possible resources have been exhausted through institutional means and it is necessary to resort to violence in order satisfy certain vital needs of a community.
(b) When there are frustrating situations that cause discontent among the inhabitants.
(c) When agreements or treaties that generate displeasure for the parties involved are broken, which, in the long run, can generate violent acts.
(d) When there are abuses of power that affect the community.
(e) When there are abuses of some natural resource such as water – when inhabitants of the upper part of the basin area use the resource to a greater extent, ignoring those inhabitants in the lower basin who also depend on the same resource.

There are occasions when disputes do not go through established institutional processes and conflict manifests as an open confrontation between the parties. These local disputes over water can be due to several factors: (1) lack of adequate hydraulic policies, (2) lack of governance (Rogers and Hall 2003), (3) market effects that encourage a failure to take care of the water resource, and (4) the absence of property rights.

Some authors stress the importance of conflict as an opportunity to implement institutional mechanisms that allow users and authorities to achieve agreements (Gleick 1993).

Although literature refers mainly to international conflicts, part of the analysis can be extrapolated to internal conflicts. According to Petzold-Bradley et al. (2001), for example, cross-border issues (degradation, scarcity, or poor distribution of resources shared by two or more countries) can create conflict and regional instability, but they can also be an opportunity for the prevention and mitigation of conflicts through mechanisms of cooperation and negotiation.

Comparative studies have shown that environmental stress (either environmental degradation or scarcity of resources) can, under certain political, economic, and social conditions, contribute to or accelerate the presence of serious conflicts in developing countries (Petzold-Bradley et al., 2001). Hence, the study of conflict is important for the prevention of conflict itself.

Therefore, analyses that make possible the identification of possible conflict scenarios (OECD 2011) and which can provide advance signals are very important in seeking to prevent conflict.

It is important to study variables that intervene in the generation and development of a conflict. A review of the body of literature on conflict in general and environmental conflict specifically indicates a large number of factors of interest: (1) water scarcity, which according to Gleick (1993) is a function of the demand and supply of the resource, which in turn is determined by climatic variables, infrastructure, the situation regarding ground and surface water, among others; and (2) political variables (the motivations of the parties involved, their needs, their capacity for organization, economic and political resources).

In Mexico, market mechanisms, prices, and tariffs have been used in a limited way by government agencies to regulate demand in recent years. Rather, strategies have been preferred to make large investments to expand supply (even with high environmental costs) and, when chronic or acute shortages persist, some rationing mechanism is implemented.

One of the reasons that the price system has not been the main instrument to regulate demand or to redistribute access to the resource is its potential to unleash conflicts. There is strong political pressure to continue providing subsidies and to avoid new tariffs.

Obviously, it is expected that any movement in this direction will generate resistance from the existing beneficiary groups. On the other hand, when property rights over the resource or its use are not well defined, conflict is one of the mechanisms that interest groups use to define property rights in their favor. As the problem of shortage grows, it will be more important to face resistance (both in the agricultural sector and in the urban sector) than to have the price of water reflect the existence of competitive uses.

There are institutional mechanisms, such as the Basin Councils (Consejos de Cuenca), to alleviate the problems. This model began to be used in Mexico at the end of the 1990s as a space for users and various levels of government to discuss the use and distribution of water in their respective basins. Currently, its function is to provide a means of participation for the diverse actors and to generate recommendations. A condition of the policy proposals is that they must be effective; otherwise, they lose their function to solve a social problem. To meet this objective, the analysis of conflict, when it advocates the identification of reactions to certain public actions, can be part of a feasibility analysis. The knowledge of the actors that intervene in conflicts can become the support for public-policy proposals. Effective implementation often depends on the anticipation of who will oppose, and with what means, public action.

Knowledge of power relations is important for understanding dynamics of conflict (Fisher 2000) and understanding why one actor would have an advantage over another. Sometimes, disadvantaged positions are generated by a lack of enough resources, both political and economic, which impedes reaching a satisfactory agreement. From this perspective, it is important to have knowledge of the interests, resources, and motivations of the actors to understand how power relations occur and anticipate possible strategies to follow in striving for a status quo that proves to be satisfactory. The motivations of actors need to be specified, and an inventory of their material and political resources must be made, which would require more information, but the way in which the groups have mobilized would provide the initial data for this.

The policy proposals related to subsidy reduction or price generation are very unpopular, but the conflict they generate is not the same in each region since the degree of tension depends on the negotiation process, the degree of organization, and the resources that those who are in opposition may have, among other variables.

7.9   Practical Proposals to Deal with Each Type of Obstacle

The specific proposals of the present study for solving the water problem in Mexico are the following:

(a) It is critically necessary to review the allocation of water for the various users within the country. To achieve this, irrigation methodologies must be updated, users must be trained with respect to various crops, and a thorough review of the efficiency of irrigation districts must be undertaken, including changing cultivation plans to convince the various stakeholders of the importance of the necessary changes. The process will optimize and take advantage of the allocation of the volumes of water that have been assigned to date.

(b) The main causes of deforestation in the country must be reviewed. Deforestation does not necessarily occur because of the change in land use due to the construction of infrastructure. Rather, deforestation occurs on innumerable occasions due to trespassing, lack of environmental care, and lack of moisture in the soil, which will cause forest fires and transcendental changes in reforestation. If changes in land use are assigned for infrastructure or mining concessions, the construction of forest nurseries and the replacement of land and habitats must be intensively promoted.

(c) Use of water for human beings must naturally be the top priority, although it is necessary to review the state of drinking water and sewerage networks, replacing old networks and significantly improving their efficiency. Disasters due to water scarcity and urban flooding should be given the highest possible priority.

(d) Sanitation of water used in the cities must also be one of the main aspects, and cities must be assigned the necessary resources so that the water is not subject to contamination when transferred from one basin to another. The same principle should apply for the sanitation of aquifers.

(e) Emphatically, it is proposed that river basins must be the planning unit and that water transfer must be established when necessary, in exchange for sanitizing and compensating for the basin water that is used.

(f) Storage of water, whether in aquifers or in dams, must have utmost priority because that is where the much-needed water reserves that society will require in the immediate short term will be obtained. It is proposed to raise the height of the dams (if the useful height is increased by 5%, then water storage can be increased by 20%) to increase the necessary transfers, as is already the case in several countries. This aspect, with the reduced costs of intermittent energies (i.e., US $20–30/MWh), will make possible the planning of the water transfer from the river mouths in areas where water is in abundance to arid zones and to cities in areas where the water supply is limited.

(g) Hydraulic research should be significantly strengthened by considering the technical, political, social, and environmental aspects of the various basins and their respective conditions to consider a more equitable and sustainable use of our hydraulic resource. This must begin immediately.

(h) There is an urgent need to discuss a clear and sustainable water-security agenda in Mexico by reviewing the various aspects of needs and uses of water where needed. It is necessary to undertake careful planning before floods occur due to extreme events by defining a detailed territorial ordering system, and it will be necessary to review cases of extreme scarcity of water resources that exists in various zones of the country.

(i) It is most urgent to make a detailed urban growth plan, to define the injection zones and absorption wells in the cities, and to obviously monitor and adjust accordingly the water quality that is delivered by the distribution networks.

(j) Finally, I propose the creation of a national hydraulic plan that is not subject to any political agenda and that is defined as a task to be fulfilled in an absolute manner for the welfare of the society once that plan has been analyzed, reviewed, discussed, and agreed upon by the various sectors of the society.

7.10 Conclusions

Sustainability of water management in Mexico implies that current consumption must be made at a rate that makes possible sufficient volumes and adequate quality of the resource for future generations. Such a pattern of consumption is possible only through a substantial change in the way in which demand is regulated, but the measures necessary to reform the current management policy are confronted by institutional and political barriers. Obstacles can be overcome using other means, with detailed analyses of the water situation in Mexico in its most critical dimensions, such as economic, institutional, and political spheres, for example. Universities and higher-education institutions should play a leading role in defining the water-related needs that Mexico presents in order resolve the inherent problems in a harmonious manner.

References

Arreguin F, Alcocer-Yamanaka VH, Marengo M H (2012) Los Retos del agua en México, Revista de Ingeniería Hidráulica

Arreguin-Cortes FI, Cervantes-Jaimes CE (2018) Water security and sustainability in Mexico. In: Raynal-Villaseñor JA (ed) Water resources of Mexico. Springer International Publishing A G, Cham

Arreguin-Cortés FI, Pérez ML, Mogollón HM (2012) Mexico's water challenges for the 21st century. In: Oswald Spring Ú (ed) Water resources in Mexico, Hexagon series on human and environmental security and peace, vol 7. Springer, Berlin, Heidelberg

CENAPRED   (2017a)   https://www2.deloitte.com/mx/es/pages/dnoticias/articles/impacto-economico-desastres-naturales.html. Accessed 5 Mar 2018

CENAPRED   (2017b)   http://www.bvsde.paho.org/cursodesastres/diplomado/curso1/tema1.html. Accessed 5 Mar 2018

CONAGUA (Comisión Nacional del Agua) (2012a) Atlas del Agua en México. Comisión Nacional del Agua, México, 142 pp.

CONAGUA (Comisión Nacional del Agua) (2012b) Compendio Estadístico de Administración del Agua (CEEA), edición 2012b. Comisión Nacional del Agua, 2012b, México, 85 pp.

CONAGUA (Comisión Nacional del Agua) (2014) Situación del Subsector Agua Potable, Alcantarillado y Saneamiento. Comisión Nacional del Agua, México

CONAGUA (Comision Nacional del Agua) (2017) Estadísticas del Agua en México, Edición 2016. http://201.116.60.25/publicaciones/EAM_2016.pdf. Accessed 5 Mar 2018

Deutsch M, Coleman MT (2000) The handbook of conflict resolution: theory and practice. Jossey-Bass Inc. Publishers, San Francisco

Fisher S (2000) Working with conflict: skills and strategies for action. Zed Books, London

Gleick PH (1993) Water and conflict: fresh water resources and international security. Int Secur 18 (1):79–112

Gonzalez VF (2018) "El Agua en México: Retos y Soluciones", 2017. Foros Universitarios "La UNAM y los desafíos de la Nación", UNAM. 496 p

Grey D, Sadoff CW (2007a) Agua para el crecimiento y el desarrollo. Documento Temático. México, D.F.: IV Foro Mundial del Agua, Comisión Nacional del Agua y Consejo Mundial del Agua, 2006, 70 pp. https://www.bing.com/news/search?q=cenapred+inversion+en+riesgos +2017&qpvt=cenapred+inversion+en+riesgos+2017&FORM=EWRE. Accessed 5 Mar 2018

Grey D, Sadoff WC (2007b) Sink or Swim? water security for growth and development. Water Policy 9(6):545–571. https://doi.org/10.2166/wp.2007.021. Published December 2007

INEGI (2015) "Estadísticas a Propósito del Día Mundial Del Agua (22 De Marzo)" Datos Nacionales. 20 de Marzo De 2015 Aguascalientes, Ags. www.inegi.org.mx/sistemas/temas/ default.aspx?s=est&c=17484

Martínez-Austria P (2007) Efectos del Cambio Climático en los Recursos Hídricos de México. IMTA, Jiutepec

Martínez-Austria P, van Hofwegen (2006) Synthesis of the 4th world water forum. National Comission of Water in Mexico. Accessed 5 Mar 2018

Montero-Martínez MJ, Martínez-Jiménez J, Castillo-Pérez NI, Espinoza-Tamarindo BE (2010) Escenarios climáticos en México proyectados para el siglo XXI. Precipitación y temperaturas máxima y mínima. In: Martínez-Austria PF, Patiño-Gómez C (eds) Atlas de vulnerabilidad hídrica de México ante el cambio climático. Semarnat-IMTA, México

Oberg J (1996) Conflict mitigation in reconstruction and development. Peace Conflict Stud 3 (2):205–236

OECD (2011) Water governance in OECD countries: A multi-level approach, OECD studies on water. OECD, París. https://www.oecd.org/governance/regional-policy/48885867.pdf. Accessed 5 Mar 2018

Petzold-Bradley E, Carius A, Vincze A (2001) Responding to environmental conflicts: implications for theory and practice. Kluwer Academic, Dordrecht

Rivas-Acosta I, Güitrón-de-los-Reyes A, Ballinas-González HA (2010) Vulnerabilidad hídrica global: aguas superficiales, En Atlas de vulnerabilidad hídrica de México ante el cambio climático.. III. Instituto Mexicano de Tecnología del Agua, Jiutepec, pp 81–113

Rogers P, Hall AW (2003) Effective water governance. Technical paper no. 7. Global Water Partnership, Stockholm

Ruiz R (2009) "Estrategia y prioridades de financiamiento de la Ciencia y la Tecnología 2009–2012", versión preliminar. Academia Mexicana de Ciencias, Los Cipreses

UNFPA (2011) Estado de la población mundial 2011. Fondo de Población de la Naciones Unidas, Nueva York, 132 pp.

Villagómez P, Bistrain C (2008) Situación demográfica nacional. Consejo Nacional del Población, México

World Economic Forum (2011a) The global information technology report 2010–2011. https:// www.ifap.ru/library/book494.pdf. Accessed 5 Mar 2018

World Economic Forum (2011b) Water security. The water-food-energy-climate nexus. The World
   Economic Forum Water Initiative, Island Press, London
WRG (2009) Charting our water future. Economic frameworks to inform decision-makers. The
   2030 Water Resources Group, Washington, DC, 198 pp

Chapter 8
Wastewater Treatment in Mexico

Cynthia G. Tabla-Vázquez, Alma C. Chávez-Mejía, María T. Orta
Ledesma, and Rosa M. Ramírez-Zamora

Abstract This chapter presents a global perspective and analysis of the current situation of the legal and regulatory framework of Mexico for wastewater treated for discharge and reuse. Also, a comparative discussion of the situation of the installed Mexican wastewater treatment trains and facilities is shown. Finally, it presents an overview of the research and development trends of new wastewater treatment processes and applications of treated wastewater in Mexico. These three issues are discussed with respect to other countries.

In the first section, a comparative and critical analysis of the normative frameworks of Mexico and several countries worldwide for wastewater treated for discharge and for reuse is presented. Specifically, three official Mexican standards will be analyzed for the regulation of the permissible maximum limits of pollutants in residual water: the NOM-001-SEMARNAT-1996 (PROY-NOM-001-SEMARNAT-2017) that establishes the permissible limits of pollutants in wastewater discharges into receiving water bodies owned by the nation; the NOM-002-ECOL-1996 that establishes the maximum permissible limits of pollutants in wastewater discharged into municipal or urban sewer systems to prevent and control the pollution of national water and its goods; and the NOM-003-SEMARNAT-1997 that establishes the permissible limits of pollutants for treated wastewater for reuse in public services.

In the second section, regarding the sanitation infrastructure, data on the number and types of wastewater treatment trains and facilities of Mexico, as well as of the effluent quality, are presented and compared in the context of some Latin American countries. The comparison is made to establish the situation of the country and to propose some recommendations to solve the problems associated with these issues. For example, of the total wastewater flowrate generated in Mexico, only 92% is collected. Nevertheless, only 63% of the collected wastewater is treated in 2536 plants with an installed capacity of 123.4 m^3/s. The processes used to treat most of

C. G. Tabla-Vázquez · A. C. Chávez-Mejía · M. T. Orta Ledesma · R. M. Ramírez-Zamora (✉)
Instituto de Ingeniería, Coordinación de Ingeniería Ambiental, Universidad Nacional Autónoma de México, Mexico City, México
e-mail: RRamirezZ@iingen.unam.mx

© Springer Nature Switzerland AG 2020
J. A. Raynal-Villasenor (ed.), *Water Resources of Mexico*, World Water Resources 6,
https://doi.org/10.1007/978-3-030-40686-8_8

this flowrate are as follows: lagoon systems, 776 wastewater treatment plants (WWTPs), and 732 activated sludge WWTPs.

Finally, in the third section "Overview of the research and development trends of new wastewater treatment processes and applications of treated wastewater in Mexico," wastewater is regarded as a source of high value-added by-products such as biohydrogen, biogas, biodiesel, bioalcohols, fertilizers, and bioplastics. This can make more sustainable the wastewater treatment processes since, historically, the Mexico treatment systems have required considerable energy and resources, which is a scenario that drains investments and is not viable in a low-carbon economy. Thus, several case studies are discussed and compared to some works reported in the literature worldwide to establish trends and the degree of development in Mexico for new wastewater processes to produce fuels and bioplastics.

Keywords Wastewater treatment · Legal framework · Disposal and reuse · Limits for pollutants

8.1 A Comparison of Mexican and International Legislation for Treated Wastewater

In recent years, guidelines to promote regulations of wastewater harnessing have been proposed. However, progress in terms of regulation is still precarious for some regions of the world, particularly those that are not fully developed economically. Mexico possesses some of the most advanced wastewater legislation in Latin America due to research and regulation, and also in terms of the reuse of this resource. It corresponds to the *Comisión Nacional del Agua* (National Water Commission, CONAGUA), a dependency of the government relying on *Secretaria del Medio Ambiente y Recursos Naturales* (Ministry of Environment and Natural Resources, SEMARNAT) for the administration and preservation of national water resources, as well as domestic water systems, wastewater reuse and exchange activities, for its construction, expansion, rehabilitation, studies, and actions. The regulations also allow that they can benefit from federal programs. For this purpose, the following considerations that promote the indirect reuse of treated water are considered, including the handling and disposal of sludge generated during the treatment of wastewater.

The Mexican Official Standard NOM-001-SEMARNAT-1996 established the maximum permissible limits of pollutants in wastewater discharges into national water and goods (rivers, streams, canals, drains, reservoirs, lakes, and lagoons); Mexican marine zones, and estuaries and soils (irrigation of green areas, infiltration, and other irrigation). This standard fixed gradual and progressive compliance according to the population range for municipal discharge and to the pollutant concentration. The compliance deadline expired for all cases on January 1, 2010, requiring the modification of the classification of the receiving water bodies and the

Table 8.1 Permissible limits according the reuse of treated wastewater (NOM-003-SEMARNAT-1997) (SEMARNAT 2003a)

Types of reuse	Monthly average				
	Fecal coliforms MPN/100 mL	Helminth eggs, eggs/L	Oils and fats mg/L	BOD$_5$ mg/L	TSS mg/L
Public services (direct contact)	240	≤ 1	15	20	20
Public services (indirect or occasional contact)	1000	≤ 5	15	30	30

MPN Most probable number

approach to subsequent use to improve their management and protection (preliminary draft PROY-NOM-001-SEMARNAT-2017, December 2017). In this new modification, a total of 22 parameters are regulated, including temperature, oils and fats, total suspended solids, chemical oxygen demand, total organic carbon, total nitrogen, total phosphorus, helminth eggs, *Escherichia coli*, *Enterococcus faecalis*, pH, true color, severe toxicity, arsenic, cadmium, cyanide, copper, chromium, lead, mercury, nickel, and zinc, all of them according to their monthly and daily average values (SEMARNAT 1996a).

The NOM-002-ECOL-1996 establishes the maximum permissible limits of pollutants in wastewater discharges to municipal or urban sewer systems to prevent and control the pollution of national water and its goods, as well as to protect their infrastructure, being mandatory for those responsible for their own waste. This ecological standard does not apply to the discharge of domestic wastewater, rainwater, and industrial wastewater where separate drainage systems are in place (SEMARNAT 1996b).

The NOM-003-SEMARNAT-1997 establishes the maximum permissible levels of pollutants for treated wastewater for reuse in services for the public with direct, indirect, or occasional contact. This standard applies to parameters of fecal coliforms, helminth eggs, biochemical oxygen demand (BOD$_5$), oils and fats, and total suspended solids (TSS), as microbiological and physical–chemical parameters, respectively (Table 8.1) (SEMARNAT 2003a). In addition to these pollutants, treated wastewater reused in services for the public shall not contain concentrations of heavy metals and cyanides exceeding the maximum limits considered in the standard NOM-001-SEMARNAT-1996.

Public authorities are responsible for the treatment of wastewater used for public services (those in which the public is directly or indirectly exposed to water). They have the obligation to monitor the wastewater quality and its conditions, establishing the frequency and the number of samples, and depending on the pollutants to evaluate. The methods of sampling and laboratory analysis are established in the standards NMX-AA-003, NMX-AA-102-1987, and the NMX-AA-42-1987. On the other hand, the sludge generated by wastewater treatment plants is classified as special handling waste and is regulated by the *Ley General para la Prevención y Gestión Integral de los Residuos* (Law for the Prevention and Integral Management of Wastes). In addition, there exists standard NOM-004-SEMARNAT-2002, which

Table 8.2 Quality of treated wastewater for artificial recharge (NOM-014-CONAGUA-2003) (CONAGUA 2009a)

Type of pollutant	Superficial/subsuperficial	Direct
Pathogenic microorganisms	Removal or inactivation of enter-pathogenic microorganisms (Enterovirus, *Legionella* sp., *Vibrio cholerae*, heterotrophic bacteria counts, *Cryptosporidium* sp. and *Giardia lamblia*)	
Pollutants regulated by standard	Permissible limits NOM-127-SSA1-1994 (Environmental health, water for human use, and consumption). Permissible limits of quality and treatments to which water must be submitted for its purification.	
Pollutants not regulated by standard	BOD$_5$ < 30 mg/L, TOC = 16 mg/L	TOC < 1 mg/L

TOC total organic carbon, *BOD* biochemical oxygen demand

establishes the minimum quality for the reuse of sludge as a fertilizer or soil amendment, although there are no regulations or guidelines related to their application or management (SEMARNAT 2003b).

Regarding the recharge of aquifers with treated water, the SEMARNAT issued Standard NOM-014-CONAGUA-2003 and NOM-015-CONAGUA-2007 (CONAGUA 2009a, b). The first aims at the regulation of artificial recharge (surface/subsurface or direct) of aquifers with treated wastewater, regarding the criteria to comply in terms of quality, operation, and monitoring (Table 8.2). Also, it is specified that it should be given information on the location of the recharging points that are projected to be built and possible sources of groundwater contamination. Additionally, the applicant must give information about the source of water recharge, the hydrogeology of the area and toxicological studies, among other data. NOM-015-CONAGUA-2007 regulates the characteristics and specifications of the work and the water for artificial infiltration into the aquifers; it is complementary to NOM-014-CONAGUA-2003 (CONAGUA 2009a).

There are other countries in Latin America making regional advances in researching, studying, analyzing, and solving problems related to wastewater reuse. The following profiles them as emerging countries developing their own regulations, such as **Bolivia** (whose current regulations are based on the EPA guidelines and Mexican standards), **Ecuador** (which has a national policy on water and sanitation (Executive Decree No. 2766); however, this policy does not include the reuse of water or sludge), **El Salvador** (a draft law is currently under discussion in the legislature. The new *Estrategia Nacional de Saneamiento Ambiental* (National Strategy of Environmental Sanitation) includes the reuse of treated wastewater as one of its strategic lines of promotion, although this is not yet articulated in detail), and **Paraguay** (the responsible institution for the control and analysis of sewage is the *Ente Regulador de Servicios Sanitarios* (Agency Regulatory of Health Services). Significant improvement of interagency collaboration is necessary for the efficient management of wastewater, as well as increasing the budget allocated to this sector. At the moment, there are no programs or policies (on the national or local governmental levels) aimed at farmers who use wastewater (treated or not) in the areas of agricultural irrigation (FAO 2017).

8.2   Analysis of the Wastewater Treatment Plants Installed in Mexico

8.2.1   Number and Flowrate of Wastewater Treatment Plants Installed in Mexico

Historically in Mexico, the number of WWTPs facilities has risen steeply. While in 1992 there were only 394 plants treating 30.55 m^3/s of wastewater, today the number of treatment plants has increased to more than 2536, providing sanitation for 123.92 m^3/s of municipal wastewater (CONAGUA 2016a, b).

The progress in the number of these treatment plants is positive, however, it is not enough. The sanitized flow still represents only a disappointingly small proportion of the total generated, since only 63% of municipal wastewater is treated throughout the country (SEMARNAT 2017). Therefore, the Mexican authorities carry out actions such as the one implemented in 2016, where the construction of 59 new plants was achieved in 2015, growing the sanitized flowrate grew from 120.9 to 123.6 m^3/s on the national level (SEMARNAT 2017). Some of the facilities put into operation were the "Hermosillo" treatment plant (Sonora) whose installed capacity is 2500 L/s and "La Marina" (Guerrero) treating 300 L/s of the Las Salinas lagoon and from the Zihuatanejo Bay (CONAGUA (2017).

In addition, necessary sanitation is considered in the continuous development of the country with infrastructure megaprojects, such as the New Mexico City International Airport. That infrastructure project involves a WWTP with an installed capacity of 76 L/s of wastewater and 21 plants to clean up 564 L/s, a flow equal to the basins of nine rivers in the region east of Lake Texcoco (SHCP 2018).

The location of the WWTPs differs from the distribution of the population throughout the territory. The northern states of the country, such as Sinaloa (2.9 million inhabitants), Durango (1.7 million inhabitants), and Chihuahua (3.5 million inhabitants) have 282, 231, and 184 WWTPs, respectively. In contrast, large regions such as the State of Mexico (16.1 million inhabitants), Mexico City (8.9 million inhabitants), and Jalisco (7.8 million inhabitants), have in operation 124, 29 and 134 WWTPs, respectively (INEGI 2015). This disparity is not due to the generated volume of municipal wastewater but to the type of treatment of the installed system.

8.2.2   Treatments Used in the Sanitation of Wastewater

According to data from CONAGUA in Mexico, two main types of municipal wastewater treatments are installed: primary and secondary (CONAGUA 2015a).

8.2.2.1   Primary Treatment

The goal of primary treatment is the removal of large-sized organic matter, sands, colloidal matter, oils, and fats. The sequential processes that are involved in this treatment are as follows : the primary process, advanced primary, Imhoff tank, and septic tank.

In Mexico, there are 198 primary treatment WWTPs of which 57.6% are septic tanks and 26% Imhoff tanks. Both systems serve small communities that have few sewerage services (up to 5000 inhabitants in the case of Imhoff tanks) or buildings, condominiums, hospitals, etc. (CONAGUA 2015b). Even with their usefulness, these are processes that generate unpleasant odors and do not eliminate the organic matter or microorganisms of the effluent for discharge, which become a potential source of infection.

8.2.2.2   Secondary Treatment

Secondary treatment degrades organic matter from food waste, human wastes, cleaning products, etc. For this purpose, technologies of treatment use biological processes with reactors containing suspended or fixed biomass. In addition, there are dual-types reactors that stimulate the increase in the overall efficiency of treatment through the combination of biological processes. In all cases, the conditions for cellular metabolism can be aerobic, anaerobic, or anoxic.

In comparison to the primary process, the secondary plants are vastly superior because they are integrated by 1912 WWTPs, essentially as stabilization lagoon systems (776) and activated sludge (732) (CONAGUA 2016a).

Stabilization ponds are not only the most used technology in secondary processes but are also the total number of treatments used in the country. These consist of depository bodies where water is stored to undergo physical, chemical, and biological transformations in a natural way (OPS 2005). Examples of stabilization lagoons are the Gómez Palacio plant (Durango) and Santa Bárbara (Chihuahua), with installed capacities of 500 L/s and 42 L/s, respectively (CONAGUA 2015a). These treatment systems require low investments for their operation and maintenance, and they also remove bacteria, pathogens, protozoa, and helminth eggs. As in septic tanks, bad odors are generated in the stabilization ponds. However, the greatest weakness of this technology are perhaps the concomitant need for large areas of land and its strong dependence on environmental conditions, which can generate degradation in the amount of effluent and an overload of pollutants (OPS 2005).

Next to WWTPs with stabilization lagoons are the technologies that have activated sludge–reactor systems. In total, 732 plants of activated sludge have been installed, equivalent to 28.9% of the total national facilities. This process basically consists of a reactor in which microorganisms perform a metabolic process on the organic matter present in the water, as long as they are supplied with sufficient oxygen. The system is simple and commonly operates with volumetric loads of 0.8–2.0 kg BOD/dm^3 in a high-rate process, consuming 54–81 m^3 of air for every kg of BOD removed. However, the technology produces a large amount of sludge in the range of 0.6–0.8 kg/sludge/kg of BOD removed that must be stabilized before its final disposal to prevent unhealthy conditions and other problems, such as odors or aerosols (CONAGUA 2015c).

On the other hand, only 5.6% of wastewater treatment corresponds to the upflow anaerobic sludge blanket (UASB). Its operation principle consists of feeding from the bottom of a tank the wastewater that flows upward through a layer of microbiological sludge. Under anaerobic conditions, the microbiological sludge degrades the organic matter of the wastewater-generating gases (mainly methane and carbon dioxide). The gases, the treated water, and some mud particles rise toward the top of the tank. In this section, the gas and sludge are separated from the treated water (López Hernández et al. 2017).

Other processes, such as ditches, bio-disks, and dual systems, are among the least used; they together make up only 2.5% of all available technologies.

8.2.2.3 Tertiary Treatment

Tertiary treatment processes include membrane filtration, adsorption with activated carbon, disinfection, etc. They remove nutrients, organic compounds or pathogens using physical, chemical, and biological processes, as well as combinations of them. This type of system is the least frequently used with only six of them installed and in operation in WWTPs belonging to tertiary treatment (CONAGUA 2015a, c).

8.2.3 Treated Flowrate According to WWTPs

Primary WWTPs processes 5.3% of the total treated flowrate of wastewater (6482 m^3/s), and WWTPs secondary processes handle the remaining 92.4%.

As shown in Fig. 8.1, having a larger number of treatment plants does not mean processing a greater quantity of water with the desired quality, depending on the type of installed technology.

While activated sludge delivers more than half of the nationwide-treated flowrate (56%), a similar number of stabilization ponds plants deliver only 11.6% (14.3 m^3/s). A similar case is for the 11 dual WWTPs that provide 13.58 m^3/s, (CONAGUA 2016a).

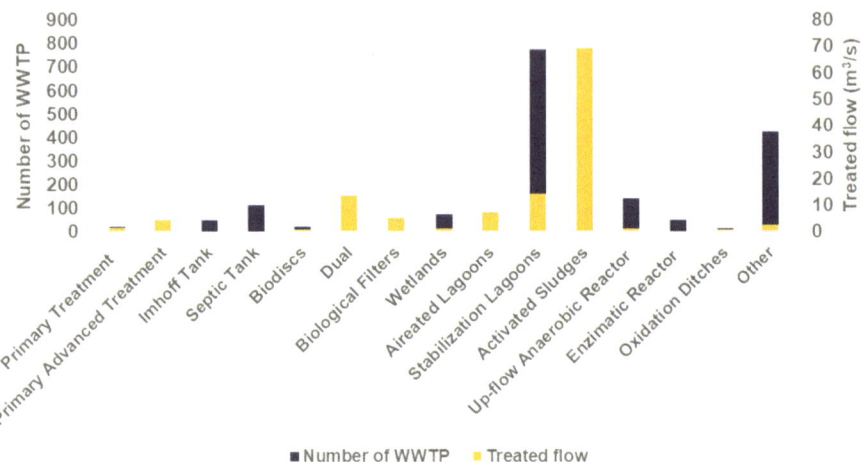

Fig. 8.1 Number of WWTPs and flowrates treated as a function of the treatment process in Mexico. (CONAGUA 2016a)

8.2.4 Parameters of Effluent Quality

Effluent quality establishes the efficiency required to select the type of process needed to achieve it. On a design basis, a primary process usually reduces around 20% if the initial values of BOD_5 and 50% TSS; meanwhile secondary treatment achieves removal of around 70% for BOD_5 and at least of 75% for COD before discharge The tertiary treatment is known to decrease at least 95% of BOD_5, 85% of COD, 70% of nitrogen removal, 80% of phosphorus removal, and more than 99% of viral inactivation (disinfection, filtration, or ozonation) microbiological removal (CONAGUA 2015c; UN 2016).

In Mexico, the current state of treatment efficiency of municipal wastewater is uncertain. All the data about quality parameters are not exhaustive, probably due to a lack stringent monitoring.

To elucidate the quality of municipal discharges after treatment, the Organization for Economic Cooperation set out data on the capacity design, incoming load and effluent of Mexican WWTPs, reported in terms of BOD_5. The organic pollution expressed as BOD_5 is an important parameter of quality because it indicates the level of negative impacts on fisheries, food security, and population, (UNESCO 2017).

According to values reported for 2010–2013, the BOD_5 design capacity improved from 2082.5 to 2467.7 ton O_2/day. Likewise, the incoming load increased from 1735.4 to 2056.5 ton O_2/day, and the effluents characteristics changed proportionally from 321.4 to 380.1 ton O_2/day. Thus, the removal percentage was around 81.5% each year (OECD 2018).On this basis, it is insufficient to conclude that the effluents achieved the standards for its release. Mainly, this is because there have been cases of anomalies in the operation of the WWTPs.

In 2012, a study assessed the quality of the treatment processes of some wastewater treatment plants in the north of Mexico. The results for all WWTPs reached concentrations of suspended solids and BOD_5 above the permissible limits of the official Standard NOM-003-SEMARNAT-1997. In particular, the concentration in the effluent of "Arenitas" WTP located in Tijuana city, ranged from 340 to 86,740 MPN/100 mL when the limit is 240 MPN/100 mL. The effluent is regularly used as a source for agricultural irrigation, but this represents environmental and health risks (Romero et al. 2014).

It is necessary to establish a current national framework for nutrient concentrations in WWTPs discharges and its impact on the eutrophication of freshwater or coastal ecosystems, as well as the contribution to the increase in groundwater pollution.

Moreover, given the technological development of the country and the new consumption patterns, it is fundamental to incorporate the emerging pollutants present in wastewater, such as pharmaceutical products, personal care products, pesticides and herbicides, and gasoline additives, among others (UNESCO 2017).

Implementing better quality control of the treated effluents would enable knowing the technological requirements of treatment, and it would assure the appropriate discharge or reuse of the treated wastewater.

8.2.5 Discharge and Reuse

Water bodies receive approximately 70% of the flowrate treated in WWTPs. Within all the localized discharge sites, approximately 53% of the water is discharged into rivers, 15% into streams, 10% into drains or collectors, 9% into channels, 4% into seas, gulf or oceans, and the rest is channeled into estuaries, aquifers, reservoirs or dams, among others (CONAGUA 2015a; Anda et al. 2016).

If the water is not then discharged, it is reused mainly for irrigating agricultural lands, green areas, golf courses, etc. This productive activity employs almost 82% of the total reused water, while 15% goes to the mining industry and cooling towers, as part of an industrial process or power generation. The remaining 3% is the provision of pipes.

The country's intention to reuse WWTPs effluent is incipient up to this date; there are still challenges to be achieved in a sustainable way. One of them is related to the scarcity of the exploitation of the WWTP process to obtain energy resources, for example, by recovering nutrients.

Energy Generation and Nutrient Recovery in a WWTP Process It is possible to recover energy from several processes used for wastewater treatment, such as sludge incineration, microbial fuel cells, aerobic granular sludge technologies, anaerobic ammonium oxidation, and biomass manipulation. Of all available technologies, the production of biogas is the most common (UNESCO 2017). The biogas generated by the anaerobic digestion of municipal wastewater has a methane content of

Table 8.3 WWTP of activated sludge with generation systems for profitable electric power (IMTA 2016)

WWTP	State	Process	Flow treated (m³/s)	Receptor of water
Villa Alvarez	COL	SA	0.850	Colima river
Principal	COAH	SA	0.900	Not available
Metropolitana San Jerónimo	GTO	SA	0.110	San Jerónimo stream
León	GTO	P	1.525	Las Mulas stream
Agua Prieta	JAL	SA	6.030	Santiago river
El Ahogado	JAL	SA	2.100	Santiago river
San Pedro Martir	QRO	DUAL	0.547	Agricultures of San Pedro Martir

Mexican states: *COL* Colima, *COAH* Coahuila, *GTO* Guanajuato, *JAL* Jalisco, *QRO* Queretaro. Treatments: *SA* Sludge activated, *P* Primary treatment

approximately 60–85% v/v. The specific calorific value of biogas with 60% methane is 21.5 MJ/Nm³, equivalent to 67% of the lower calorific capacity of natural gas (López Hernández et al. 2017). Therefore, those processes provide sanitation of municipal wastewater while they can simultaneously tap sources of power generation.

Table 8.3 shows the treatment plants with an aerobic stabilization system for sewage sludge in Mexico. Only a few WWTPs take advantage of the biogas generated in the activated-sludge process. Nevertheless, there are 28 WWTPs with activated sludge with the capability to use biogas for the generation of profitable electric power (IMTA 2016).

The "*Metropolitana San Jerónimo* WWTP" which handles wastewater from the municipalities of *San Francisco* and *Purísima del Rincón* in the state of Guanajuato carries out the cogeneration of electric power. The main process consists of an activated-sludge reactor where the biogas results in an energy supply. This WWTPs was one of the first biogas cogeneration plant installed in the country and has two motor generators (750 kilowatts); at the same time, the sludge serves as an agricultural soil amendment, and the water benefits 114,000 inhabitants (Gobierno de Guanajuato 2018).

Nowadays, akin to energy generation from biogas, there is special interest in recovering phosphorus because of the potential depletion of this resource in the next century (Vaccari and Strigul 2011). This is a promising alternative for obtaining the nutrients in wastewater, mostly notable in decentralized systems. Until now, the technologies to extract phosphorus from wastewater are in an early stage and they are very sophisticated (UNESCO 2017).

8.2.6   Wastewater Generation and Treatment in Latin America

Latin America is one of the regions with the world's greatest biodiversity; although up to now it has had available water resources, it does not handle well the generated wastewater. According to data from the World Bank (2013), 70% of Latin America's wastewater returns to the rivers without treatment. The situation has generated pathogenic, organic, and saline contamination of important rivers, putting at risk the inhabitants of contracting cholera, dysentery, typhoid, and polio (UNEP 2016).

On average, only 50% of the population in Latin America is connected to sanitary drainage, of which 30% receives a type of treatment using stabilization ponds, sludge-activated reactors or UASBs (World Bank 2017; Noyola and Morgan 2013). Data compiled by the OECD for 16 countries in Latin America in the 2008–2012 period show that Brazil, Mexico, and Argentina are the countries that generate the largest volumes of wastewater. In this period, Chile treated 69% of the wastewater generated, Mexico 41%, Nicaragua 37%, Brazil 30%, Peru 28%, Cuba 27%, and Argentina 12%. Regarding reuse, data are only available for four countries: Mexico, Argentina, Peru, and Brazil (OECD 2018).

Recently, sanitation has been improved due to the efforts of countries, such as Chile and Mexico that currently treat 90% and 63% of their wastewater, respectively. However, other countries such as Costa Rica processes only 4% (World Bank 2017).

Evidently, pollution impacts the environment negatively, affects the economy, and represents a health risk to the country's population. Therefore, it is essential to subject wastewater to sanitation operations.

8.2.7   The Placement of Mexico in the Worldwide Wastewater Treatment Field

Mexico has a municipal WWTPs infrastructure greater than several high-income countries, such as the UK (1856 WWTPs), Japan (2184 WWTPs), Poland (1574 WWTPs), and Canada (1265 WWTPs), among others. However, the Mexican systems treats only 63% while most of these nations sanitize at least 69%, and, in the case of the UK, 98%. More stringently installed treatment is the main reason for this. While the Mexican wastewater systems are primary and secondary treatment, disinfection, sand filtration, and microfiltration are extensively employed in high-income countries. For example, there are 3729 wastewater treatment plants in France, and of this total 3018 have advanced treatment operations and the remaining 711 are secondary systems. Other similar cases are the UK, Denmark, Germany, Poland, and Italy (EEA 2018).

Table 8.4 Removal efficiencies and nutrient discharges of WWTP located in European Union countries and Mexico (EEA 2018)

	Germany	Italy	Denmark	Czech Republic	Mexico
Treatment	Sand filter	Disinfection	Sand filter	Microfiltration	Dual[a]
WWTP	Ska Römerstein-Böhringen	Lagosanto	Søholt	Dobrichovice	WWTP AGS City
Volume WWT (m³/day)	599,755	520,550	4,872,432	402	172,800
%BOD removal	–	98	99	97	59
%COD removal	–	89	96	94	79
% N removal	89	56	95	–	37
% P removal	100	100	97	75	–
Discharged N (kg/day)	4	4	12	0	101[b]
Discharged P (kg/day)	0	0	1	1	–

[a]Discharge to San Pedro River in Aguascalientes, Mexico
[b]Nitrogen discharged (mg/L) in the 2006 rainy season

As a result of the application of the best treatment technologies in these countries, the removal percentages of BOD, COD, nitrogen, and phosphorus are higher than those of the Mexican WWTPs, as shown in Table 8.4. This table indicates that the total nitrogen discharged into the San Pedro River in Mexico in the 2006 rainy season was above limits established by NOM-001-SEMARNAT-1996 (60 mg/L). This value was of 17,453 kg/d and did not comply with the Mexican standard and Urban Waste Water Treatment Directive (UWWTD) (Torres-Guzmán et al. 2010).

In recent years, Mexico and Chile have become the countries with the highest flowrate of wastewater treatment in Latin America. In addition, unlike Nicaragua, El Salvador, Panama, and Haiti where around 25% of the population does not have access to sanitation facilities, Mexico has 98% coverage in sewerage and basic sanitation services (FAO 2017; CONAGUA 2016a).

An increase in the number of treatment plants in Mexico does not guarantee the sanitation of the country's municipal wastewater because the correct operation of the wastewater treatment plants must also be insured. Without complete knowledge of the pollutants in the treated water, it is difficult to determine the needs of the various types of WWTPs. Therefore, strict compliance measures and strong monitoring must be enforced more seriously. The authorities must enable adequate sanitation and must ensure the sustainable development of the country only through environmental programs with documented investment (International Trade Administration 2016). The research results about current issues in treatment technologies are imperative to be able to make suitable decisions.

8.3 Overview of the Research and Development of Wastewater Treatment Processes and Applications of Treated Wastewater in Mexico

Since 1966, investigations of wastewater treatment technologies have produced around of 18,458 documents worldwide. More than a half (52.5%) have taken place in China, the US, Spain, and India. For its part, Mexico has published only 172 papers, within which 126 appeared during the period from 2009 to 2018. The *Universidad Nacional Autonoma de Mexico, Instituto Mexicano de Tecnologia del Agua* and *Universidad Autonoma Metropolitana* are the three major contributors of the national research production in Mexico. Primary research areas include environmental science, chemical engineering, engineering, chemistry, biochemistry, energy, immunology and microbiology, agricultural and biological sciences, earth and planetary sciences, and computer science, among others.

For each research area, the objectives are mainly the efficient removal, reuse of alternatives, and recovery of value-added products such as some fuels. To explicate this, several applications are briefly described in the next section.

8.3.1 Microplastics

Microplastics are plastic particles smaller than 5 mm that can absorb persistent organic pollutants and heavy metals from water sources. As emerging pollutants, they were are not considered in the design of conventional treatment plants and therefore not necessarily eliminated.

Talvitie et al. (2017) studied the removal of microplastics in four different municipal wastewater treatment plants utilizing various advanced final-stage treatment technology. They found that primary and secondary treatment in WWTPs are able to remove microplastics however, small amounts of pollutants could be discharged in effluents, and thus advanced final-stage wastewater treatment technology must be installed. The MBR process removes 99.9% of microplastics, rapid sand filters 97%, dissolved air flotation 95%. and disc filters 40–98.5% during the treatment (Talvitie et al. 2017).

Investigations have examined the microbial biodegradation of plastics, such as polyethylene (PE), polystyrene (PS), polyethylene terephthalate (PET), and polyurethane (PUR). Some technologies involve fungi that biodegrade PE, PP, and PS with heat or UV irradiation (Wu et al. 2017), or *Acinetobacter sp* to degrade PE and an Actinomycete strain for PS (Tokiwa et al. 2009).

Microplastics are a concern because their presence can affect various aquatic species and enter the food chain of the population. To solve the problem, governments should consider decreasing the amount of materials delivered to the environment, the enforcement of satisfactory technology, improving the removal efficiency of systems, and updating the WWTPs installed.

Unfortunately, in Mexico, the presence of polyacrylic, polyacrylamide, polyethylene terephthalate, polyester, and nylon microplastics has been identified in national assessments (Piñon-Colin et al. 2018). The data reported covers no greater a period than five years, and thus research on treatment technologies is at an early stage of development.

8.3.2   Bioplastics

Plastics occupy a large part of the volumes of sanitary landfills, so their replacement with biodegradable plastics such as polyhydroxyalkanoates (PHA) has recently become a goal. Biodegradable plastics or bioplastics originate from renewable resources and are entirely biodegraded by microorganisms in the natural medium (Dias et al. 2006). The generation of PHA is possible in WWTPs. The process has of two stages, one involving feeding the microorganisms and the other starving the microorganisms. Due to the dynamic of feeding, the microorganisms will produce PHA as a carbon reserve (Pittmann and Steinmetz 2017).

Pittman and Steinmetz (2017) evaluated the production of PHA using sludge generated from various WWTPs. They achieved a PHA production of up to 28.4% of cell dry weight (CDW) at a low substrate concentration, 20 °C, neutral pH-value and at 24 h (Pittman and Steinmetz 2017). From these results, the authors estimated that German WWTPs could provide more than 19% of the worldwide production in 2016. In the case of the European Union, the WWTPs installed could generate about 120% of the worldwide production.

In Mexico, research is being carried out to produce PHA and establish systems that make possible obtaining the best yields. Vargas et al. (2014) developed a control strategy that estimates the concentration of the carbon source and the rate of biodegradation to elucidate the optimal degradation time. Through this strategy, they achieved a substrate conversion rate of 1.370 ± 0.598 mg PHA/mg COD/d, (Vargas et al. 2014).

Due to the commercial value of PHAs, their production using wastewater as raw material is an interesting opportunity as a novel investment. Thus, Mexican scientists and authorities should consider further developing this research area in future programs to cover sanitation and sustainable development.

8.3.3   Pharmaceuticals

Pharmaceutical products are incorporated into wastewater by human excretion. They are usually difficult to eliminate in the conventional treatment plants, so they are discharged with the effluents (Amouzgar et al. 2016). Like other countries in the world, in Mexico, there is evidence of the presence of pharmaceuticals in water sources and treated effluents. For example, José Abraham Rivera-Jaimes et al.

(2018) studied the occurrence of pharmaceuticals in wastewater in Cuernavaca, (Morelos, Mexico). The analysis of effluent of the main treatment plant Acapantzingo revealed 12 compounds, such as Sulfamethoxazole, Diclofenac, Bezafibrate, Trimethoprim, Naproxen, and Indomethacin, among others. As a result, a great part of Mexican research is focusing on the degradation of those and other pollutants by employing advanced oxidation processes (AOPs).

– Acosta-Rangel et al. (2018) assessed with a photocatalysis system the degradation of Tinidazole, as a model compound. The results achieved around a 68% removal of Tinidazole in water after 1 h of treatment using silica xerogel doped with iron.
– Arzate-Salgado et al. (2016) achieved the complete removal of Diclofenac in 90 min using two metallurgical slags in heterogeneous Fenton-like reactions.
– Durán-Alvárez et al. (2016) tested the degradation of ciprofloxacin by heterogeneous photocatalysis. The photocatalysts were bi-metallic Au–Ag and Au–Cu nanoparticles deposited on TiO_2. They distinguished ciprofloxacin by-products showed residual antibiotic activity. Total degradation was achieved after 360 min of irradiation at the conditions given.

While it is true that it is essential to implement systems for removing pharmaceuticals, the technologies investigated are not economically competitive yet. Therefore, an important challenge for the scientific community is to link the technical efficiency to an economic benefit.

8.3.4 Pesticides

In 2016 in Mexico, around 27,761 tons of insecticides were produced, as well as 31,298 tons of herbicides and defoliants (INEGI 2014). These compounds are very relevant to the agricultural sector because they prevent, destroy, or repel pests in the fields. However, they are also a potential risk of contamination of surface and groundwater intended for human consumption. Pesticides such as DDT, Aldrin, and Atrazine, among others, are poorly removed in WWTPs, so they can become introduced into fields by irrigation (Ochoa-Rivero et al. 2017). Some investigations for the treatment of pesticides are as follows:

– Aquino et al. (2017) evaluated the degradation of Atrazine, Simazine, Hydroxyl Triazine, and Propazine using a photo-assisted electrochemical process. The electrochemical degradation experiments were carried out in a tubular flow reactor, six tubular anodes, a Ti-mesh cathode, and a UV germicidal lamp. Atrazine and Propazine were efficiently removed under laminar flow conditions (330 L/h), while 90% of the Simazine content was removed under turbulent flow conditions (20 kWh m^{-3}) (Aquino et al. 2017).
– Bandala et al. (2006) examined the removal of Aldrin, Dieldrin, Heptachlor, and Heptachlor epoxide by activated carbon. The data showed a degradation tendency

correlated with the chemical structure of the pesticide. The removal efficiency ranged from 95.45% to 97.18%, varying according to the dose and the type of pesticide (Bandala et al. 2006).

The use of these pesticides is a delicate balance between risk and benefit for the agricultural sector. It is possible that, in addition to implementing technologies like the previous ones, chemical formulations that reduce the risk should also be better developed.

8.3.5 Fuels

Greenhouse gas emissions and their effect on climate change force countries to look for an alternative energy to replace fossil fuels. One of the goals to which Mexico has committed is to supply 35% of the national electricity generation by renewable-energy sources (SENER 2015). To achieve this, efforts have been made to increase the number of investigations that provide long- and medium-term solutions. The energetics mainly investigated so far are hydrogen, methane, bio-diesel, and bio-alcohols.

– Arzate-Salgado et al. (2016) investigated hydrogen production capacity on a pilot-plant scale through the photocatalytic treatment of a synthetic formic acid effluent. The hydrogen obtained was 6500 mmolH_2/L using Au/TiO$_2$ after 5 h of irradiation (Arzate Salgado 2016).

Other researchers have shown that successful sewage treatment is possible when combining microalgae with bacteria, micro-organisms which act together to remove a variety of water pollutants.

– Currently, Orta et al. (2017) are developing a project called "Novel pilot-scale system for wastewater/leachates treatment and carbon dioxide capture using microalgae and ozone-flotation." They have proven at pilot scale (1.2 m^3 raceway reactor) that this technology reduces carbon emissions and energy needed to clean water, producing microalgae. Microalgae readily absorb nutrients and other pollutants present in wastewater using carbon dioxide and sunlight (Orta Ledesma et al. 2017). The algal biomass obtained can be converted into ecologically attractive products such as bioplastics, biofuels, or biofertilizers. Additional microalgae advantages over land crops or plants for biodiesel production include the use of wastewater or seawater for microalgae cultivation to reduce clean water requirements and carbon footprints associated with water treatment (Rodolfi et al. 2009); nutrients recycled into microalgae biomasses when they are cultured in wastewater, eliminating requirements for fertilizers and the pollution effects associated with them (Mata et al. 2010); CO_2 recycling based on microalgae (Rodolfi et al. 2009) and the use of residual algae biomass, after oil extraction, as fertilizer or as a feedstock to produce bio-alcohol (Mata et al. 2010). For the process to produce biodiesel sustainably from microalgae, it is necessary that the microalgae grow in wastewater, which significantly reduces wastewater nutrients and total coliforms. In the second part of this technology, they use the ozone flotation process, which does not require the use of flocculants or change in the pH of the system (Alves Oliveira et al. 2018). It had been reported that ozone could induce cell lysis, the release of intracellular organic matter (IOM) (Henderson et al. 2008). This IOM could act as a natural surfactant for the flotation process, (Valeriano González et al. 2016). On the other hand, most of the existing harvesting processes are either expensive or time intensive, thus making microalgae biodiesel generation counterproductive. Ozone flotation could be a good solution because it improves the efficiency of biomass recovery, the lipid productivity and save ozone dose (Valeriano et al. 2020). Furthermore, research into ozone-flotation harvesting of the resultant biomass has shown it to be viable and resulting in oils (FAMEs) of greater oxidation stability (Velasquez-Orta et al. 2014). Another advantage is the water-quality improvement because of the ozone action involving organic and inorganic pollutants (Langlais et al. 1991).

Mexico is aware that fossil fuels are no longer a good option for the energy sector. The damages are greater than the benefits. The prices of fuels, such as gasoline, increase day by day and are not as efficient as the technologies just presented.

8.3.6   Other Components

Within the country's commercial activities, tanning and fabric dyeing generate wastewater that cannot be treated by conventional WWTPs. The dyes that they release generally require advanced technologies for removal. Advanced technologies, such as photocatalysis and Fenton's reagent, among others, are also being investigated.

Bilal et al. (2018) published research on bioremediation using horseradish peroxidase and cross-linked polyacrylamide to remove methyl orange. Moreover, they concluded that the maximum efficiency achieved to biodegrade this dye was over 90%.

8.3.7   Mexican Research from a Global Perspective

We assessed information of a database of professional journals to place Mexican investigation within a global perspective. The database used was Scopus, the Title-Abs-Key (wastewater AND treatment AND technologies) and the data generated in May 2018. During the period 1966–2018, the results indicate China as the country with the highest number of publications. Then the US, Spain, India, and Germany follow in that order. Mexico contributed just 4.5% of the total number of documents published by China, 5.3% with regard to the US, and 16.7% compared to Spain (Elsevier Scopus 2018).

The research areas differ from country to country, but environmental science is the largest field in which the most publications are categorized. Mexico has the lowest number of publications in all research areas: environmental science, engineering, chemical engineering, chemistry, energy, material science and agricultural, and biological sciences.

Even though the topical areas listed above are the apparent trend among all countries analyzed, the research in Mexico follows this order:

Environmental Science > Chemical Engineering > Chemistry > Biochemistry, Genetics and Molecular Biology > Energy > Agricultural and Biological Sciences > Material Science

China invests more effort in energy and biochemistry, genetics and molecular biology than in agricultural science. Such a field of study has the lowest number of publications reported for all the fields analyzed in this chapter, except for that of Mexico. These data show that energy and material science are fields of opportunity. Energy from wastewater technologies is related to alternatives fuels, such as hydrogen and methane. Finally, material science is related to the recovery of value-added products such as bioplastics.

8.4   Summary

In Mexico, the quality of effluent discharges, and wastewater reuse are regulated through the Official Mexican Standards. NOM-001-SEMANARNAT-1996 established the limits of pollutants that can be legally discharged into national water bodies. To improve the regulation of this standard, the quality parameters have been modified to culminate in 22 regulatory parameters stipulated in PROY-NOM-001-SEMARNAT-2017. Other standards such as NOM-002-ECOL-1996 and NOM-003-SEMARNAT-1997 define the maximum permissible limits of pollutants in wastewater for municipal discharge and reuse, respectively.

On the regional level, these Mexican standards are the model for other countries in Latin America that have neglected regulating wastewater reuse.

The wastewater reuse in Mexico is performed in WWTPs by implementing two main types of municipal wastewater treatment processes: primary and secondary. The latter is the most common with 1912 WWTPs that deliver 92.4% of the total treated flowrate of wastewater, mainly via lagoon stabilization systems and activated sludge. According to CONAGUA data, a very small portion of WWTPs employ tertiary treatment.

While Mexican wastewater treatment systems are primary and secondary in nature, in high-income countries, disinfection, sand filtration, and microfiltration are extensively employed. More specifically, in the UK, Denmark, Germany, Poland, and Italy, the effluent qualities are therefore superior to Mexican effluents with respect to the removal of BOD, COD, nitrogen, and phosphorus.

Even though in recent years there are more WWTPs, the qualities of Mexican WWTPs effluents are not published in national reports and, as a consequence, the characteristics of effluents are uncertain, especially for emerging pollutants.

On the other hand, it should be noted that Mexico, is one of few countries in Latin America that reports data on wastewater reuse. The reuse activities not only incorporate treated wastewater but also by-products generated in the WWTPs, such as is the case of the production of biogas by activated sludge. Also, Mexican research has far yet to go to explore new treatment systems in areas as environmental science, engineering, chemical engineering, chemistry, energy, material science, and agricultural, and biological sciences.

8.5   Conclusions

The development of regulations is directly proportional to the country's development, and, in Latin America, it is clear that Mexico has shown the greatest progress in the area. In other countries of Latin America such as Chile, Brazil, Argentina, and Colombia, studies are motivated to take advantage of wastewater reuse, resulting in the existence of a standard or in consideration of a regulation in this area.

The standards are essential for the control of contamination of water bodies and national goods; however, stringent monitoring of effluents is also necessary to avoid irregular compliance with the law.

On the other hand, a greater incorporation of tertiary treatments should be evaluated and considered as treatment models, such as those applied in some high-income countries.

Mexico is on a very good track to implement programs to increase wastewater treatment to cover 100% of the country's sanitation. It excels in the results presented in Latin American countries; however, there are many challenges that still have to be met. For example, the challenges include increasing wastewater reuse in productive processes in the country and changing the perspective of wastewater use to add areas of opportunity such as the generation of electricity, the recovery of phosphorus, and nitrogen. It also requires more research and scaling of technologies that involve emerging pollutants, and, like conventional processes, determine the recovery of valuable products such as the bioplastics, biofuels, and bio-fertilizers. These systems can be the beginning of investigations that can economically support part of the operation of treatment plants.

References

Abraham R-JJ, Cristina P, María M-AR, Jaume A, Damia B, de Alda Miren L (2018) Study of pharmaceuticals in surface and wastewater from Cuernavaca, Morelos, Mexico: Occurrence and environmental risk assessment. Sci Total Environ 613–614(1):1263–1274. https://doi.org/10.1016/j.scitotenv.2017.09.134

Acosta-Rangel A, Sánchez-Polo M, Polo AMS, Rivera-Utrilla J, Berber-Mendoza MS, Acosta-Rangel et al (2018) Tinidazole degradation assisted by solar radiation and iron-doped silica xerogels. Chem Eng J 344:21–33

Alves Oliveira G, Carissimi E, Monje-Ramírez I, Velasquez-Orta SB, Teixeira Rodrigues R, Orta Ledesma MT (2018) Comparison between coagulation-flocculation and ozone-flotation for Scenedesmus microalgal biomolecule recovery and nutrient removal from wastewater in a high-rate algal pond. Bioresour Technol 259:334–342. https://doi.org/10.1016/j.biotech.2018.03.072

Amouzgar P, Wong MY, Horri BA, Salamatinia B (2016) Advanced material for pharmaceutical removal from wastewater. In: Smart materials for waste water applications. Scrivener Publishing/Wiley, Hoboken

Anda J De, Alberto L, Villegas-garc E et al (2016) Manual de Agua Potable, Alcantarillado y Saneamiento Manual de Agua Potable , Alcantarillado y Saneamiento Diseño de Plantas de Tratamiento Pretratamiento y Tratamiento Primario Comisión Nacional del Agua. Publicaciones Estadísticas y Geográficas SINA 130:92. https://doi.org/10.3390/w10020099

Aquino JM, Miwa DW, Rodrigo MA, Motheo AJ (2017) Treatment of actual effluents produced in the manufacturing of atrazine by a photo-electrolytic process. Chemosphere 172:185–192. https://doi.org/10.1016/j.chemosphere.2016.12.154

Arzate Salgado SY, Ramírez Zamora RM, Zanella R et al (2016) Photocatalytic hydrogen production in a solar pilot plant using a Au/TiO2photo catalyst. Int J Hydrog Energy 41:11933–11940. https://doi.org/10.1016/j.ijhydene.2016.05.039

Bandala E, Andres-Octaviano J, Pastrana P, Torres L (2006) Removal of aldrin, dieldrin, heptachlor, and heptachlor epoxide using activated carbon and/or Pseudomonas fluorescens free cell cultures. J Environ Sci Health – Part B Pestic Food Contam Agric Wastes 41:553–569. https://doi.org/10.1080/03601230600701700

Bilal M, Rasheed T, Iqbal HMN et al (2018) Horseradish peroxidase immobilization by copolymerization into cross-linked polyacrylamide gel and its dye degradation and detoxification potential. Int J Biol Macromol 113:983–990. https://doi.org/10.1016/J.IJBIOMAC.2018.02.062

CONAGUA (2009a) Norma Oficial Mexicana NOM-014-CONAGUA-2003 Requisitos para la recarga artificial de acuíferos con agua residual tratada

CONAGUA (2009b) Norma Oficial Mexicana NOM-015-CONAGUA-2007, Infiltración artificial de agua a los acuíferos.- Características y especificaciones de las obras y del agua

CONAGUA (2015a) Inventario Nacional de Plantas Municipales de Potabilización y de Tratamiento de Aguas Residuales en Operación. Diciembre 2014. Mexico

CONAGUA (2015b) Manual de Agua Potable , Alcantarillado y Saneamiento Manual de Agua Potable, Alcantarillado y Saneamiento Diseño de Plantas de Tratamiento Pretratamiento y Tratamiento Primario Comisión Nacional del Agua. CONAGUA, Mexico

CONAGUA (2015c) Manual de Agua Potable, Alcantarillado y Saneamiento Manual de Agua Potable, Alcantarillado y Saneamiento Introducción al Tratamiento de Aguas. CONAGUA, Mexico

CONAGUA (2016a) Procesos de tratamiento de aguas residuales municipales por caudal tratado (2016). http://sina.conagua.gob.mx/sina/tema.php?tema=plantasTratamiento&ver=reporte&o=0&n=nacional#ui-state=dialog. Accessed 25 Mar 2018

CONAGUA (2016b) Numeragua México. Publicaciones Estadísticas y Geográficas SINA 9133

CONAGUA (2017) Reporte de Acciones del Programa Nacional Hídrico 2014–2018. Mexico

Dias JML, Lemos PC, Serafim LS et al (2006) Recent advances in polyhydroxyalkanoate production by mixed aerobic cultures: from the substrate to the final product. Macromol Biosci
6:885–906. https://doi.org/10.1002/mabi.200600112

Durán-Álvarez JC, Edwin A, Ramírez-Zamora RM, Zanella R (2016) Photocatalytic degradation of
ciprofloxacin using mono- (Au, Ag and Cu) and bi- (Au–Ag and Au–Cu) metallic nanoparticles
supported on TiO2 under UV-C and simulated sunlight. Catal Today 266:175–187. https://doi.
org/10.1016/j.cej.2018.03.051

EEA (2018) Urban wastewater treatment. https://www.eea.europa.eu/themes/water/water-pollu
tion/uwwtd/interactive-maps/urban-waste-water-treatment-maps#tab-based-on-data. Accessed
29 Apr 2018

Elsevier Scopus (2018) https://www.scopus.com/term/analyzer.uri?
sid=9e245c6d0f8568ca9c3bf5e9f7a11142&origin=resultslist&src=s&s=TITLE-ABS-KEY%
28wastewater+treatment+technologies%29&sort=plf-f&sdt=cl&sot=b&sl=48&count=793&
analyzeResults=Analyze+results&cluster=scoaffilctry%2C%22. Accessed 1 May 2018

FAO (2017) Reutilización de aguas para agricultura en América Latina y el Caribe: Estado,
principios y necesidades. http://www.fao.org/3/a-i7748s.pdf. Accessed 30 Apr 2018

Gobierno de Guanajuato (2018) Planta de Tratamiento Metropolitana "San Jerónimo" para los
municipios de San Francisco y Purísima del Rincón. In: Comunidad Sustentabilidad. http://
www.guanajuato.gob.mx/ceag/plantas.php. Accessed 1 Apr 2018

Henderson R, Parsons SA, Jefferson B (2008) The impact of algal properties and pre-oxidation on
solid-liquid separation of algae. Water Res 42:1827–1845. https://doi.org/10.1016/j.watres.
2007.11.039

IMTA (2016) Revisión y actualización del potencial de biomasa para generación de energía
eléctrica a partir de plantas de tratamiento de aguas residuales presentado en el Inventario
Nacional de Energías Renovables (INERE). México

INEGI (2014) Volumen de producción de fertilizantes, insecticidas y plaguicidas por tipo de
producto, 2005 a 2016

INEGI (2015) Información por entidad.Población. http://cuentame.inegi.org.mx/monografias/
informacion/sin/poblacion/default.aspx?tema=me&e=25. Accessed 2 Apr 2018

International Trade Administration (2016) Top markets report environmental technologies, USA

Langlais B, Reckhow DA, Brink RD (1991) Ozone in water treatment: application and engineering,
1st edn. Am Water Works R.F. CRC Press. CRC Press. ISBN 9780873714747

López Hernández JE, Ramírez Higareda BL, Bayer Gomes Cabral C, Morgan Sagastume JM
(2017) Guía técnica para el manejo y aprovechamiento de biogás en plantas de tratamiento de
aguas residuales, GIZ. Programa Aprovechamiento Energético de Residuos Urbanos en México

Mata TM, Martins AA, Caetano NS (2010) Microalgae for biodiesel production and other applications: a review. Renew Sust Energ Rev 14:217–232. https://doi.org/10.1016/j.rser.2009.07.
020

Noyola A, Morgan JGL (2013) Selección de tecnologías para el tratamiento de aguas residuales
municipales, 1st edn. Universidad Nacional Autónoma de México, Mexico

Ochoa-Rivero JM, Reyes-Fierro AV, Peralta-Pérez MDR et al (2017) Levels and distribution of
pollutants in the waters of an aquatic ecosystem in Northern Mexico. Int J Environ Res Public
Health 14. https://doi.org/10.3390/ijerph14050456

OECD (2018) OECD statistics. In: Wastewater treat. http://stats.oecd.org/index.aspx?
DataSetCode=water_treat. Accessed 10 Apr 2018

OPS (2005) Guía Para el Desempeño de Tanques Séptico, Tanques Imhoff y Lagunas de
Estabilización. Lima, Perú

Orta Ledesma MT, Monje-Ramírez I, Velasquez-Orta S, Rodriguez-Muñiz V, Yáñez-Noguez I
(2017) Ozone for Microalgae Biomass Harvesting from Wastewater. Ozone Sci Eng
39:264–272. https://doi.org/10.1080/01919512.2017.1322488

Piñon-Colin T de J, Rodriguez-Jimenez R, Pastrana-Corral MA et al (2018) Microplastics on sandy
beaches of the Baja California Peninsula, Mexico. Mar Pollut Bull 131:63–71. https://doi.org/
10.1016/j.marpolbul.2018.03.055

Pittmann T, Steinmetz H (2017) Polyhydroxyalkanoate production on waste water treatment plants: process scheme, operating conditions and potential analysis for German and European municipal waste water treatment plants. Bioengineering 4:54. https://doi.org/10.3390/bioengineering4020054

Rodolfi L, Zittelli GC, Bassi N et al (2009) Microalgae for oil: strain selection, induction of lipid synthesis and outdoor mass cultivation in a low-cost photobioreactor. Biotechnol Bioeng 102:100–112. https://doi.org/10.1002/bit.22033

Romero S, Villagomez A, Trasviña D et al (2014) Treatment and use of wastewater in Mexicali, Baja California, Mexico. WIT Trans Ecol Environ 181:591–601. https://doi.org/10.2495/EID140501

SEMARNAT (1996a) Norma Oficial Mexicana NOM-001-ECOL-1996 que establece los Límites Máximos Permisibles de Contaminantes en las Descargas de Aguas Residuales en Aguas y Bienes Nacionales

SEMARNAT (1996b) Norma Oficial Mexicana NOM-002-SEMARNAT-1996 que establece los límites máximos permisibles de contaminantes en las descargas de aguas resiudales de alcantarillado urbano o municipal

SEMARNAT (2003a) Norma Oficial Mexicana NOM-003-SEMARNAT-1997 que establece los límites máximos permisibles de contaminantes para las aguas residuales tratadas que se reúsen en servicios al público. D Of la Fed México, pp 2–6

SEMARNAT (2003b) Norma Oficial Mexicana NOM-004-SEMARNAT-2002. Lodos y biosólidos. Especificaciones y límites máximos permisibles de contaminantes para su aprovechamiento y disposición final. D Of la Fed México, 15 de agosto de 2003

SEMARNAT (2017) Logros 2016 Programa Nacional Hídrico 2013–2018. México

SENER (2015) Prospectiva de Energías Renovables 2016–2030

SHCP (2018) Construcción, operación y mantenimiento de 21 plantas de tratamiento de aguas residuales cercanas al nuevo aeropuerto internacional de la ciudad de méxico. In: Proy Mex. http://www.proyectosmexico.gob.mx/proyecto_inversion/019-naicm-23-plantas-de-tratamiento-de-aguas-residuales/. Accessed 1 Apr 2018

Talvitie J, Mikola A, Koistinen A, Setälä O (2017) Solutions to microplastic pollution—removal of microplastics from wastewater effluent with advanced wastewater treatment technologies. Water Res 123:401–407. https://doi.org/10.1016/J.WATRES.2017.07.005

Tokiwa Y, Calabia BP, Ugwu CU, Aiba S (2009) Biodegradability of plastics. Int J Mol Sci 10:3722–3742. https://doi.org/10.3390/ijms10093722

Torres-Guzmán F, Avelar-González FJ, Rico-Martínez R (2010) An assessment of chemical and physical parameters, several contaminants including metals, and toxicity in the seven major wastewater treatment plants in the state of Aguascalientes, Mexico. J Environ Sci Health – Part A Toxic/Hazard Subst Environ Eng 45:2–13. https://doi.org/10.1080/10934520903388517

UN (2016) AQUASTAT. In: Treated/generated municipal wastewater. http://www.fao.org/nr/water/aquastat/main/index.stm. Accessed 22 Apr 2018

UNEP (2016) A snapshot of the world's water quality: towards a global assessment. United Nations Environment Programme, Nairobi

UNESCO (2017) Informe Mundial de las Naciones Unidas sobre el Desarrollo de los Recursos Hídricos 2017.Aguas Residuales.El recurso desaprovechado. Francia

Vaccari DA, Strigul N (2011) Extrapolating phosphorus production to estimate resource reserves. Chemosphere 84:792–797. https://doi.org/10.1016/j.chemosphere.2011.01.052

Valeriano González MT, Monje-Ramírez I, Orta Ledesma MT et al (2016) Harvesting microalgae using ozoflotation releases surfactant proteins, facilitates biomass recovery and lipid extraction. Biomass Bioenergy 95:109–115. https://doi.org/10.1016/j.biombioe.2016.09.020

Valeriano MT, Orta Ledesma MT, Velasquez-Orta S, Monje-Ramirez I (2020) Harvesting of microalgae using ozone-air flotation for recovery of biomass, lipids, carbohydrates, and proteins. Environ Technol. In press

Velasquez-Orta SB, García-Estrada R, Monje-Ramirez I, Harvey A, Orta LMT (2014) Microalgae harvesting using ozoflotation: effect on lipid and FAME recoveries. Biomass Bioenergy 70:356–363. https://doi.org/10.1016/j.biombioe.2014.08.022

Vargas A, Montaño L, Amaya R (2014) Enhanced polyhydroxyalkanoate production from organic wastes via process control. Bioresour Technol 156:248–255. https://doi.org/10.1016/j.biortech. 2014.01.045

World Bank (2013) Río Matanza-Riachuelo Un 70% de las aguas residuales de Latinoamérica vuelven a los ríos sin ser tratadas

World Bank (2017) Investing in wastewater in Latin America can pay off. http://blogs.worldbank. org/water/how-can-we-make-wastewater-investments-sustainable-latin-america. Accessed 20 Apr 2018

Wu W-M, Yang J, Criddle CS (2017) Microplastics pollution and reduction strategies. Front Environ Sci Eng 11:6. https://doi.org/10.1007/s11783-017-0897-7

Chapter 9
Climate Change and Water Resources in Mexico

Polioptro F. Martinez-Austria

Abstract Achieving water security has become the strategic objective of water management, from the point of view of international organizations, governments, and experts. However, at the global, national, and regional levels, water security is a growing concern. In this context, climate change poses enormous additional challenges that need to be addressed. There are expectations of lower precipitation and therefore availability in large regions of the planet, particularly those located in midlatitudes, as well as an increase in the occurrence and severity of droughts and extreme storms. Climate-change research on aspects of vulnerability and impacts should be a guide to decision-making that increases resilience. Mexico, due to its geographical location and climatic conditions, will be one of the most affected countries both concerning its water resources and extreme weather events. In this chapter, the general perspective of the observed climate-change trends and expected effects, especially that of Mexico, is analyzed within the context of current knowledge, including the effects on water availability and extreme phenomena such as droughts, floods, and severe storms, as well as heat waves and their impacts on human health.

9.1 Overview

In the recent global risk report of the World Economic Forum (2017), water crisis is ranked third as a global risk with the greatest impact and is among those crises with the highest probability of occurrence. The water crisis, moreover, is associated with two other major global risks: the occurrence of extreme climatic events and the failure to mitigate and adapt to climate change. These risks, all of them of great impact and probability of occurrence, are in a feedback loop with each other, so that the presence of any of them increases the probability of the occurrence of the rest.

P. F. Martinez-Austria (✉)
Department of Civil and Environmental Engineering, Universidad de las Americas Puebla, Cholula, Puebla, Mexico
e-mail: polioptro.martinez@udlap.mx

© Springer Nature Switzerland AG 2020   157
J. A. Raynal-Villasenor (ed.), *Water Resources of Mexico*, World Water Resources 6,
https://doi.org/10.1007/978-3-030-40686-8_9

Already in 2011, the water initiative of the World Economic Forum stated, "We simply cannot manage water in the future as we have in the past, or else the economic web will collapse (World Economic Forum 2011)."

In many countries, water security has not been achieved and, in fact, it is increasingly threatened. Population growth, economic development, urbanization, and natural climatic variability, as well as the effects of global climate change, continue to increase the pressure on water resources, in such a way that conditions of scarcity, permanent or recurrent, have already been recorded in some regions. The number and impact of natural water-related disasters continue to grow.

Without considering the effects of climate change, the Water Resources Group (2009) analyzed the global water demand by 2030 and found that, if measures are not adopted to increase efficiency in the use of water and to reduce consumption, the demand will be 40% greater than the supply of water. However, these global figures hide the huge differences in regional scarcities.

There are various water scarcity indexes (i.e., Florke and Alcamo 2007), among which it is important to use one that considers environmental needs. In the Human Development Report 2006 that was dedicated to water (UNDP 2006), the Water Stress Index (WSI) proposed by Smakhtin et al. (2004) was used and is as follows:

$$WSI = \frac{\text{Withdrawals}}{\text{Average Water Available} - \text{Environmental needs}} \quad (9.1)$$

From the results of Smakhtin et al. (2004), Rekacewicsz (2006) produced the map in Fig. 9.1. As can be seen, there are many large basins in which the water resource is

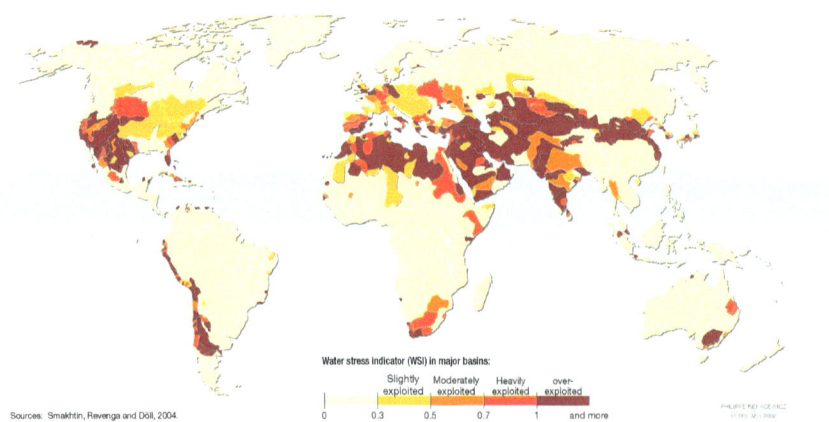

Fig. 9.1 Water Stress Index (Rekacewicsz 2006)

overexploited, particularly in Europe, North Africa, the Middle East, India, China, and North America. In the case of Mexico, practically all the basins in the center and north of the country, in addition to the basin of the Lerma River and the Valley of Mexico, have levels of high exploitation or overexploitation.

Increases in water demand, mainly in agriculture, caused by population growth, urbanization, and economic development will be exacerbated by global warming, while in many regions, natural water availability will be depleted and the number of water-related natural disasters will grow. This scenario poses a daunting challenge for water management.

9.2   Climate-Change Scenarios

The Representative Concentration Pathways (RCPs) describe four twenty-first century scenarios that involve increases in the solar radiative force, going from 2.6 W/m^2 up to 8.5 W/m^2, which may be easily reached if the actual trends continue. The difference in the resulting temperature, precipitation, and sea-level increase, between both extremes, is illustrated in Fig. 9.2 showing the expected changes by the 2081–2100 period, in comparison with average of the 1986–2005 period.

Temperature increases will not be uniform but concentrate more in the northern hemisphere and in some specific regions. As can be seen in Fig. 9.2, the expected temperature increase in Mexico is among the highest on the continent and will affect areas that already suffer from high temperatures, further increasing heat-wave dangers.

From the point of view of water resources, the temperature increase will mean a greater demand for water to meet the most critical uses: the environment, agriculture, and cities. This increase in water demand will be aggravated by the foreseen changes in precipitation because forecasts indicate that there will be a decrease in the midlatitudes, where Mexico is located. The result will be an expansion in demand along with a decrease in the supply of water, which will lead to greater water stress and, in some areas, extreme scarcity.

Because the rainfall–runoff relationship is not linear, it is to be expected that changes in precipitation, by percentage, will lead to even greater changes in runoff. This decrease will have important effects on the availability of water for environmental and human uses. In an extensive study, using an ensemble of 11 global hydrological models, Schewe et al. (2014) developed the map shown in Fig. 9.3, describing the expected variations in annual runoff with a temperature increase of 2 °C. The intensity of each color indicates the level of agreement between the different models used. In North America, with a high level of agreement between the models, reductions of 10–30% in runoff are to be expected. If the temperature increase were 3 °C, the decrease in runoff could reach up to 50% in some places.

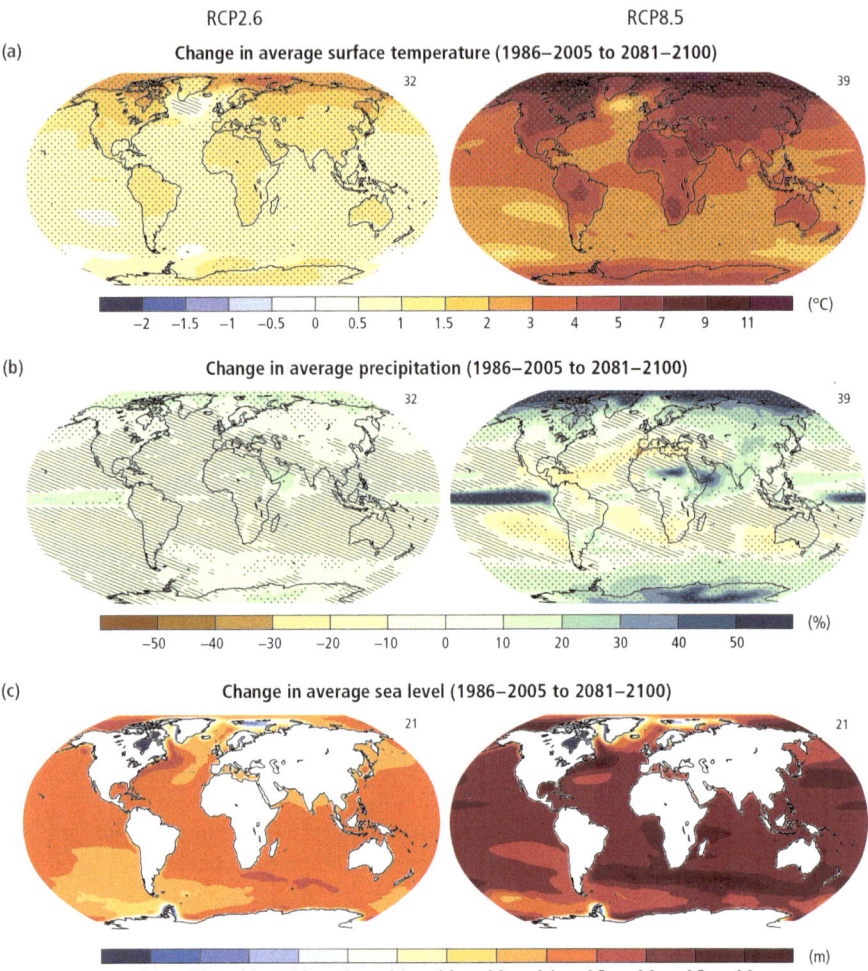

Fig. 9.2 Coupled Model Intercomparison for 2081–2100 under RCP2.6 (left) and RCP8.5 scenarios (IPCC 2014)

9.3   Water Resources in Mexico

In a global analysis, Mexico's water balance is already negative: more water is used than naturally available on a sustainable basis. In an analysis conducted by the National Water Commission (CONAGUA 2010), it was found that the 2010 gap between availability and demand was already 11,500 million m^3 and, if the current trend continues, the gap will rise to 23,000 million m^3 in 2030 (Fig. 9.4). These figures are generated without considering the climate change.

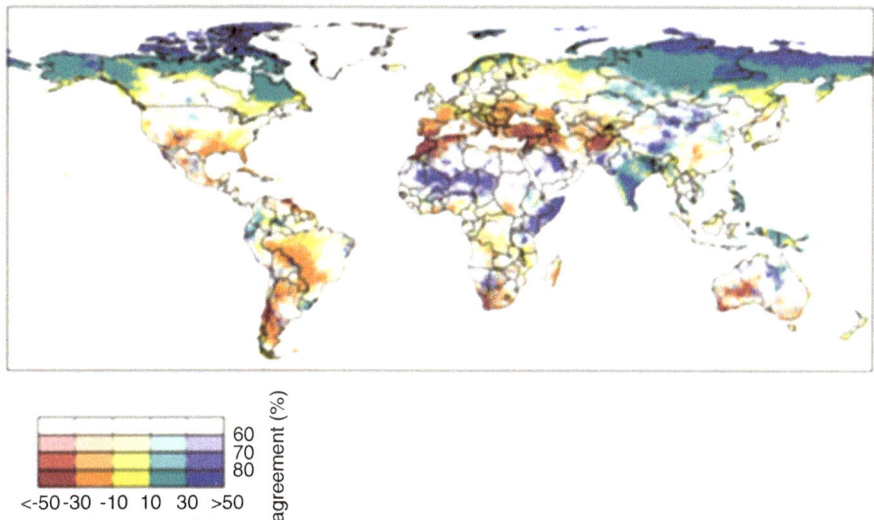

Fig. 9.3 Decrease in runoff with an increase in temperature of 2 °C in scenario RCP8.5, compared to the current situation (Schewe et al. 2014)

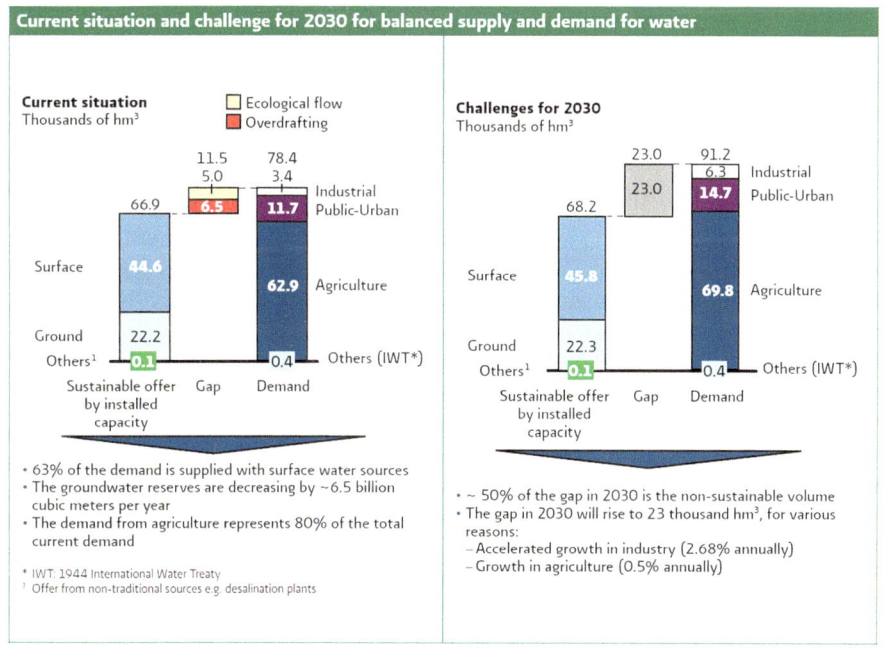

Fig. 9.4 Global water balance in Mexico between 2010 and 2030 (CONAGUA 2010)

Table 9.1 Per capita removable water in Mexican hydrological regions, 2014 and 2030 (with data from CONAGUA 2015)

Number	Name	Per capita renewable water 2014 (m³/year)	Per capita renewable water 2030 (m³/year)
I	Peninsula of Baja California	1135	899
II	Northeast	2951	2465
III	North Pacific	5730	5062
IV	Balsas	1896	1644
V	South Pacific	6084	5660
VI	Bravo River	1014	857
VII	Northern Central Basins	1738	1532
VIII	Lerma-Santiago-Pacifico	1469	1267
IX	North Gulf	5366	4710
X	Central Gulf	9075	8196
XI	Southern Border	19,078	16,334
XII	Yucatan Peninsula	6494	5026
XIII	Valley of Mexico	150	136

For water-management purposes, the Mexican territory has been divided into 13 hydrological–administrative regions that group various basins. Table 9.1 shows the main characteristics of these hydrological regions, which can be identified in Fig. 9.5, including the per capita available water in 2014 and the one for 2030.

The water situation in Mexico has already reached critical levels in some regions. Without considering the effects of climate change, but only the demographic ones and according to the Falkenmark water stress criterion, by the year 2030 most of the Mexican territory will be in conditions of water stress, scarcity, or absolute scarcity, as can be clearly seen in Fig. 9.5. The situation in the Valle de Mexico region, home to more than 23 million inhabitants, and with a per capita availability of less than 150 m³/inhabitant/year, is particularly worrisome. This hydrological region already imports large volumes of water from other nearby basins and overexploits its aquifers. This situation worsens over time, and no appropriate solutions have been implemented.

With respect to groundwater, 38% of the allocated water for consumptive uses in Mexico comes from this source (CONAGUA 2015) and, more importantly, 75% of the water used in cities comes from aquifers.

In this regard, of the 653 aquifers registered in Mexico, 195 (almost 30%) have not more availability, that is, natural recharge is completely allocated. Likewise, 106 aquifers are under conditions of overexploitation. Of special concern are those aquifers located in large urban centers and agricultural production in the Mexican highlands, such as the Valley of Mexico, the Bajío Region, the La Laguna Region, and northern Chihuahua. Additionally, there are 31 aquifers with brackish water and 15 coastal aquifers with saline intrusion (CONAGUA 2015). Figure 9.6 shows the aquifers without water availability or those that are already overexploited.

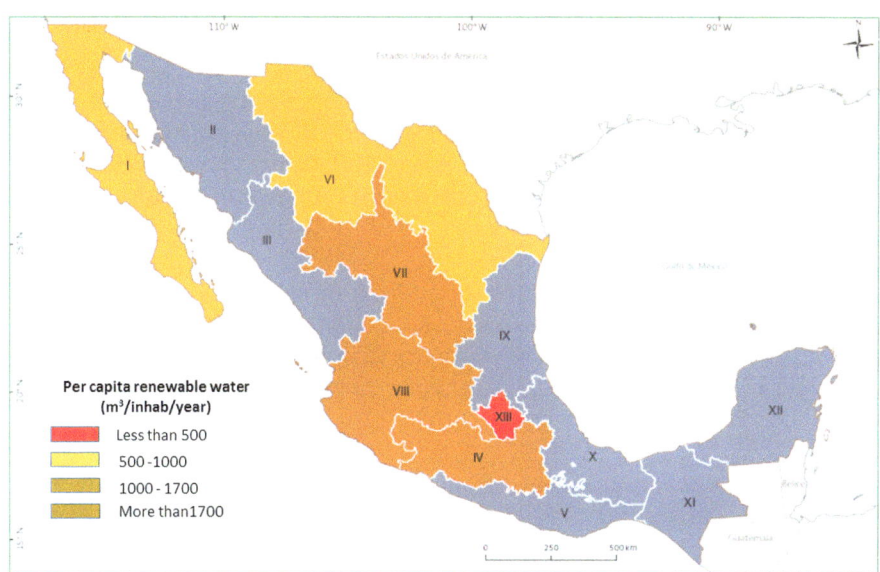

Fig. 9.5   Renewable per capita water in 2030 (CONAGUA 2015)

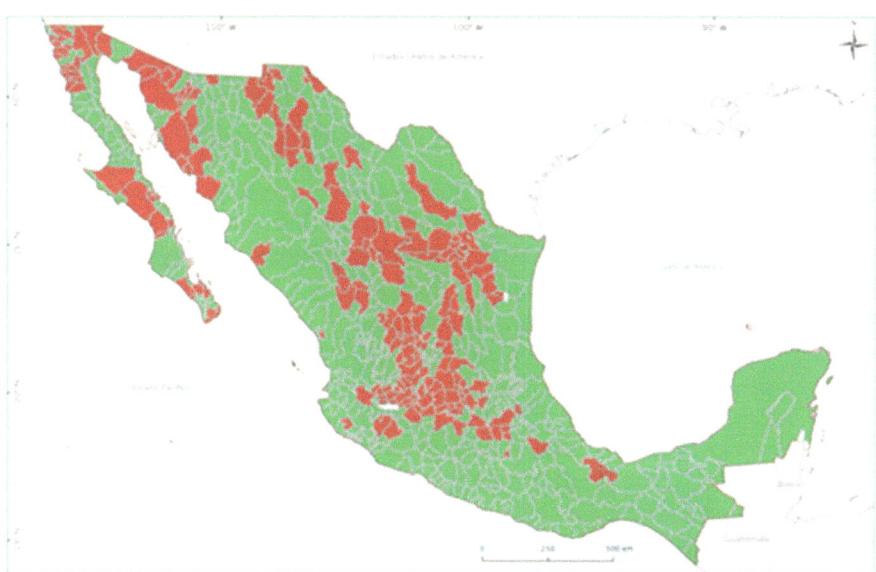

Fig. 9.6   Aquifers without water availability or overexploited, in red (CONAGUA 2015)

9.4 Effects of Climate Change on Mexico

The hydrological cycle is closely linked to the temperature of the planet, both atmospheric and oceanic, so that global warming will have important effects on precipitation, runoff, aquifer recharge, water quality, and hydro-meteorological extremes. In general, in midlatitudes and subtropical zones, significant decreases in precipitation and runoff are expected, which will cause scarcity and greater pressure on water resources in these regions.

In the case of Mexico, both extremes are to be expected, in various parts of its territory. Thus, increases in the severity and frequency of droughts are expected, particularly in its north-central zone, as well as an increase in the intensity of storms and floods in the center-south region and in the coastal areas.

9.5 Temperature

The results of global circulation models are not enough to estimate the effects of climate change on local scales. The Intergovernmental Panel on Climate Change (IPCC) estimated the performance of global models, making a comparison between their results and the climate observed during the period 1980–1999. Regarding temperature, when multimodel results are analyzed (the average of 23 general circulation models), the estimation error (i.e., the difference between the observations and the model) is rarely greater than 2 °C, although individual models can show errors close to 3 °C (Randall and Word 2007). Consequently, the IPCC notes that the characteristics at large scales are simulated with greater accuracy than the regional scales. The local analysis of impact and vulnerability to climate change, therefore, should be strongly based on observational evidence, and it is crucial to study the observed regional trends.

There are already reliable indicators that allow us to affirm that we have begun to observe the effects of climate change in Mexico. According to the 2016 weather report of the American Meteorological Society (Blunden and Arndt 2017), there is already a clear trend of temperature increase in the last decades, as shown in Fig. 9.7, in which one can notice that, compared to the 1981–2010 base period, an anomaly of almost two degrees was recorded between the years 1970 and 2016. This increase in temperature has serious implications for the climate of Mexico, although with differentiated effects on the regional level.

Future temperature projections are no less worrisome. According to the most recent calculations for Mexico, made by the National Institute of Ecology and Climate Change (INECC), in the RCP6.0 scenario at the end of this century, the temperature in the country would increase 2.5–3 °C in the center-south region of the country and up to 5 °C in the center-north region (Fig. 9.8). In scenario RCP8.5, the most unfavorable, the increase in temperature, practically in the whole territory, would be 5 °C and even more in the central-northern region (INECC 2016).

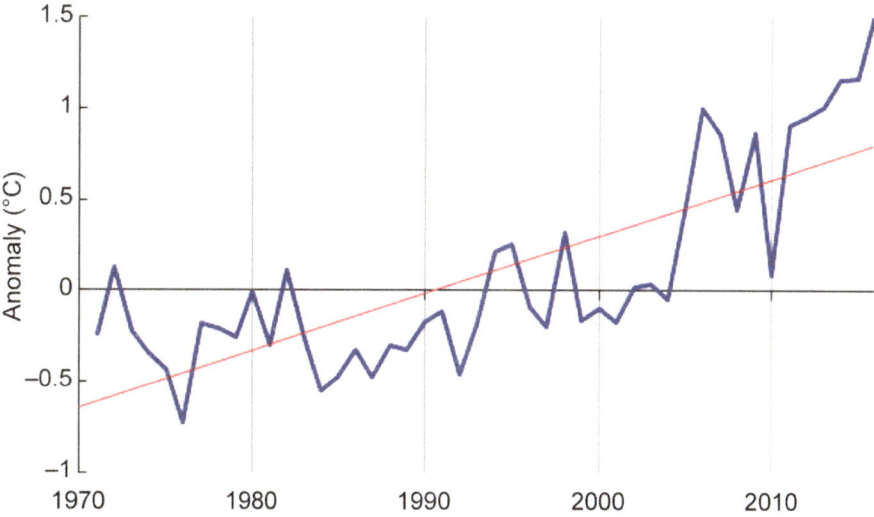

Fig. 9.7 Annual mean temperature anomalies (°C, blue line, 1981–2010 base period) for Mexico. Red line represents the linear trend over this period (Mekonnen et al. 2017)

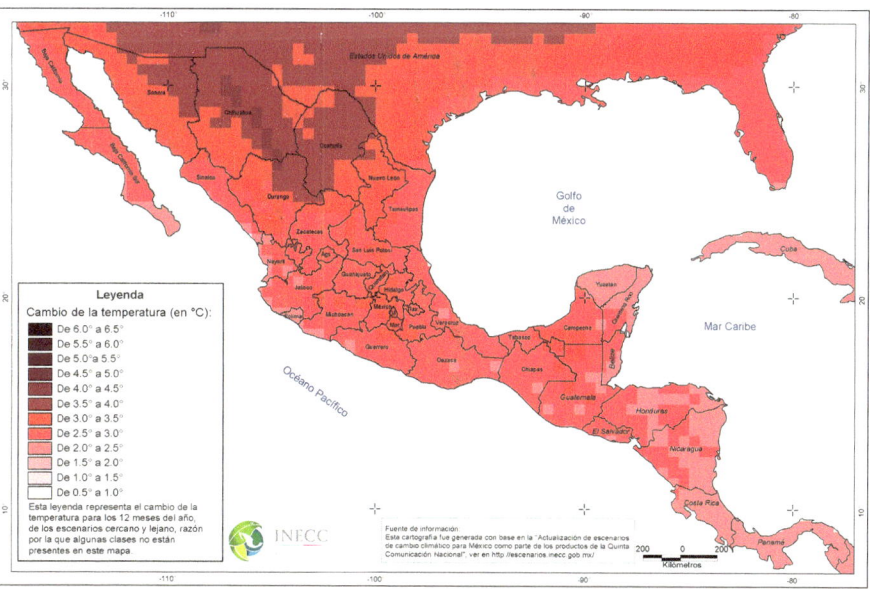

Fig. 9.8 Expected anomaly in the average annual temperature in Mexico at the end of the twenty-first century (2075–2099) in CPR6.5 Scenario (INECC 2016)

The effects of these temperature changes on water management will be very large. They will be observed in an increase in demand for water for food production, because crops—in such varieties that can withstand these new temperatures—will require greater water consumption. Ojeda Bustamante et al. (2015) conducted a vulnerability analysis of irrigated agriculture in Mexico, finding that the regions where the largest irrigation districts are located will have high or very high levels of vulnerability.

The increase in average temperature is accompanied by a significant increase in maximum temperatures, which produce heat waves that have proven to be extremely dangerous to health.

A heat wave can be defined qualitatively as a period, usually lasting several days, of temperatures significantly higher than average. Their importance lies mainly in their effects on human health, producing disorders that cause minor alterations in or even the collapse of the body's capacity to regulate its temperature by means of changes in blood circulation or sweating. People with preexisting illnesses, such as respiratory or cardiovascular diseases, can experience negative health consequences during extreme heat episodes. In extreme cases, these health effects can lead to death. The elderly and small children are particularly vulnerable to heat waves.

In the last few decades, there have been records of particularly dangerous heat waves that have caused at least several deaths, even in developed countries with good public-health services. It is worth mentioning here the heat wave that hit Chicago in 1995, causing 514 heat-related deaths (Whitman et al. 1997), the heat wave of 2003 in Europe, which affected mainly France and caused almost 15,000 deaths (Hémon and Jougla 2003, cited by Le Tertre et al. 2006), and the one that occurred in the Russian Federation in 2010, which caused 55,736 deaths (CRED-UNISDR 2016).

The change in maximum temperature is clearly observable in the northeastern region of Mexico and southeastern US, which are in the same climatic region (BWh in the Köppen–Geiger classification), in the Sonora–Mojave Desert area. In a study of six major cities of the region, it was reported that during the warmer months, there occurred an increase of more than two degrees in the maximum temperatures in the period 1960–2011 (Fig. 9.9) and an increase in the number of days per month in which the temperature exceeds the threshold of heat waves. This corresponds to the 90th percentile of the maximum temperatures observed (Fig. 9.10) and had extended from five to 15 days on average. This behavior of the temperature influences the increase in mortality in the hottest periods (Martinez-Austria and Bandala 2017).

In the city of Mexicali, which has the highest mortality rates due to heat stroke in Mexico, a correlation can be identified between maximum temperatures and the general mortality rate. At and above a maximum temperature of 47 °C, the mortality rate per 10,000 inhabitants increases drastically, as can be seen in Fig. 9.11.

9.6   Water Availability

While changes in temperature will increase the demand for water, mainly for the environment and food production, the expected changes in precipitation and runoff point to a decrease in water availability. Figure 9.12 shows the expected changes in

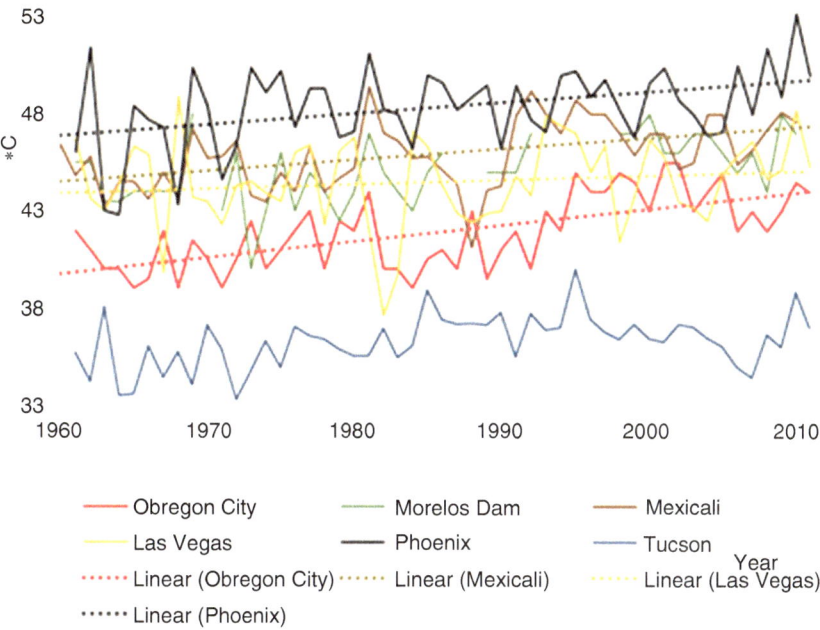

Fig. 9.9 Maximum monthly temperature variations and linear trend lines for August, in six cities of the Sonora–Mojave Desert region in the period 1960–2011 (Martinez-Austria and Bandala 2017)

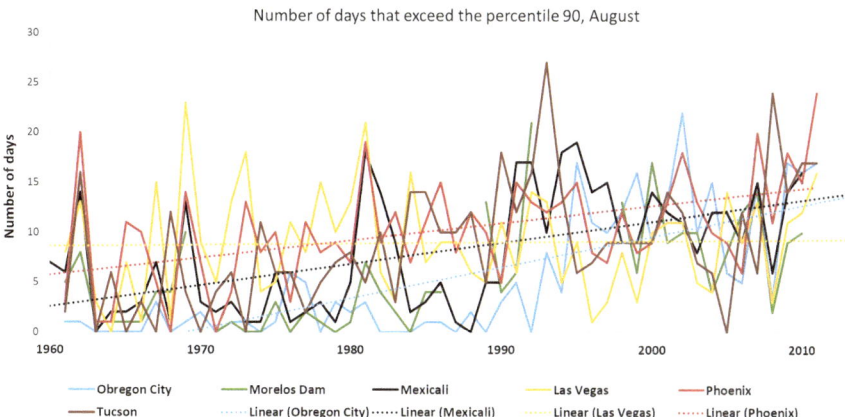

Fig. 9.10 Number of days exceeding the 90th percentile of the maximum temperature during the month of August in the period 1960–2011 in six cities of the Sonora–Mojave Desert region (Martinez-Austria and Bandala 2017)

rainfall in the RCP6.0 scenario, for spring–summer, which is the period of highest rainfall in the country, calculated by Salinas Prieto et al. (2015), with respect to the base period 1971–2000. Decreases in precipitation are expected throughout Mexico, with reductions of between −9% and −15% in the north and northwest of Mexico,

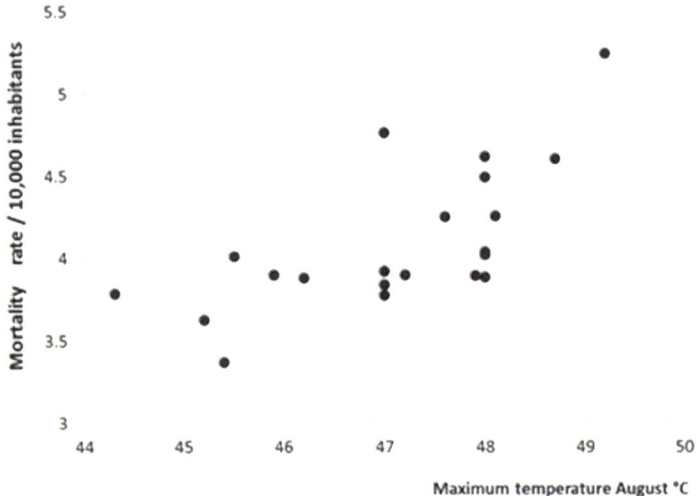

Fig. 9.11 Total mortality rate (per 10,000 inhabitants) versus maximum temperature during August (1990–2010) in Mexicali, Mexico (Martinez-Austria and Bandala 2017)

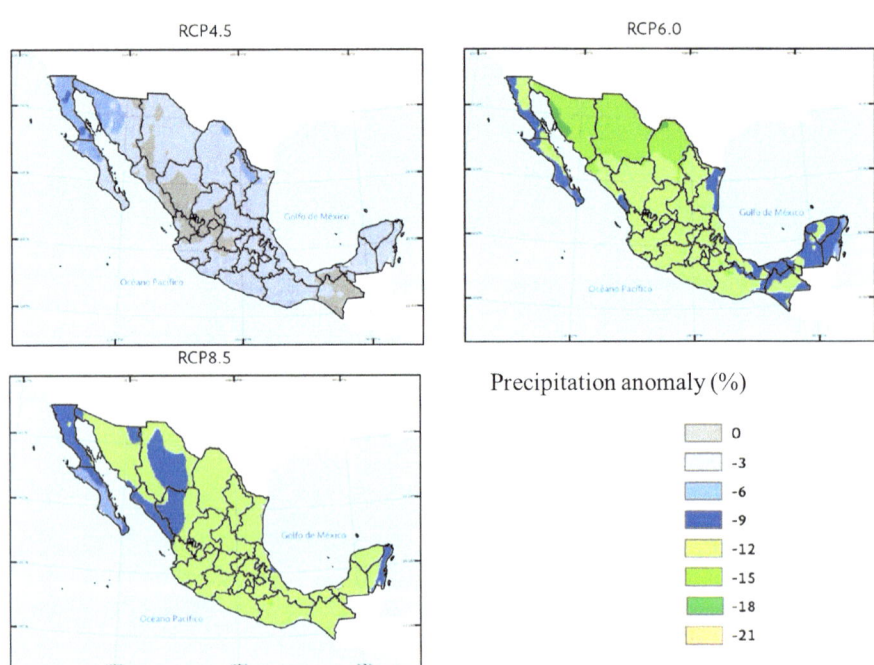

Fig. 9.12 Changes in spring–summer rainfall for the period 2075–2099 in Mexico, for several climate-change scenarios, compared to the period 1971–2000 (Salinas Prieto et al. 2015)

the region where the main irrigation systems are located, and which includes watersheds where water stress and scarcity are already expected, caused by growing demand, which will be aggravated by this decrease in precipitation.

Due to factors such as higher evapotranspiration of natural vegetation, drier soils, and higher evaporation, the expected reduction in runoff will be greater than that estimated in precipitation. As mentioned before (see Fig. 9.3), in Mexico, significant runoff reductions between 10% and 30% are expected. However, these results come from models of global circulation and have great uncertainty at local scales. Determination of its regional magnitudes requires analysis by basin, through the application of rain–runoff models that, although the simplest ones are used, require a significant quantity of local information of the basin under study. Probably for this reason, there are few studies on runoff vulnerability in Mexico.

Rivas Acosta et al. (2010) conducted studies on changes in runoff for some of the major basins in Mexico. In the Conchos River basin, for example, using the balance method of the Mexican official standard for water-balance calculations, and applying the estimated decrease in scenario A2 precipitation, they determined reductions of 25% in the runoff for 2100. These results, in a basin that already has extractions that keep new water allocations practically unavailable, represent a very important challenge for the future management of water and make expensive adaptive measures necessary in the agricultural sector.

9.7   Extreme Events

Most of the research works on the effects of climate change have focused mainly on the averages of climatic variables, that is, for example, the change in medium temperature or average annual precipitation. However, in an atmosphere and in an ocean with greater energy, more intense extreme phenomena are expected, mainly, more frequent heat waves with higher maximums and extreme rainfall or droughts with greater recurrence and intensity, as well as larger storms and floods. These phenomena are no longer a matter only for the future, since in fact they are recorded year after year with greater severity in various parts of the world. As stated: "The signal of climate change is emerging from the "noise"—the huge amount of natural variability in weather" (Carey 2012).

The number and cost of natural disasters related to water have registered a continuous rise in recent decades (UNESCO 2009), as shown in Fig. 9.13. As can be seen, especially since 1990, the number of disasters, such as floods and damaging winds, has continuously increased. Glokany (2009) has conducted a study of trends in the number of climate-related disasters over the last century. The result is not very encouraging, as shown in Fig. 9.14, which makes evident the exponential growth.

Extreme rainfall, with its harmful effects such as floods and landslides, causes loss of human life and damage to infrastructure and the productive sector and often reverses years of progress. Consequently, flood protection is among the greatest challenges to water security in Mexico. In 2010 alone, the cost of damage caused by extreme hydro-meteorological phenomena amounted to US$6600 million. In Nuevo

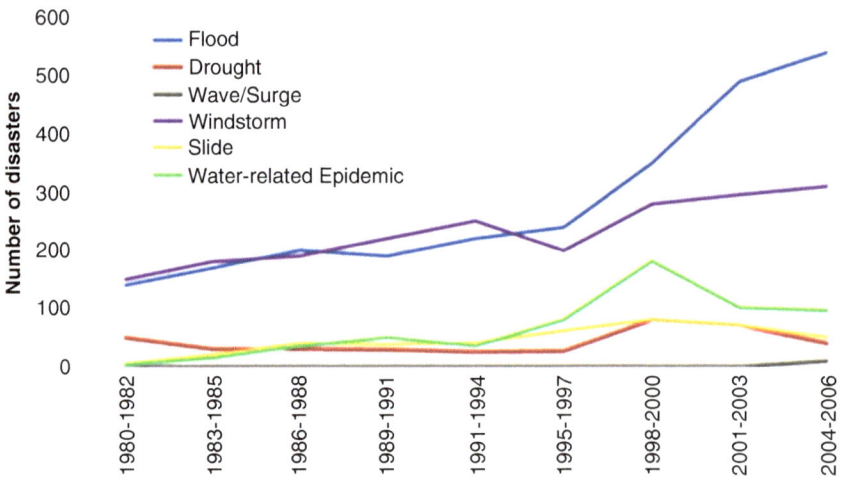

Fig. 9.13 Water-related disasters, 1980–2006 (UNESCO 2009)

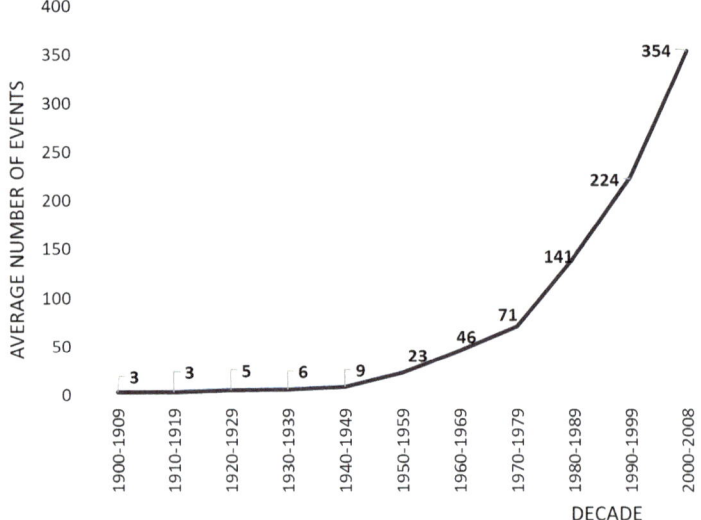

Fig. 9.14 Water-related disasters, recorded in the period 1980–2006. Average by decade. (With data from Glokany 2009)

León, the damage from Hurricane Alex in 2010 represented 2.45% of the state's Gross Domestic Product (GDP), and in Veracruz, in 2016, the floods caused by the storms Karl and Matthew inflicted damage equivalent to 4.8% of the state GDP. In 2010, 739 municipalities in the country received a declaration of natural disaster due to hydro-meteorological events (CENAPRED 2012). In 2007, most of the state of

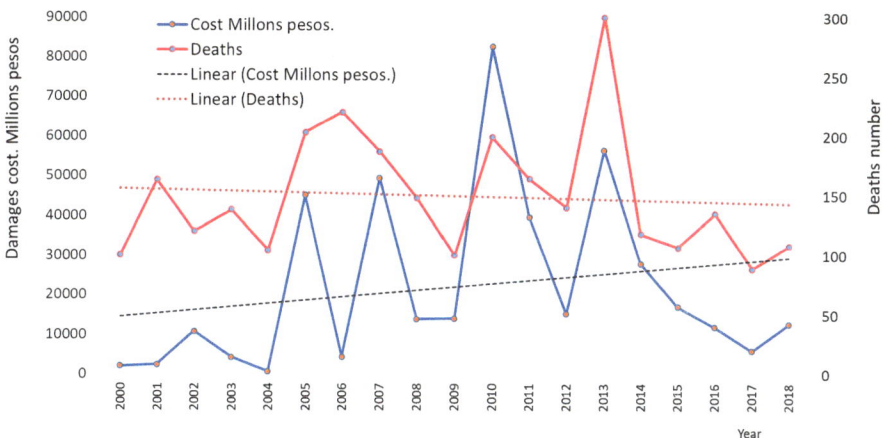

Fig. 9.15 Damage costs and deaths in Mexico due to hydro-meteorological disasters. (Prepared with data from CENAPRED's reports)

Tabasco suffered floods, with huge economic and social costs. Unfortunately, as can be seen in Fig. 9.15, the number of deaths and the damage costs in Mexico have been increasing continuously in recent years.

On the other hand, droughts are the hydro-meteorological events that cause the most extensive economic and social damage. Drought is associated with phenomena, such as climate migration and damage to the environment, including permanent ones such as desertification. Its frequency and intensity, presumably because of climate change, have also increased since 1970 to date, and are the cause of the greatest number of deaths among extreme climatic events, as is evident from the list of the 10 largest hydro-climatological disasters recorded in the period 1970–2012 (World Meteorological Organization 2014) and shown in Table 9.2.

In Mexico, recurrent droughts of great intensity occur that cover almost the entire national territory. Figure 9.16 shows the percentage of Mexican territory suffering drought of varying magnitude, from 2003 to 2017, according to the national drought monitor. The drought of 2012 covered more than 90% of the territory.

On the other hand, there is already a statistically significant trend of decreased precipitation in the northern region of the country, as will be subsequently shown.

One of the most accepted methods for estimating droughts is the Standardized Precipitation Index (SPI) (Hayes et al. 2011), which represents the number of standard deviations in which the transformed value of precipitation deviates from the historical average, which therefore represents the value zero. The SPI is used in most countries due to the availability of data, for its easiness of interpretation, as well as for its ability to be calculated for short or very long periods of time. Another advantage of SPI is that it enables observing not only the abnormally dry periods, but also the extremely humid ones, and the variability between them. Table 9.3 shows the humidity ranges of the SPI and their significance.

Table 9.2 Disasters ranked according to reported deaths, globally in the period 1970–2012 (World Meteorological Organization 2014)

Rank	Disaster type	Year	Country	Number of deaths
1	Drought	1983	Ethiopia	300,000
2	Storm (TC)	1970	Bangladesh	300,000
3	Drought	1984	Sudan	150,000
4	Storm (TC)	1991	Bangladesh	138,866
5	Storm (*Nargis*)	2008	Myanmar	138,366
6	Drought	1975	Ethiopia	100,000
7	Drought	1983	Mozambique	100,000
8	Extreme temperature	2010	Russian Federation	55,736
9	Flood	1999	Venezuela	30,000
10	Flood	1974	Bangladesh	28,700

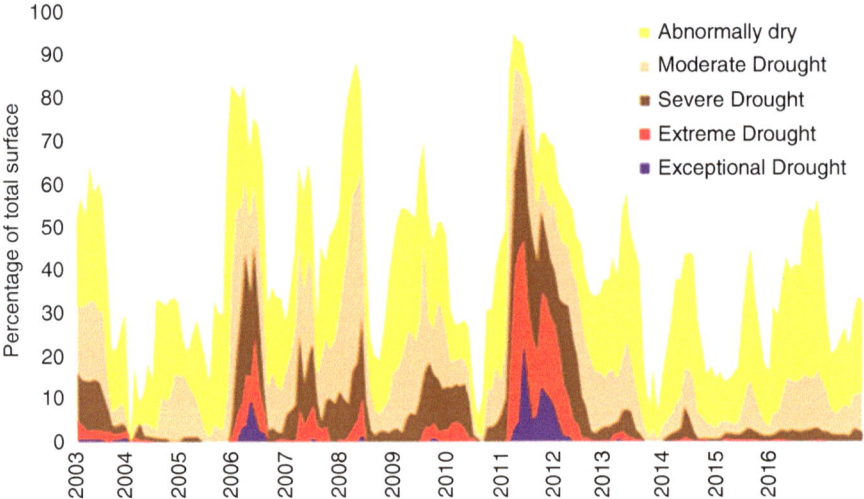

Fig. 9.16 Area of Mexico, by percentage, under various conditions of drought in the period 2004–2017. (With data from National Meteorological Service, SMN 2017)

As an example of precipitation trends in northern Mexico, consider the Conchos Basin, which is the main tributary of the Rio Grande, and is mainly in the state of Chihuahua. This basin is of great importance in Mexico because it supplies large irrigation districts, the capital of the state of Chihuahua, and other cities. In addition, it is of international importance because a proportion of its runoff must be delivered to the US, according to a treaty signed between both countries in 1944.

In Fig. 9.17, the calculated variation of the SPI at the La Mesa weather station, located in the state of Chihuahua, is shown. The SPI values for 6, 9, and 12 months are shown, as well as the linear adjustment for this last case. As can be seen, there is a clearly identifiable tendency to lower amounts of precipitation, especially since the 1970s. While in previous decades the precipitation was characterized by alternating

Table 9.3 Standard Precipitation Index (SPI) and weather conditions

SPI	Classification
2.0 or more	Extremely wet
1.5–1.99	Very wet
1.0–1.49	Moderately wet
−0.99 to 0.99	Near normal
−1.0 to −1.49	Moderate drought
−1.5 to −1.99	Severe drought
−2 or less	Extreme drought

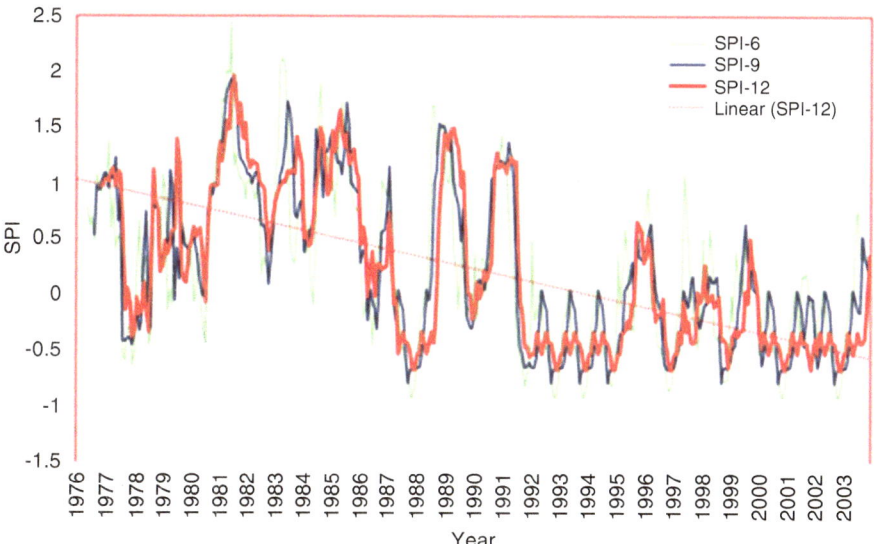

Fig. 9.17 Variation of the Standardized Precipitation Index in the La Mesa climatological station, in the Conchos River basin (Martinez-Austria and Irula-Luztow 2016)

wet and dry periods, there have been no wet periods since the beginning of the 1990s. This situation has caused the area cultivated in the irrigation districts of the region to rarely receive the needed water in all sectors, in addition to difficulties complying with the commitments of Mexico to the US, as established in the Water Distribution Treaty between both countries.

9.8   Concluding Remarks

Climate change will have great repercussions on Mexico's water resources. The combination of higher temperature and the current demographic processes will cause increasing growth in demand. Simultaneously, decreases in precipitation are expected in most of the continental territory, which will cause a decrease in the naturally available water. Thus, on the one hand, the demand for water will increase, and on the other, its natural supply will decrease. Together, both trends will cause huge pressure on Mexico's water resources.

The extreme weather phenomena, such as storms, floods, and droughts, have been increasing, in such a way that the cost of the damages and the loss of human lives have swelled. These extreme phenomena, under climate change, will continue to grow in intensity and frequency. Another extreme phenomenon of great importance for the people's health will be the maximum temperatures and their effects on human health, especially during heat waves.

These events are not only in the distant future. Many of the effects are already evident in obvious trends observed during recent decades. It is very important to be aware of this situation and not to postpone further the design and application of adaptive actions to increase resilience and reduce the vulnerability of cities, communities, agriculture, and the environment.

References

Blunden J, Arndt DS (2017) State of the climate in 2016. Bull Am Metorol Soc 98(8):Si–S277. https://doi.org/10.1175/2017BAMSStateoftheClimate.I

Carey J (2012) Extreme weather is a product of climate change. In: Storm warnings. Climate change and extreme weather. Scientific American, New York

CENAPRED (2012) Características e impacto socioecnómico de los principales desastres ocurridos en la República Mexicana en el año 2010. México, Centro Nacional de Prevencion de Desastres (National Center of Disasters Prevention)

CONAGUA (2010) 2030 water Agenda. Comision Nacional del Agua, México

CONAGUA (2015) Atlas del Agua en México 2015. Comisión Nacional del Agua, Ciudad de México

CRED-UNISDR (2016) Poverty & death: disaster mortality 1996–2015. Centre for Research on Epidemiology of Disasters (CRED) Université Catholique de Lovain, Belgium

Florke M, Alcamo J (2007) Assessment of global scale water stress indicators. In: Lozán J, Grassl H, Hupfer P, Schönwiese ML, Schönwiese C (eds) Wissensechaftliche Auswertungen. Hamburg, p 384

Glokany IM (2009) Deaths and death rates from extreme weather events: 1900–2008. J Am Physicians Surg 14(4):102–109

Hayes M, Svoboda M, Wall N, Widhal M (2011) The Lincoln declaration on drought indices. Am Meterol Soc (April):485–488

Hémon D, Jougla E (2003) Surmortalité liée á la canicule d'aout 2003. Rapport détape (1/3). Estimation de la surmortalité et principales caractéristiques épidémiologiques. INSERM, Paris

INECC (2016) Escenarios de cambio climático, November 13. From INECC Cambio Climático. http://escenarios.inecc.gob.mx/. Accessed 7 June 2017

IPCC (2014) Climate change 2014: synthesis report. Contribution of Working Groups I, II and III to the Fifth Assessment Report of the Intergovernmental Panel on Climate Change. IPCC, Geneva

Le Tertre A, Lefranc A, Eilstein D, Iercq C, Medina S, Blanchard M, . . . Ledrans M (2006) Impact of the 2003 heatwave on all-cause mortality in 9 French cities (Wilkins LW (ed)). Epidemiology 17(1):75–79

Martinez-Austria PF, Bandala ER (2017) Temperature and heat-related mortality trends in the Sonoran and Mojave Desert region. Atmosphere 8(53):13. https://doi.org/10.3390/atmos8030053

Martinez-Austria PF, Irula-Luztow F (2016) Trends of precipitation and climate change in Rio Conchos Basin, Mexico. Aqua-LAC 8(2):79–88

Mekonnen A, Renwick JA, Sánchez Lugo A (2017) Regional climates. In: Blunden J, Arndt DS (eds) State of the climate in 2016, vol 98. Bulletin of American Meteorological Society. American Meteorological Society, Boston, pp 173–228. https://doi.org/10.1175/2017BAMSStateoftheClimate.i

Ojeda Bustamante W, Flores Velázquez J, Ontiveros Capurata R, Íñiguez Covarrubias M (2015) Vulnerabilidad de la agricultura de riego en México ante el cambio climático. actualización 2014. En: Agua IM (ed) Atlas de vulnerabilidad hídrica de México ante el cambio climático. Instituto Mexicano de Tecnologia del Agua, Jiutepec, pp 101–130

Randall DA, Word RA (2007) Climate models and their evaluation. In: IPCC (ed) Climate 2007: impacts, adaptation and vulnerability. IPCC, Cambridge, pp 589–662

Rekacewicsz P (2006) Water stress index. https://www.grida.no/resources/5586. Accessed 18 Aug 2017

Rivas Acosta I, Güitron de los Reyes A, Ballinas González H (2010) Vulnerabilidad hídrica global: aguas superficiales. In: Austria PM, Gomez CP (eds) Atlas de vulnerabilidad hídrica de México ante el cambio climático. Instituto Mexicano de Tecnologia del Agua, Jiutepec, pp 81–114

Salinas Prieto JA, Colorado Ruiz G, Maya Magaña ME (2015) Escenarios de Cambio Climático para México. In: Atlas de vulnerabilidad hídrica de México ante el cambio climático. Instituto Mexicano de Tecnologia del Agua, Jiutepec, pp 41–70

Schewe J, Heineke J, Gerten D, Haddelhand I, Arnell N, Clark D et al (2014) Multimodel assessment of water scarcity under climate change. Proc Natl Acad Sci U S A 111 (9):3245–3250. https://doi.org/10.1073/pnas.1222460110

Smakhtin V, Revenga C, Döll P (2004) In: CGIAR (ed) Taking into account environmental water requirements in global scale-water resources assessment. Comprehensive Assessment Secretariat, CGIAR, Colombo

SMN (2017) Monitor de Sequía. On line in National Meteorological Service, Mexico. http://smncnagobmx/es/climatologia/monitor-de-sequia/monitor-de-sequia-en-mexico.   Accessed 8 Dec 2017

UNDP (2006) Human development report 2006. Beyond scarcity: power, poverty and the global water crisis. United Nations Development Programme, New York

UNESCO (2009) Global trends in water-related disasters: an insight for policymakers. United Nations Educational, Scientific and Cultural Organization, Paris

Water Resources Group (2009) Charting our water future. Economic frameworks to inform decision-making. 2030 Water Resources Group, New Delhi

Whitman S, Good S, Donoghue ER, Benbow N, Shou W, Mou S (1997) Mortality in Chicago attributed to the July 1995 heat wave. Am J Public Health 87(9):1515–1518

World Economic Forum (2011) Water security. The water-food-energy climate nexus. Island Press, London

World Economic Forum (2017) The global risks report 2017. World Economic Forum, Geneva

World Meteorological Organization (2014) Atlas of mortality and economic losses from weather, climate and water extremes (1970–2012). WMO, Geneva

Chapter 10
Water Security and Sustainability in Mexico

Felipe I. Arreguin-Cortes and Claudia Elizabeth Cervantes-Jaimes

Abstract This chapter provides an analysis of the water risk in Mexico based upon its geographic location, the current situation of its water resources, its population growth, and the impact of economic globalization and of current and future technological changes, as well as the effects of global climate change, all of which place several regions of the country at high water risk and lead to the conclusion that Mexico needs a sound water-security program.

It is acknowledged that, since 1976, upon developing the National Water Plan and setting it as a guiding instrument for water-policy actions that must be carried out by the federal administration every 6 years—through cross-cutting, sectorial, regional, special, and institutional programs—Mexico has had the elements for establishing a water-security program based on three guidelines: efficiency, sustainability, and a cross-cutting approach, with a multidimensional application of at least six guiding principles: governance, social, economic, environmental, human development, and global.

Finally, the point is made that scientific and technological knowledge is crucial for bolstering a water-security program.

10.1 Introduction

Since the UN Bretton Woods Agreement of 1944, remarks have been made about the risk of living in a world without water and with part of nature in danger of extinction.

In the 1970s, and especially at the United Nations Water Conference held in Mar del Plata, Argentina, in 1977, a second global warning was presented, derived from the Club of Rome's prospective study *The Limits to Growth*.

F. I. Arreguin-Cortes (✉)
Instituto de Ingeniería UNAM, Ciudad de Mexico, Mexico
e-mail: farreguin2011@gmail.com

C. E. Cervantes-Jaimes
Tecnológico de Monterrey, ITESM, Monterrey, Mexico
e-mail: celizacervantes@gmail.com

© Springer Nature Switzerland AG 2020   177
J. A. Raynal-Villasenor (ed.), *Water Resources of Mexico*, World Water Resources 6,
https://doi.org/10.1007/978-3-030-40686-8_10

In the Rio de Janeiro Earth Summit, Brazil, 1992, there was a third warning due to the unsustainable use of natural resources, and in the 2nd World Water Forum, held in The Hague in 2000, it was concluded that water insecurity was growing in many regions of the world.

At the World Summit on Sustainable Development, held in Johannesburg, South Africa in 2002, continuous warnings were made about the insecurity of water resources because there had been no significant advances by that date. In the 6th World Water Forum, held in Marseille in 2012, member states were called upon to take on formal commitments in relation to water and environmental conservation, and in the 7th World Water Forum in Daegu-Gyeongbuk, the Republic of Korea, 2015, it was concluded that the solutions agreed upon in Marseille had been insufficiently adopted, so it was requested that a group of experts work on water security and a global strategy be defined.

The World Bank report entitled *High and Dry: Climate Change, Water, and the Economy*, presented at the World Economic Forum in Davos, Switzerland, in January 2016, warns for the first time that the demand for water will rise exponentially, due to the combined effects of growing populations, rising incomes, and expanding cities. This will lead to increasing water scarcity, exacerbated by climate change.

And, since 1988, the Intergovernmental Panel on Climate Change (IPCC) has issued assessment reports; the sixth of them is currently being drafted. In the fifth Assessment Report (IPCC 2013), the effects of climate change on the water cycle were explicitly pointed out:

Changes in the global water cycle in response to warming over the twenty-first century will not be uniform. IPCC (2013) stated, "The contrast in precipitation between wet and dry regions and between wet and dry seasons will increase, although there may be regional exceptions. The global ocean will continue to warm during the 21st century. Heat will penetrate from the surface to the deep ocean and affect ocean circulation."

The Arctic sea ice cover will very likely continue to shrink and thin, and the Northern Hemisphere spring snow cover will decrease during the twenty-first century as the global mean surface temperature rises (IPCC 2013). The global glacier volume will further decrease, too. What's more, the global mean sea level will continue to rise during the twenty-first century (IPCC 2013).

IPCC (2013) stated, "Under all RCP scenarios, the rate of sea level rise will very likely exceed that observed during 1971 to 2010 due to increased ocean warming and increased loss of mass from glaciers and ice sheets."

In December 2015, the Paris Agreement was signed, whose goal was to prevent the average global temperature from increasing more than 2 °C over preindustrial levels and to make additional efforts, so that global warming did not exceed 1.5 °C.

But despite all these warnings supported by sound scientific studies, and the various commitments established by nations, water resources are still being negatively affected in many regions of the world.

10.2   Water Security

The concept of water security has been defined by various institutions, including the Global Water Partnership (GWP 2000), the United Nations Water Group (UN-Water 2013), the International Hydrological Program of the United Nations Educational, Scientific and Cultural Organization (IHP-UNESCO 2012), and the United Nations Economic Commission for Latin America and the Caribbean (Peña 2016).

In Mexico, the Mexican Institute of Water Technology (IMTA) has adopted the concept of water security proposed by the Economic Commission for Latin America and the Caribbean (ECLAC) and has adapted it to the national context as follows:

To ensure the capacity to access, utilize, use, and harness water in a sustainable way, as well as to manage, plan, handle, and administer, in an integrated fashion, the interrelations and interventions among the various sectors associated with water resources.

To ensure the availability of water in an adequate quantity and of acceptable quality to preserve a climate of peace and political stability, to sustain and protect water supplies for all living beings and for all social, economic, and environmental activities.

To ensure the capacity to mitigate and adapt to acceptable and manageable levels in the face of natural and anthropogenic phenomena related to the quantity and quality of water that endanger the population, the economy, and the environment.

Climate change will have noticeable effects on temperature and rain patterns in many regions of the planet, with hydrological impact reflected in sea-level rise, more intense and frequent heat waves, reduced or lost permafrost or periodic snow, growing destructiveness of tropical cyclones, the movement of cyclone and/or tornado-prone zones, and the quicker reentry of precipitation water to the atmosphere due to increasing evaporation, among other factors, all of which will increase water insecurity and will multiply the risk of conflicts. In conclusion, water insecurity is an important threat to economic growth and stability worldwide.

Due to its geographic location, the current situation of its water resources, its accelerated population growth, the impact of economic globalization and of current and future technological changes, as well as the effects of global change, Mexico needs a sound water-security program.

10.3   Mexico: A Vulnerable Country

The north of Mexico is close to 30° north latitude, where antitrade winds descend and preclude the development of important cloud groups that may produce rain. This region is also known as the strip containing the world's major deserts; therefore, this part of the territory has an arid climate and scant precipitation (see Fig. 10.1).

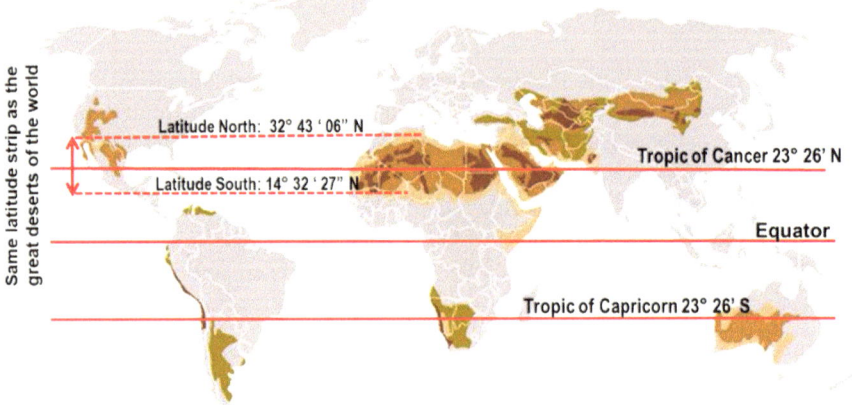

Fig. 10.1 Location of Mexico in the strip of the great deserts

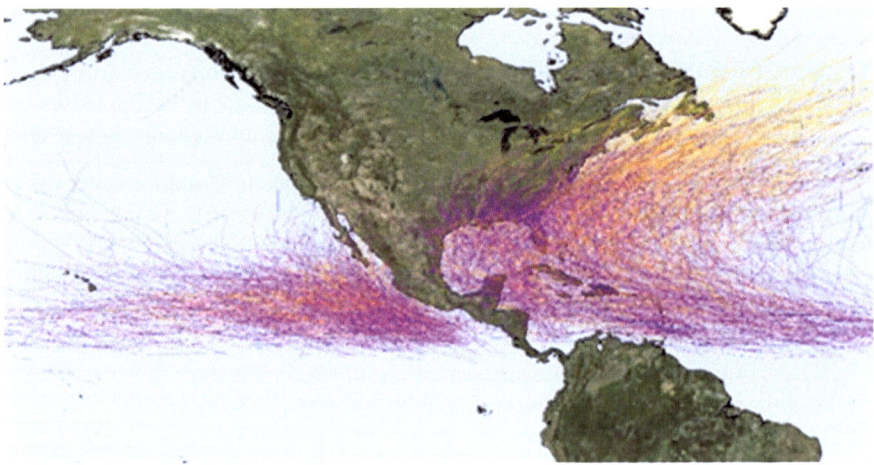

Fig. 10.2 Historical paths of the tropical cyclones of both regions. (Proprietary development, with data from NOAA 2016)

The difference in climate in the Mexican Republic with respect to other regions located in the strip of the major deserts is due to its location close to two very active cyclonic regions: those of the Atlantic and North Pacific (see Fig. 10.2). Atmospheric global circulation patterns, and its proximity to the Intertropical Convergence Zone, generate a nearly permanent provision of humidity; thus, the arid conditions are balanced out by the presence of hurricanes and tropical storms, among other meteorological phenomena.

Historical records of global temperature made by NASA began in 1880 (see Fig. 10.3). According to these data, the planet has warmed 1.1 °C since the late

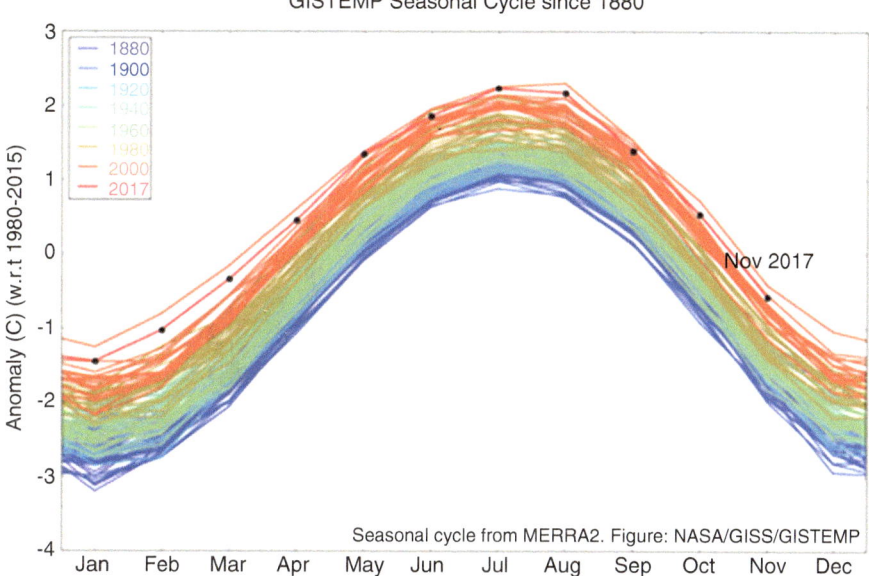

Fig. 10.3 Monthly temperature anomalies, 1880–2015. Goddard Institute for Space Studies Surface Temperature Analysis (GISTEMP Team 2018)

nineteenth century, although most of the global warming has occurred in the last 35 years, with 16 of the 17 warmest years recorded since 2001. The year 2016 was the third consecutive year in which record global temperatures were recorded (NASA 2017). According to NASA, global surface temperatures in 2017 ranked as the second warmest since 1880, and the National Oceanic and Atmospheric Administration (NOAA), in an independent analysis, concluded that 2017 was the third-warmest year in their records (NASA 2018).

Some of the effects of this global temperature increase are precipitation events of shorter duration and higher intensity, greater evaporation, longer dry seasons, and more intense and frequent heat waves; these factors render many regions of the country vulnerable. CNA (2017) estimated the vulnerability to drought as a function of the following:

- Exposure level: the quantification of the difficulty of a planning cell (a group of municipalities that belong to the same state, within the limits of a hydrological subregion) to satisfy its water demand by 2030
- Sensitivity: an estimate of the impact on economic, commercial, industrial, and agricultural activities of the population by 2030
- Adaptation capacity: the degree of aquifer exploitation (see Fig. 10.4)

Another worrisome effect is the risk of floods in many regions of Mexico, since 162,000 km^2 of its territory is susceptible to being affected by this phenomenon (see Fig. 10.5).

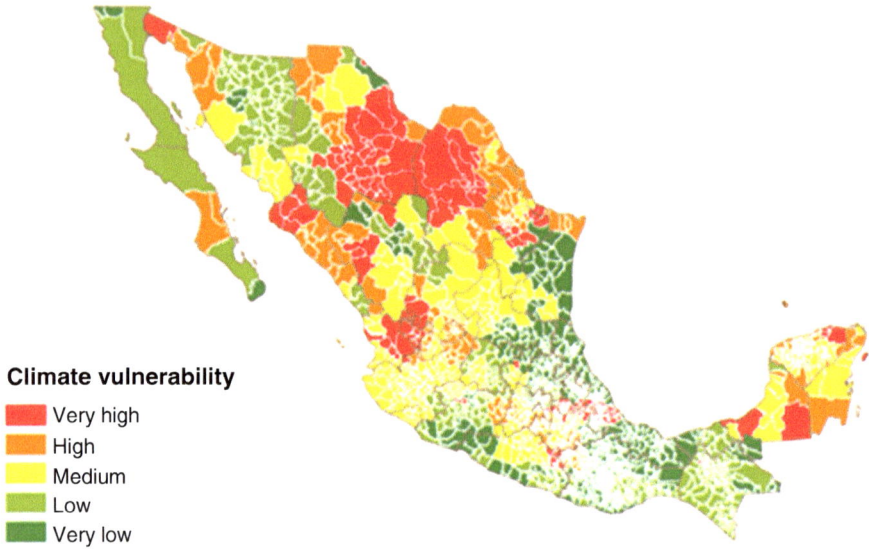

Fig. 10.4 Climate vulnerability at the municipal level in 2012 (CNA 2017)

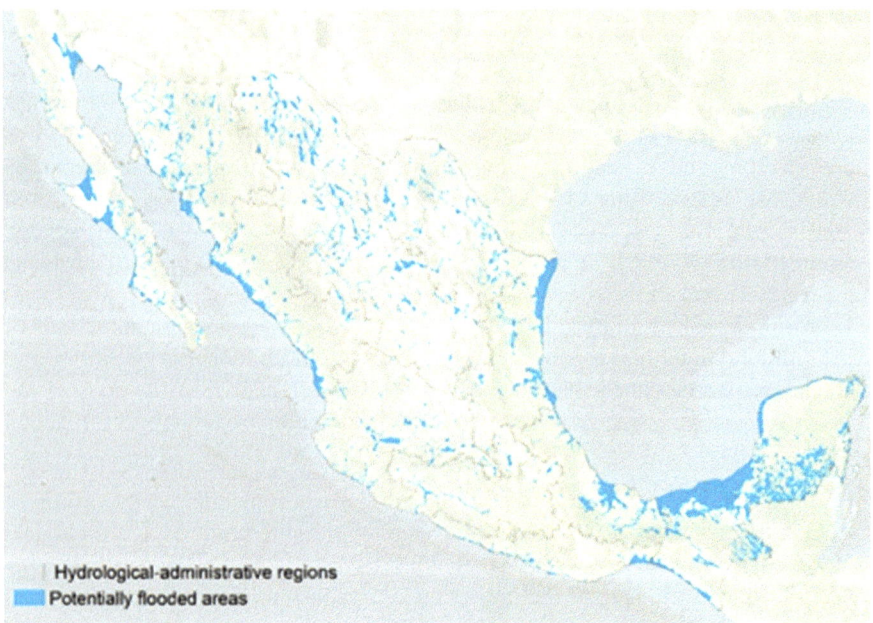

Fig. 10.5 Potential inundation zones. (Proprietary development, with data from INEGI 2012)

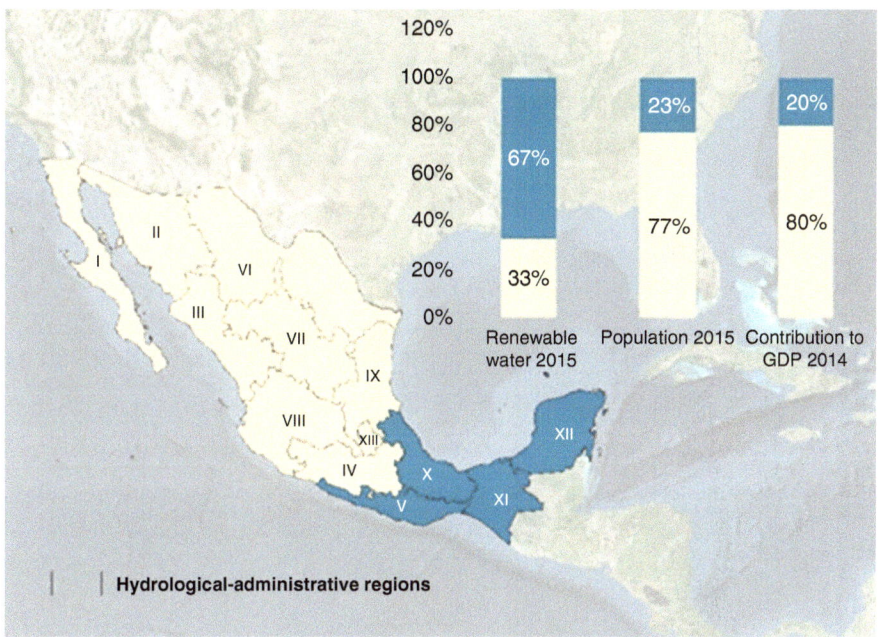

Fig. 10.6 Per capita availability of renewable water, population, and Gross Domestic Product (GDP). (Proprietary development, with data from CNA 2017)

Population growth, besides being directly related to increasing demand, has an unfavorable distribution regarding the availability of water resources. Seventy-seven percent of the population is settled where there is only 33% of total water resources available (see Fig. 10.6).

Figure 10.7 shows the water cycle at a national level. Normal precipitation in the country (1981–2010) is 740 mm. Considering the extent of the territory, it means that Mexico receives 1449.5 km^3 of water annually, 1050 km^3/year of which is lost to evapotranspiration.

From that annual volume, 307 km^3/year runs off through the national territory and 91.8 km^3/year infiltrates into aquifers. It is important to note that the national runoff includes the inflows that Mexico receives from the USA, established by the Treaty on International Water Distribution between the United Mexican States and the USA (Water Treaty, IBWC 1944), through the Colorado River, as well as those water inflows coming from Guatemala, with whom there is no treaty, through the Grijalva and Usumacinta rivers. For its part, Mexico delivers to the USA, from the Rio Bravo watershed, an average of 431.70 hm^3 annually within 5-year cycles, complying with the 1944 Treaty.

For various uses, 33.3 km^3/year and 52.3 km^3/year are extracted from aquifers and surface sources, respectively. The agro-livestock sector is the main consumer of this resource, using 65.4 km^3/year.

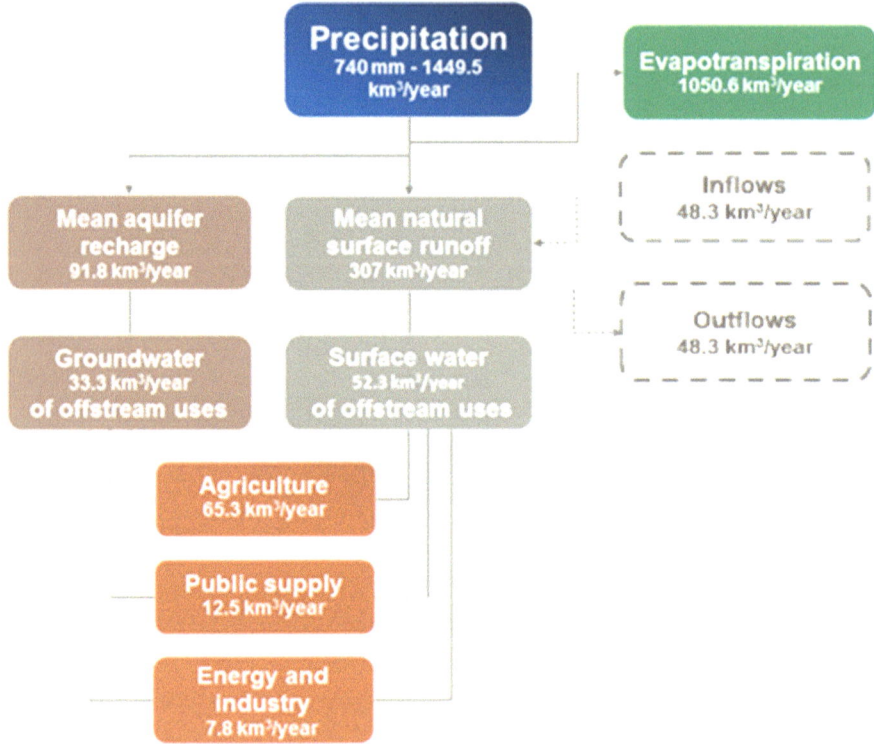

Fig. 10.7 Water cycle at the national level. (Proprietary development, with data from CNA 2017)

10.4   Water Availability

In the national territory, surface water is distributed throughout 731 watersheds. Eight are transboundary watersheds, and 627 had water availability, according to the information published December 31, 2015 (Fig. 10.8).

Groundwater availability is monitored in 653 hydrogeological units, which, for practical purposes, are called aquifers, although they are actually groups of them (Fig. 10.9). The number of overexploited aquifers as of December 31, 2015, is 105 (CNA 2017).

Water quality is measured by means of several parameters, three of which are: Five-Day Biochemical Oxygen Demand (BOD5), Chemical Oxygen Demand (COD), and Total Suspended Solids (TSS). TSS measures the amount of sedimentation solids, suspended solids, and organic and/or colloidal material (Fig. 10.10). They originate in wastewater and from soil erosion (CNA 2017).

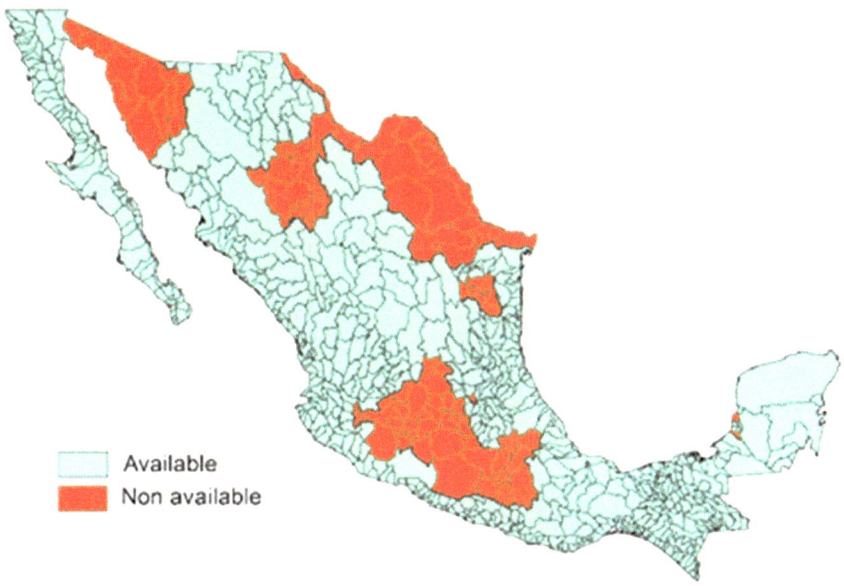

Fig. 10.8 Surface-water availability in Mexico. (Proprietary development, with data from SINA 2017)

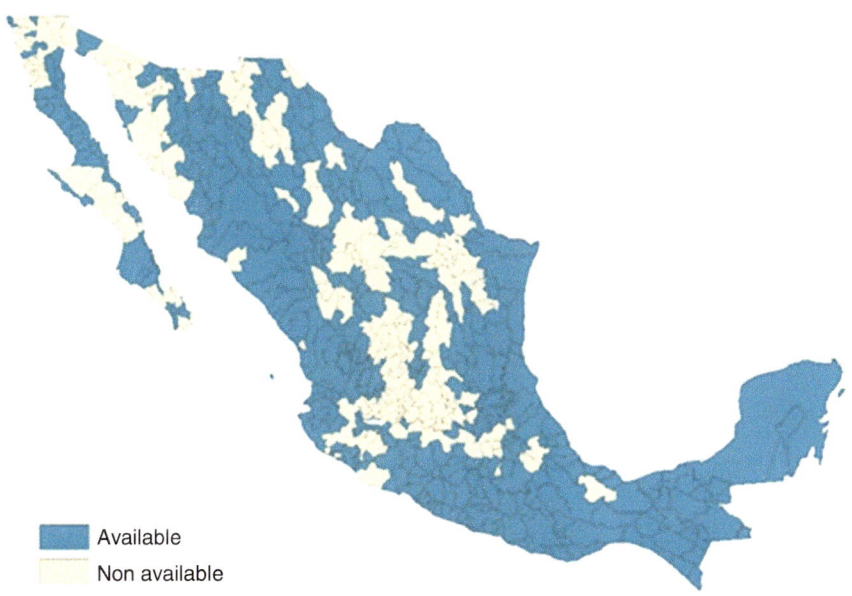

Fig. 10.9 Groundwater availability at the national level. (Proprietary development, with data from SINA 2017)

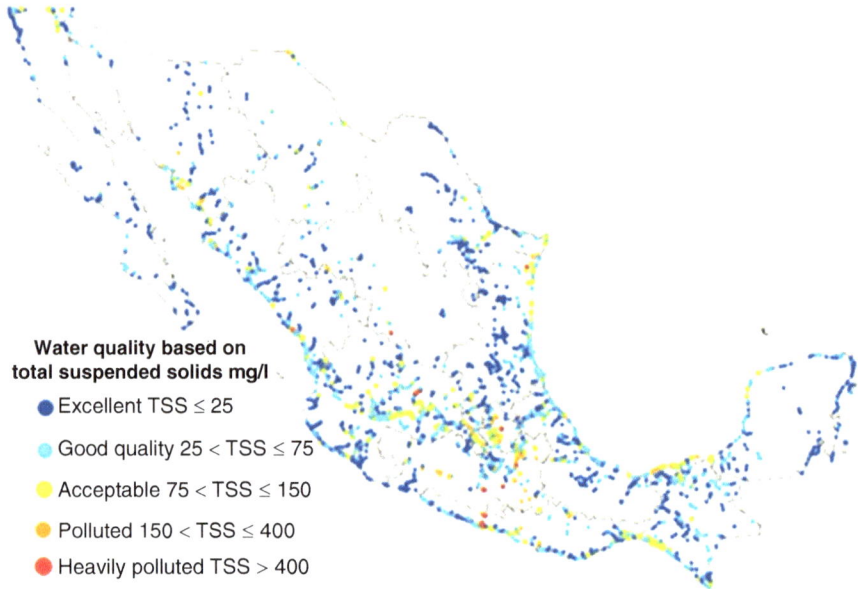

Water quality based on
total suspended solids mg/l

● Excellent TSS ≤ 25

● Good quality 25 < TSS ≤ 75

● Acceptable 75 < TSS ≤ 150

● Polluted 150 < TSS ≤ 400

● Heavily polluted TSS > 400

Fig. 10.10 Monitoring sites and interpretation of quality according to TSS levels. (Proprietary development, with data from SINA 2017)

10.5 The Main Challenges of Water Management

Efficient water management in Mexico involves finding solutions to factors such as scarcity, contamination, impact of global change on the water cycle, lack of land-use planning, the need for better regulations, increasing financial resources, and greater investments in technological development and research.

An example of how sea-level rise affects the national territory is the Yucatan Peninsula, one of the most vulnerable regions to this effect. Here, the main water supply is the aquifer, whose natural current situation is that of a reduced hydraulic head above sea level, high transmissivity, the presence of saline interface at shallow depth, and a reduced volume of freshwater. Due to these conditions, the sea-level rise could translate into a considerable depletion of freshwater in that region (see Fig. 10.11).

On the other hand, many of the problems related to natural disasters are due to poor land-use planning. The invasion of water into federal and protected zones, as well as of streams, lakes, and reservoirs, has been the cause of tragedies that have occurred with floods that many times do not even come close to probable maximum flows.

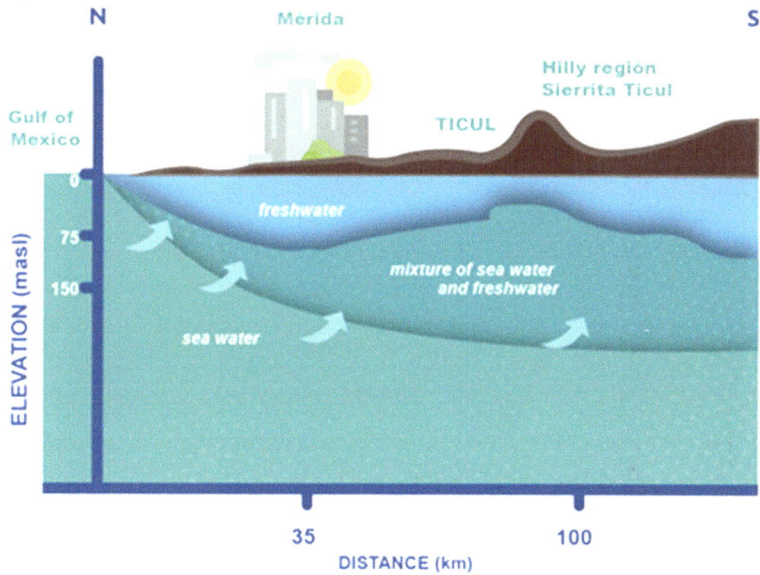

Fig. 10.11 Vulnerability of the Yucatan Peninsula

An example of this happened in the municipality of Motozintla, Chiapas, in 2005, which suffered serious economic and human losses during the passing of Hurricane Stan. All this was due to the combined threat or phenomenon and the exposure of a population in vulnerable conditions because most of the houses in that zone did not have the necessary structural characteristics to endure intense precipitations or flooding, much less that inflicted by the concomitant dragging of debris (see Fig. 10.12).

Several theses state that the problem of supplying water to a population to meet their needs is not that of scarcity, but of management. Without taking a stand on any of these two extremes, Mexico requires better water management. This can only be achieved with a new General Water Law.

In our country, investment in technological research and development amounted to 0.57% of the Gross National Product (GNP) in 2016, which represents the lowest level among members of the Organization for Economic Cooperation and Development (OECD).

An increase in investment in technological research and development would result in the expansion and enhancement of critical aspects such as measurement, instrumentation, and monitoring, modeling, weather forecasting, and training of experts and decision makers, as well as the creation of tools for outreach to society, creating awareness of the situation and explaining how people can contribute to face the main challenges.

Fig. 10.12 Population settled on the riverbed in Motozintla, Chiapas, after the passing of Hurricane Stan, October 2005

10.6   Technology and Water

Hydroinformatics is the application of information and communications technologies to water-resources management. Some of these technologies had their origins in the third industrial revolution in the second half of the twentieth century and include tools such as mathematical modeling.

However, disruptive technologies, belonging to the fourth industrial revolution, such as The Internet of Things, Big Data, and Cloud Computing, are not being used to the fullest in water management, and although the impact of revolution 4.0 in all fields is unknown, its potential for improving the efficiency of processes is well established; therefore, there are reasons to have high expectations for their inclusion (see Fig. 10.13).

Some of the advantages that technology offers for the management of water resources are as follows:

- Remote hydrometeorological and hydrometric data acquisition in watersheds and aquifers
- Storage and management of large amounts of heterogeneous data on every watershed and event (big data)
- Automated monitoring systems based on remote sensing (Internet of Things) to speed up processing and forecasting
- Greater productivity in less time and with fewer resources

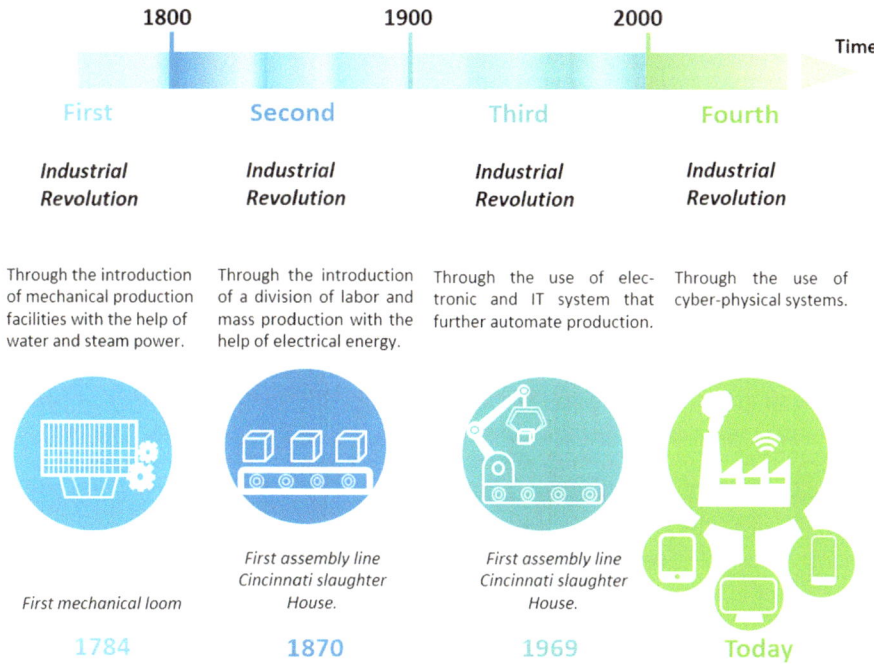

Fig. 10.13 Industrial revolutions

- Generation of scenarios that consider several factors and variables by using calibrated models with updated and reliable information
- Development and implementation of Early Warning Systems
- Synthesis of relevant information for supporting decision-making

Satellites have contributed to the exponential growth of the amount of available data in the hydrometeorological field. In hydrology and geohydrology, they are used for the measurement and transmission of data and the identification of flooded zones and aquifers; in meteorology and climatology, they are an essential tool for modeling, the follow-up of tropical cyclones, nowcasting, the monitoring of the surface temperature of the sea, and the seasonal movement of the Intertropical Convergence Zone (ITCZ). In November 2017, the GOES-16 satellite (initially called GOES-R) began operating. This new generation of satellites will enable better observation of environmental phenomena.

LIDAR (Light Detection and Ranging or Laser Imaging Detection and Ranging) enables calculating the distance from a laser emitter to an object, by measuring the time it takes from the emission of the pulse to the detection on the reflected signal. Its use in engineering is widespread in the field of cartography. It is used in topographical, hydrological, plant cover, and infrastructure (roads, dams, bridges, buildings, etc.) and location surveys, among other uses.

Unmanned Aerial Vehicles or drones can be equipped with Global Positioning System (GPS), infrared sensors, high-resolution cameras, and satellite or ground-station information transmission systems, depending on their cargo capacity. Some of LIDAR's applications in water-resources management are the following: level monitoring, saline-concentration monitoring, temperature monitoring, discharge estimation, leak detection in pipes, and soil modeling.

In Mexico, robots have been used in drinking water and sewerage systems to inspect pipes, manholes, and interconnections of home intakes or discharges to mains. There are more sophisticated submarine robots that have been used for the inspection of dam walls, lakebeds, and well networks and oil platforms.

One of the tools that is already being used is the Internet of Things (IoT). This is a network of interconnected objects with exclusive access, based on standard communication protocols (Bassi and Horn 2008). From this interconnection, it is possible to obtain information on their status by monitoring variables such as temperature, movement, humidity, and level—to mention just a few—by means of sensors that are interconnected with a wireless network.

Big Data, for its part, refers not only to a large amount of data, but also includes specific techniques for their management because traditional techniques are not sufficient or effective to handle massive amounts of data.

Big Data tools have a great potential for the management of hydrometeorological data, since their sources of information and formats vary greatly. Among these, we can mention data from conventional stations, from automatic stations, satellite images, radar images, and results from atmospheric, hydrological, and hydraulic models.

Machine learning consists of algorithms that perform automatic learning through tasks such as pattern recognition, category classification, and trial-and-error testing, where a response is used as feedback. Neuronal Networks—a type of algorithm to generate artificial intelligence that consists in developing a relationship between input and output variables to extract the complex behavior among variables—have been used for modeling and forecast of runoff, groundwater levels, and estimations of the shape of a hydrogram from meteorological and physiographical parameters.

Using cloud computing, it is possible to perform tasks that require great computational capacity without purchasing costly hardware and software. The National Institute of Standards and Technology (NIST) defines it as "a model for enabling convenient, on-demand network access from any place (internet) to a shared pool of configurable computing resources (e.g., networks, servers, storage, applications, and services) that can be rapidly provisioned and released with minimal management effort or service provider interaction" (Zhang et al. 2010).

One of the needs in water-resources modeling is high computational capacity and the use of Big Data, which is why cloud computing offers opportunities for sharing information, as well as software and platforms for the execution of models.

10.7 Water Security in Mexico

In Mexico, the relevance of water issues was laid out in the political agenda for the first time in the mid-1970s, when the National Hydraulic Plan was presented in 1976. From this point forward, and once the Planning Law was issued in 1982, the National Development Plan was established as a guiding instrument for water-policy actions that must be carried out by the federal administration during every 6-year term, and with it the planning of sectorial policies to be carried out during the term by means of cross-cutting, sectorial, regional, special, and institutional programs.

Within this regulatory and guiding framework, water has been instituted as a national security issue by considering the need to protect our country from external and internal threats and risks, as well as to preserve our democracy through sustained economic, social, and political developments. In this sense, at the sector level and as a national premise, the National Water Plan aims to achieve water security and sustainability.

As previously noted, Mexico is vulnerable due to its geographic location, to a natural water scarcity, coupled with various factors, including a management that has led to consequences that are now significant. For instance, in 1975, with a population of 62 million inhabitants, there were 36 overexploited aquifers, no watersheds with a water deficit, and only a few cases of contamination. On the other hand, in 2015, with 119 million inhabitants, there were 105 overexploited aquifers, 104 watersheds with a water deficit, and 60% of wastewater (municipal and industrial) that was generated in the main urban centers and irrigation zones in the country and that was not treated, contaminating rivers or other water bodies in the country.

Water-use practices in Mexico show that demand is greater than available supply in many regions, which has created an overexploitation of our water resources, putting them at a high risk. If water continues to be used and exploited without considering a sustainable strategy, with this model of exploitation and the impact of climate change on the hydrological cycle, the deficit volume will double in the next 15 years.

Some indicators of this situation are the reduced capacity of dams, the contamination of underground and surface-water bodies, the number of deforested watersheds, the overurbanized cities that affect aquifer recharge and increase runoffs, the more intense rains and more severe droughts, a larger number of zones that are vulnerable to climate variability, potential conflicts in watersheds and aquifers, inefficiency in the use of water, insufficient institutional capacities and qualified personnel, a complex and difficult-to-implement legal framework, and limited resources for cutting-edge research, among others.

Water issues must be analyzed in a multidimensional manner, since the causes and effects that limit development come from various different sources. The problems are multifactorial, and, therefore, highly complex. Thus, Mexico requires scientific and technological knowledge and innovation to face the changes of global change and climate variability, to enhance and generate efficient water governance, and to build a future with sufficient water, in terms of both quantity and quality, for social, economic, and environmental development.

It is necessary to create capacities to adapt and protect ourselves against the natural and anthropogenic risks that generate the destructive force of water, so that communities are better protected from floods, droughts, landslides, erosion, and waterborne diseases.

It is urgent for decision makers to have the capacities and instruments for assuring sustainable access to water in a sufficient, salubrious, acceptable, and affordable manner to lead a full, healthy, and productive life; for reducing poverty, promoting education, and enhancing quality of life, especially for the most vulnerable—mainly women and children—who are the ones who benefit the most from good water governance.

The development of a Water Security Program must be based on three guidelines: efficiency, sustainability, and a crosscutting approach, with a multidimensional application based on at least six guiding principles:

- Governance principle. Should be efficient for preserving a climate of peace, political stability, and protection.
- Social principle. Should ensure water supply for social well-being and quality of life.
- Economic principle. Should generate water productivity for economic growth.
- Environmental principle. Should ensure environmental sustainability and protection.
- Human development principle. Should guarantee the capacities for addressing water security.
- Global principle. Should ensure a global responsibility for water security.

  Figure 10.14 shows a general overview of the proposed water-security program.

10.8   The Role of Research, Science, Technology, and Innovation for Water Security

Viable solutions to face water issues depend, to a large extent, on scientific and technological knowledge. Thus, it is crucial to foster research, development, and innovation related to water security. In this sense, the participation of the academic and scientific sectors is of highly relevant importance.

The demand for water will continue to increase in the future, and it will be necessary to address the new needs in a framework of sustainability that will guarantee water for future users. Technological research and development are the instruments that will be used for solving the problems that preclude water from

Fig. 10.14 General
overview of the proposed
water security program

reaching the entire population. The full utilization of Mexico's water resources will
depend on the developments of the Research, Development, Technology, and
Innovation (RDTI) sector in the future.

In Mexico, the water sector requires scientific and technological elements that
support the water-security strategy, as well as the formulation of public policies that
lead to integration of a national program with a view to the future. With the aim of
enhancing the National Water Plan, the Mexican Institute of Water Technology
(IMTA) proposes the premise that water in our country be considered an element of
national security. IMTA's Institutional Strategic Program of Scientific and Technological Innovation for Water Security, according to the Planning Law, is aligned
with all the corresponding objectives of the National Development Plan and with the
programs that derive from it.

10.9   Conclusions

Due to its geographic location, the current situation of its water resources, its
accelerated population growth, the impact of economic globalization and of current
and future technological changes, as well as the effects of global change in the water
cycle, Mexico has many regions with a high level of water risk and therefore requires
a sound water-security program.

Since 1976, upon developing the National Water Plan and setting it as a guiding instrument for water-policy actions that must be carried out by the federal administration in every 6-year term—through cross-cutting, sectorial, regional, special, and institutional programs—Mexico has the elements for establishing a water-security program based on three guidelines: efficiency, sustainability, and a cross-cutting approach, with a multidimensional application of at least six guiding principles: governance, social, economic, environmental, human development, and a global perspective.

Viable solutions for meeting water-security issues depend on scientific and technological knowledge. Thus, it is crucial to foster research, development, and innovation related to water security. In this sense, the participation of the academic and scientific sectors is of critical, relevant importance.

References

Bassi A, Horn G (2008) Internet of things in 2020: a roadmap for the future. Euro Comm Inf Soc Media 22:97–114

CNA (Comision Nacional del Agua) (2017) Statistics on water in Mexico, 2016 edition. http://sina. conagua.gob.mx/publicaciones/EAM_2016.pdf. Accessed 6 Feb 2018.

GISTEMP Team (2018) GISS Surface Temperature Analysis (GISTEMP). NASA Goddard Institute for Space Studies. https://data.giss.nasa.gov/gistemp/news/20171218/. Accessed 6 Feb 2018

GWP (2000) Towards water security: a framework for action. GWP, Stockholm

INEGI (2012) INEGI Recursos Naturales [Internet]. http://www.inegi.org.mx/geo/contenidos/recnat/humedales/. Accessed 6 Feb 2018

IPCC (2013) Summary for policymakers. In: Stocker TF, Qin D, Plattner GK, Tignor M, Allen SK, Boschung J, Nauels A, Xia Y, Bex V, Midgley PM (eds) Climate change 2013: the physical science basis, Contribution of Working Group I to the Fifth Assessment Report of the Intergovernmental Panel on Climate Change. Cambridge University Press, Cambridge/New York

NASA (2017) NASA news & feature releases. NASA, NOAA data show (2016) warmest year on record globally. Posted January 18, 2017. https://climate.nasa.gov/news/2537/nasa-noaa-data-show-2016-warmest-year-on-record-globally/. Accessed 6 Feb 2018

NASA (2018) Long-term warming trend continued in 2017: NASA, NOAA. Global climate change. Vital signs of the planet. https://climate.nasa.gov/news/2671/long-term-warming-trend-continued-in-2017-nasa-noaa/. Accessed 6 Feb 2018

NOAA (2016) TROPICAL cyclone tracks. Available at: http://www.nhc.noaa.gov/climo/ Accessed 10 Dec 2017

Peña H (2016) Desafios de la seguridad hidrica en America Latina y el Caribe. CEPAL

SINA (2017) Sistema Nacional de Informacion del Agua. http://sinaconaguagobmx/sina/. Accessed 15 Dec 2017

UN (United Nations) (2003) Johannesburg summit, 2002. World summit on sustainable development. Johannesburg, South Africa. Report

UNESCO (2012) International Hydrological Programme (IHP) eighth phase: water security: responses to local, regional and global challenges, strategic plan, IHP-VIII (2014–2021)

United Nations (1944) Monetary and financial conference, Bretton Woods, New Hampshire, United States

UN-Water (2013) Water security and the global water agenda. A UN-water analytical brief, United Nations University

World Health Assembly (1977) United Nations conference on water
World Water Council (2000) Declaracion Ministerial de La Haya sobre la seguridad del agua en el
    siglo XXI. March 22, 2000, The Hague, Netherlands
World Water Forum (2015) Water action for sustainable cities and regions. https://www.google.
    com.mx/search?q=VII+Foro+Mundial+del+Agua+en+Daegu-Gyeongbuk&ei=tvg7WpPMo
    Xz0wLz6mgCw&start=10&sa=N&biw=1523&bih=871. Accessed 6 Feb 2018
Zhang Q, Cheng L, Boutaba R (2010) Cloud computing: state-of-the-art and research challenges.
    J Internet Serv Appl 1(1):7–18

Chapter 11
Expected Impacts on Agriculture Due to Climate Change in Northern Mexico

Carlos Escalante-Sandoval

Abstract In general, lack of water affects agriculture, livestock, public supply, industry, electric power generation, and flora and fauna.

In northern Mexico, lack of rainfall mainly affects the crops of rain-fed agriculture and as a result the annual average affected area is 1,100,000 hectares; however, this surface will increase due to meteorological conditions related to climate change. In addition, this region has 38 irrigation districts covering a surface area of 2,212,800 hectares (68% of the surface of all irrigation districts in the country), and it is expected that climate change will affect the levels of the crops' evaporation (ET).

The food requirements of a growing population will increase the demand for water and sowing surface, altering soil use and the environment, and, because of this, will increase water stress in the region.

The aim of this study is to estimate the long-term water availability required to satisfy the future demands of the agricultural sector and to determine the water stress generated in the region, by considering the future effects of climate change. The simulations consider the prevailing conditions up to 2013 and the rainfall anomalies for the Representative Concentration Pathways (RCP) 4.5 and (RCP) 8.5 scenarios in this part of the country.

The results indicate that effects of climate change will reduce the total renewable water resources from 76,554 hm^3/year (2015) to 55,367 hm^3/year for the RCP 4.5 (2039) or 36,356 hm^3/year for the RCP 8.5 (2099). Moreover, the irrigation requirements will increase from 34,048 hm^3/year (RCP 4.5) or 57,226 hm^3/year (RCP 8.5).

Keywords Climate change · Evaporation · Water demand · Sowing surface · Water stress · Rainfall anomalies

C. Escalante-Sandoval (✉)
Faculty of Engineering, National Autonomous University of Mexico, Mexico City, Mexico
e-mail: caes@unam.mx

© Springer Nature Switzerland AG 2020
J. A. Raynal-Villasenor (ed.), *Water Resources of Mexico*, World Water Resources 6,
https://doi.org/10.1007/978-3-030-40686-8_11

11.1 Introduction

The Mexican territory extends over a surface area of 1964 million km^2 and presents a significant seasonal variation in rainfall that has a great impact on the availability of water resources. The average per capita renewable freshwater resources are 3692 m^3/year; however, this availability reduces to 2331 m^3/year in the north of the country, and it is estimated that 76.3% of the water is utilized for agriculture.

The many factors influencing agricultural production are classified as internal or external. The increase in crop yields can be related to the genetic make-up of crop (internal factors), such as the tolerance to drought, flood, salinity, pest and diseases. The external factors able to increase or decrease the agricultural productivity include:

(a) Edaphic Factors—Crops' growth depends on soil characteristics on which they grow. For instance, aeration of soil is essential for absorption of water by roots, the soil temperature controls the processes involved in the nutrient availability and the soil's organic matter supplies all nutrients to crops.
(b) Biotic Factors—A harmful effect can be caused when there is competition between plants for nutrients and moisture.
(c) Physiographic Factors—The altitude plays a very important role in crop growth because, when it increases, it causes a decrease in temperature and an increase in precipitation and wind velocity.
(d) Climatic Factors—Higher temperatures often aggravate stress on water resources that are essential for crop growth, and they will lead to higher evapotranspiration (ET) rates, which tend to dry out the soil.

Climate change is a global phenomenon that is expected to modify agricultural productivity worldwide, threatening food security and increasing food prices. These impacts could increase due to the occurrence of droughts.

Many research studies have assessed the impacts of climate change on rain-fed and irrigated agriculture and proposed some adaptation options (Al-Bakri et al. 2010; Mongui et al. 2010; Asha et al. 2012; Guoju et al. 2013; Calzadilla et al. 2014; Valverde et al. 2015a, b). The results have shown that rain-fed and irrigated crop yields will decrease in the simulated agricultural scenarios.

In northern Mexico, deforestation and inappropriate agricultural practices are rapidly modifying the water cycle because they affect the amount of water in the soil and groundwater and the moisture in the atmosphere, constituting an important factor in global climate change.

In the case of Mexico, Hellin et al. (2014) proposed an adaptation strategy that includes using a corn variety with improved tolerance to heat stress and combined heat and drought stress. Arredondo and Huber (2011) evaluated the impacts of drought on agriculture in northern Mexico and proposed some measures of mitigation. Mundo-Molina (2015) analyzed the climate-change effects on the levels of evapotranspiration, concluding that in northern Mexico the value of ET will be 8% greater than in the rest of the country with a raise in the average temperature of 3 °C.

This chapter predicts the long-term water availability needed to satisfy the future demands of agricultural sector and the water stress generated in the region by considering the future effects of climate change.

11.2   Study Area

The states comprising northern Mexico are Aguascalientes, Baja California, Baja California Sur, Coahuila, Chihuahua, Durango, Nuevo Leon, San Luis Potosí, Sinaloa, Sonora, Tamaulipas, and Zacatecas (Fig. 11.1). These states extend over a surface area of 1,191,252 km² (details in Table 11.1), and they had a total population of 32,847,428 inhabitants in 2015 (CONAGUA 2016).

The region has a rugged topography, composed of a central plateau and two mountain chains, the Sierra Madre Oriental in the east and the Sierra Madre Occidental in the west, both of which are separated from the nearest coast by plains. There are several types of climates: the dry arid and dry semiarid climates cover 77% of the region, mainly the California Peninsula, and the lowland valleys of the west coast and central plateau, where rainfall reaches an average of only 420 mm per year. The remaining 23% of the territory has a tropical climate, mainly along the east coast, and a temperate climate in the mountain highlands. These climatological conditions introduce high levels of water stress in some states of northern Mexico.

Fig. 11.1   States in northern Mexico

Table 11.1 Surface area and population for each state of northern Mexico (2015)

State	Surface area (km^2)	Population (inhabitants)		
		Total	Urban	Rural
Aguascalientes	5618	1287,661	1,042,859	244,802
Baja California	71,446	3,484,150	3,202,534	281,616
Baja California Sur	73,922	763,928	664,805	99,123
Coahuila	151,563	2,960,681	2,672,183	288,498
Chihuahua	247,455	3710,129	3,160,350	549,779
Durango	123,451	1,764,727	1229,998	534,729
Nuevo Leon	64,220	5,085,848	4,809,752	276,096
San Luis Potosí	60,983	2,753,478	1,790,013	963,465
Sinaloa	57,377	2,984,572	2,216,210	768,362
Sonora	179,503	2,932,821	2,525,775	407,046
Tamaulipas	80,175	3,543,365	3,103,421	439,944
Zacatecas	75,539	1,576,068	953,692	622,376
Northern Mexico	1,191,252	32,847,428	27,371,592	5,475,836

11.3   Agriculture in Northern Mexico

Mexico has a robust agricultural tradition because agriculture was the basis of the major Mesoamerican civilizations such as the Aztecs, Mayas, and Olmecs, with crops such as corn, tomatoes, and beans (Pope et al. 2001). During the Spaniard conquest period, there were more people working in agriculture than in other economic activities, not only producing subsistence crop, but also in commercial agriculture to supply Spanish cities. Spaniards introduced new crops such as wheat, sugar, pears, apples, and bananas. In the nineteenth century, despite the war of independence, the colonial agricultural system retained the same role for farmers as toilers and the benefits going only to the elite. As a result, these unacceptable conditions brought on the Mexican Revolution. After this period of war (1910–1920), it was necessary to define a new distribution of lands, under a system of common tenure called *ejido*, which could be worked individually or collectively by its members. Until 1992, the system of *ejidos* remained unchanged, after which year the Mexican Constitution was reformed, permitting the *ejido* property rights to be leased and to be sold. Today, 73% of landowners are smallholders who own five or fewer hectares, 22% own up to 20 hectares, and only 5% of landholders own more than 20 hectares. Until the late 1990s, Mexico was a net exporter of agricultural products, but nowadays is a net importer, mostly from the USA. At the beginning of the twenty-first century, Mexico's main agriculture products included corn, beans, wheat, coffee, bananas, pineapples, mangos, cacao, vegetables, beef, milk, pork, and eggs. At constant prices, the contribution of the agriculture, livestock, forest use, fishing and hunting sector to the country's Gross Domestic Product (GDP) was 3.8% in 2015 (INEGI 2016). The national percentages of the agricultural and livestock production value for each state in this region are shown in Table 11.2 (INEGI 2016).

Table 11.2 Contribution to the value of national agricultural and livestock production

State	Value of the agricultural production		Value of the livestock production	
	National percentage	Place	National percentage	Place
Aguascalientes	0.6%	30°	3.2%	11°
Baja California	3.0%	14°	1.7%	19°
Baja California Sur	0.2%	30°	7.0%	4°
Chihuahua	6.3%	6°	2.9%	14°
Coahuila	1.4%	22°	4.2%	7°
Durango	1.9%	19°	5.7%	4°
Nuevo Leon	1.1%	24°	2.2%	16°
Sinaloa	8.5%	2°	3.2%	12°
San Luis Potosi	2.6%	16°	1.9%	18°
Sonora	6.9%	5°	4.4%	6°
Tamaulipas	3.5%	10°	0.9%	24°
Zacatecas	3.2%	13°	1.2%	23°
Northern Mexico	39.2%		38.5%	

Table 11.3 Average annual surface devoted to agriculture in northern Mexico (2005–2015)

State	Irrigated Surface (ha)		Rain-fed Surface (ha)	
	Sown	Harvested	Sown	Harvested
Aguascalientes	49,890	49,440	100,040	64,673
Baja California	186,426	181,397	35,209	19,608
Baja California Sur	38,956	36,546	0	0
Coahuila	151,940	147,108	127,992	107,548
Chihuahua	584,155	573,520	576,471	518,665
Durango	153,824	151,436	566,343	502,502
Nuevo Leon	95,272	92,495	258,448	247,824
San Luis Potosi	132,180	126,933	624,516	478,335
Sinaloa	877,674	798,161	427,770	350,242
Sonora	545,343	534,091	38,913	34,575
Tamaulipas	491,555	472,147	925,916	803,269
Zacatecas	156,298	152,094	1,099,674	877,584
Northern Mexico	3,463,513	3,315,366	4,781,292	4,004,826

The average area devoted to agriculture in northern Mexico during the period 2005–2015 was approximately 8.2 million hectares (SIAP 2017), of which 3.4 million are irrigated and 4.8 million are rain-fed (Tables 11.3 and 11.4). In the same period, the 15 most important crops produced 66.5 million tons per year (Table 11.5). In the case of corn, the average yield was 4.1 tons per hectare, 28.1% more than the average national yield of 3.2 tons per hectare. These 15 crops occupied an average surface of 7.2 million hectares per year (considering the agricultural year and perennial crops, under both irrigated and rain-fed regimes), representing 87.1% of the area sown in the region (SIAP 2017).

Table 11.4 Percentages of surface sown considering the autumn–winter and spring–summer seasons and perennial crops, under both irrigated and rain-fed regimes

State	Irrigated surface			Rain-fed surface		
	Autumn–winter	Spring–summer	Perennial	Autumn–winter	Spring-summer	Perennial
Aguascalientes	100.0%	22.2%	71.3%	0.0%	77.8%	28.7%
Baja California	75.0%	97.5%	96.3%	25.0%	2.5%	3.7%
Baja California Sur	100.0%	100.0%	0.0%	0.0%	0.0%	0.0%
Chihuahua	96.5%	34.4%	99.2%	3.5%	65.6%	0.8%
Coahuila	94.8%	60.6%	42.7%	5.2%	39.4%	57.3%
Durango	96.1%	13.2%	83.6%	3.9%	86.8%	16.4%
Nuevo Leon	51.3%	12.4%	24.9%	48.7%	87.6%	75.1%
Sinaloa	91.0%	11.6%	47.2%	9.0%	88.4%	52.8%
San Luis Potosi	50.2%	13.1%	22.8%	49.8%	86.9%	77.2%
Sonora	99.6%	59.8%	99.7%	0.4%	40.2%	0.3%
Tamaulipas	38.0%	9.1%	75.7%	62.0%	90.9%	24.3%
Zacatecas	100.0%	9.1%	36.1%	0.0%	90.9%	63.9%
Northern Mexico	82.7%	36.9%	58.3%	17.3%	63.1%	33.4%

The agricultural subsistence production comes from rain-fed areas. The communities settled on these regions are the most vulnerable to climate anomalies, and it is expected these conditions will intensify based on future scenarios of climate change.

The adoption of new irrigation technologies in rain-fed lands has not increased in the last 40 years, and the infrastructure has deteriorated, generating irrigation efficiencies between 33% and 55% (Pedroza-Gonzalez and Hinojosa-Cuellar 2014).

Agriculture production generally has negative impacts, such as deforestation, on natural resources. Forests are vital to life on Earth, since they play a crucial role in the water cycle, also absorbing greenhouse gases, reducing soil erosion, producing oxygen, and absorbing carbon dioxide.

Deforestation is the conversion of forests into another land use. The biggest threat to forests is the demand for arable farmland and the building of more houses to accommodate more people, which means that huge areas of forests have to be completely cut down. In northern Mexico, deforestation and inappropriate agricultural practices, along with other factors, are rapidly modifying the water cycle because they affect the amount of water in the soil and groundwater and the moisture in the atmosphere, constituting an important factor in global climate change.

In the period 1979–2012, agricultural activities increased the demand for land in northern Mexico at a rate of 151,413 hectares per year and reduced the forest cover at a rate of 30,078 hectares per year (Table 11.6). Moreover, the land-cover rate of change in the scrublands of 112,942 hectares per year is added to the former rate because the country's arid and sub-humid scrublands are also regarded as forest vegetation. Thus, the final rate due to deforestation amounts to 143,020 hectares per year (INEGI 2016). This amount implies that the region has lost 4% of the area that

Table 11.5 Characteristics of the 15 most important crops of northern Mexico (2005–2015)

		Irrigated surface			Rain-fed surface		
Crop		Surface sown (hectares)	Volume (metric tons)	Yield ton/ha	Surface sown (hectares)	Volume (metric tons)	Yield ton/ha
1	Maize grain	823,381	6,480,865	7.9	916,506	659,073	0.7
2	Sorghum grain	345,498	1,647,381	4.8	951,022	1807,685	1.9
3	Bean	151,621	237,568	1.6	1,022,183	461,125	0.5
4	Fodder oats	72,512	1,835,474	25.3	550,614	4,982,886	9.0
5	Wheat grain	447,505	2,598,469	5.8	49,608	68,854	1.4
6	Grass	128,934	3,038,808	23.6	351,745	3,975,473	11.3
7	Forage maize	70,058	3,316,500	47.3	209,934	2,014,642	9.6
8	Alfalfa	225,274	17,137,935	76.1	6	240	38.3
9	Forage sorghum	64,473	2,292,900	35.6	127,682	1,383,206	10.8
10	Sugar cane	87,527	5,680,829	64.9	69,800	3,201,610	45.9
11	Orange	68,527	1,121,957	16.4	44,095	338,133	7.7
12	Green chilli	109,311	1,865,831	17.1	90	921	10.3
13	Soy	18,496	31,271	1.7	86,032	106,630	1.2
14	Safflower	42,590	82,161	1.9	61,714	21,608	0.4
15	Nut	86,909	91,815	1.1	1443	424	0.3
	Total	2,742,617	47,459,764	17.3	4,442,475	19,022,508	4.3

Table 11.6 Land use and land cover in northern Mexico in 1979 and 2012

	1979		2012	
Land use/Land cover	Area (km^2)	Percentage	Area (km^2)	Percentage
Grassland	107,206	9.0%	107,748	9.0%
Desert scrub	609,369	51.2%	572,098	48.0%
Forest	267,603	22.5%	257,677	21.6%
Agriculture	71,017	6.0%	120,984	10.2%
Irrigated	21,470	1.8%	32,938	2.8%
Rain-fed	32,204	2.7%	48,919	4.1%
Livestock	17,343	1.5%	39,127	3.3%
Urban	715	0.1%	7860	0.7%
Other	135,342	11.4%	124,884	10.5%
Total	1,191,252	100.0%	1,191,252	100.0%

had been covered by forest and scrub. Overall, the urban areas grew from 71,521 to 786,023 hectares during the same period (or 21,652 hectares per year).

A lack of rainfall mainly affects the crops of rain-fed agriculture, which without an irrigation infrastructure cannot retain water for future use. In northern Mexico, the rain-fed agriculture surface is almost two-fold greater than that of irrigated agriculture, and its losses represent over 91% of the damaged surface (SIAP 2017). In 2011, rain-fed bean and maize production were 224,000 and 505,000 tons lower than expected, respectively, and economic damage was as high as $363 million (SIAP 2017). In the same year, some other northern states did not experience significant damage to irrigated croplands as a result of the lack of rain, but their groundwater extractions were 54% and 40% higher than the averages from 2006 to 2010, respectively (INEGI 2016).

The study area has 38 irrigation districts covering a surface area of 2,212,800 hectares (68% of the surface of all irrigation districts in the country). The planting and harvesting of corn in these districts represents 26% of the surface and 22% of the production of all irrigation districts in the country.

To demonstrate how climate change will affect the values of the maize crop evapotranspiration (ETc), an example is useful. Weather station 25,116 was analyzed, which is located in irrigation district 75 in Sinaloa, Mexico. The crop-planting seasons were autumn–winter (date of planting October 15) and spring–summer (date of planting April 1). The crop evapotranspiration, ETc, was calculated by multiplying the reference crop evapotranspiration, ETo, by a crop coefficient, Kc. The reference ETo was calculated using the FAO Penman-Monteith equation (Allen et al. 2006). In Table 11.7, the climatic variables required to compute the values of ETo for station 25,116 are shown.

Columns 2 and 3 of Table 11.8 present the values of the maize crop evaporation, ETc, for the following cases: (a) available data (1950–2013), (b) RCP 4.5 (2015–2039) and (c) RCP 8.5 (2075–2099). As just noted, the irrigation efficiencies in the agricultural sector are very low, and irrigation district 75, with an irrigation efficiency of 47%, is no exception. In columns 4 and 5, the required volumes per hectare are shown. These volumes already take into account the irrigation efficiency. Moreover, in irrigation district 75, an average of 134,300 and 35,200 hectares are usually sown in the autumn–winter and spring–summer cycles, respectively. Finally, columns 6 and 7 present the total water volume necessary to satisfy the average demand in the irrigation district. In scenario 4.5 (2015–2039), these volumes will increase by 14% (autumn–winter) and 3% (spring–summer) with respect to the current conditions (period 1950–2013), and by 30% and 18%, with respect to scenario 8.5 (2075–2099).

Other negative impacts of agriculture production on natural resources include soil erosion and salinization, overexploitation of aquifers and greenhouse gas emissions. In this sense, the agricultural sector contributes 12.3% of the total greenhouse gas (GHG) emissions in the country and the land-use change contributes 6.3%. The main sources of agricultural GHG emissions are enteric fermentation, manure left on pasture and synthetic fertilizers.

Table 11.7 Monthly temperature and rainfall for weather station 25,116 located in Sinaloa, Mexico

Case	Jan	Feb	Mar	Apr	May	Jun	Jul	Aug	Sep	Oct	Nov	Dec	Average
1950–2013													
Tmin (°C)	11.7	12.1	13.3	15.5	18.4	23.2	25.4	25.2	24.8	21.6	16.1	12.5	18.3
Tmax (°C)	26.1	27.7	29.8	32.5	35.2	37.1	37.6	37.5	36.7	35.3	30.7	26.4	32.7
Tmed (°C)	18.9	19.9	21.6	24	26.8	30.1	31.5	31.3	30.7	28.4	23.4	19.5	25.5
Hp (mm)	16.8	7.1	3.7	0.7	1.2	13.8	72.5	115.6	98.5	34.4	15.8	19.1	399.4
RCP 4.5													
Tmin (°C)	12.7	13.0	14.3	16.6	19.7	24.5	26.6	26.4	26.1	23.0	17.3	13.4	19.5
Tmax (°C)	27.3	28.9	31.3	34.0	36.8	38.5	39.2	38.9	38.1	36.8	32.0	27.6	34.1
Tmed (°C)	20	20.9	22.8	25.3	28.2	31.5	32.9	32.6	32.1	29.9	24.7	20.5	26.8
Hp (mm)	9.9	1.1	2.6	0.4	0.8	12.4	62.8	113.1	90.2	28.6	10.0	12.5	344.5
RCP 8.5													
Tmin (°C)	15.7	16	17.4	19.7	23.1	27.6	29.7	29.4	29	26.3	20.9	16.6	22.6
Tmax (°C)	30.4	32.4	34.8	37.5	40.5	42.2	42.8	41.9	40.8	40.3	35.5	30.8	37.5
Tmed (°C)	23.1	24.2	26.1	28.6	31.8	34.9	36.2	35.6	34.9	33.3	28.2	23.7	30.1
Hp (mm)	5.7	0	1.3	0	0	12.1	48.1	100.1	108.7	25.7	7.9	10.2	319.9

Table 11.8 Crop evapotranspiration and volumes necessary to satisfy the water demand of corn in irrigation district 75

Period	ETc (mm)		Volume (m³/ha)		Water Demand (m³)	
	Autumn–winter	Spring–summer	Autumn–winter	Spring–summer	Autumn–winter	Spring–summer
1950–2013	266.7	544.6	5674	11,587	762,081,064	407,870,638
RCP 4.5	304.2	561.2	6472	11,940	869,235,319	420,302,979
RCP 8.5	347.5	641.3	7394	13,645	992,962,766	480,292,766

Table 11.9 Contribution to GDP and population occupied in the primary sector in northern Mexico in 2015

State	% Country GDP	% GDP from agriculture	Economically active population	Population-occupied primary sector
Aguascalientes	1.2	1.5	574,193	25,992
Baja California	3.1	3.0	1,684,210	84,174
Baja California Sur	0.8	0.9	396,512	27,357
Coahuila	3.5	2.3	1376,024	48,719
Chihuahua	3.1	6.3	1,617,998	145,820
Durango	1.2	3.5	769,768	96,021
Nuevo Leon	7.3	1.2	2,436,540	36,375
San Luis Potosi	2.0	2.4	1,199,278	195,814
Sinaloa	2.2	7.8	1401,111	214,328
Sonora	3.3	6.4	1,414,120	156,667
Tamaulipas	3.0	3.0	1625,475	101,959
Zacatecas	1.0	2.6	619,925	114,462
Northern Mexico	31.5	40.9	15,115,154	1,247,688

Moreover, this region plays an important role in the country's economy (INEGI 2016), accounting for 31.5% of its total GDP and 40.9% of the GDP from agriculture. The primary sector employs 8.2% of the economically active population of northern Mexico (Table 11.9).

11.4   Water Availability

In Mexico, the average annual rainfall is 750 mm, and varies greatly from that in the regions in the north and northwest (average annual rainfall < 500 mm) to that in the regions in the south and southeast (average annual rainfall > 2000 mm). In addition, 68% of the normal monthly precipitation falls between June and September, which accentuates the problems related to the availability of water resources.

The country receives annually 1449,471 million m^3 of water in the form of precipitation, and it is estimated that 72.5% evapotranspirates, 21.2% runs off into rivers and streams and the remaining 6.3% naturally filters through to the subsoil and recharges aquifers. Taking into account the water outflows to and inflows from neighboring countries, the country has 446,777 million m^3 of renewable freshwater resources, an amount that is equal to 3692 m^3/inhabitant/year; however, this average availability varies significantly among the various regions of the country, particularly in the north, where the average is 2331 m^3/inhabitant/year. In the case of northern Mexico, Table 11.10 shows the total renewable water resources along with the available volumes for various sources in the period 1950–2013 (CONAGUA 2016), along with the estimated per capita volume for the years 2015, 2039 and 2099.

Table 11.10 Total and per capita renewable water resources in northern Mexico (1950–2013)

State	(1)	(2)	(3)	(4)	(5)	(6)
Aguascalientes	514	202	312	399	318	266
Baja California	2989	2062	928	858	657	499
Baja California Sur	1264	770	494	1655	841	172
Coahuila	3151	1229	1922	1064	856	689
Chihuahua	11,888	8188	3701	3204	2685	2260
Durango	13,370	12,431	939	7576	6406	5677
Nuevo Leon	4285	3416	869	843	640	459
San Luis Potosi	10,597	8564	2033	3849	3321	3022
Sinaloa	8682	6174	2508	2909	2511	2237
Sonora	7018	3739	3280	2393	1855	1409
Tamaulipas	8928	7919	1009	2520	2053	1688
Zacatecas	3868	2842	1026	2454	2155	1976
Northern Mexico	76,554	57,534	19,020	2331	1861	1356

(1) Total renewable water resources in the period 1950–2013 (in hm^3/year)
(2) Total mean natural surface runoff in the period 1950–2013 (in hm^3/year)
(3) Total mean aquifer recharge in the period 1950–2013 (in hm^3/year)
(4) Per capita renewable water resources estimated for the year 2015 (in m^3/year)
(5) Per capita renewable water resources estimated for the year 2039 (in m^3/year)
(6) Per capita renewable water resources estimated for the year 2099 (in m^3/year)

11.5   Uses of Water

In Mexico, 76.3% of the water is utilized for agriculture, 14.6% for public supply, 4.8% for energy generation excluding hydropower and 4.3% for self-supplying industry (CONAGUA 2016); however, due to the naturally arid conditions in northern Mexico, these percentages are redistributed as follows: 87.5% (Table 11.11), 9.7% (Table 11.12), 1.1% (Table 11.13) and 1.7% (Table 11.13),

Table 11.11 Average water volume utilized at irrigation areas in northern Mexico (period 2005–2015)

| State | Irrigation water (hm^3) | | | | |
	Total	Surface	Percentage	Groundwater	Percentage
Aguascalientes	487.5	175.0	35.9	312.5	64.1
Baja California	2565.3	1504.5	58.6	1060.8	41.4
Baja California Sur	330.5	29.0	8.8	301.5	91.2
Coahuila	1624.0	836.1	51.5	787.9	48.5
Chihuahua	4584.6	1916.6	41.8	2668.0	58.2
Durango	1323.8	733.2	55.4	590.6	44.6
Nuevo Leon	1446.3	804.7	55.6	641.6	44.4
San Luis Potosi	1190.9	635.4	53.4	555.5	46.6
Sinaloa	8799.6	8035.0	91.3	764.5	8.7
Sonora	6045.9	3765.5	62.3	2280.4	37.7
Tamaulipas	3478.5	3118.6	89.7	359.9	10.3
Zacatecas	1337.3	330.7	24.7	1006.5	75.3
Northern Mexico	33,214.2	21,884.4	65.9	11,329.9	34.1

Table 11.12 Average water volume utilized for public supply in northern Mexico (period 2005–2015)

| State | Drinking water (hm^3) | | | Water supply |
	Total	Surface	Groundwater	m^3/inhabitant/year
Aguascalientes	120.8	0.1	120.8	103.8
Baja California	214.3	106.7	107.5	70.5
Baja California Sur	62.3	3.0	59.3	108.8
Coahuila	205.9	18.0	187.9	78.1
Chihuahua	480.3	51.0	429.3	147.1
Durango	154.4	13.6	140.8	101.2
Nuevo Leon	512.0	356.0	156.0	114.3
San Luis Potosi	174.9	28.8	146.0	79.3
Sinaloa	410.9	181.6	229.4	156.7
Sonora	919.1	489.5	429.6	361.4
Tamaulipas	320.8	278.9	41.9	102.1
Zacatecas	115.6	7.0	108.6	83.7
Northern Mexico	3691.3	1534.1	2157.2	129.1

respectively (CONAGUA 2016). Of the 37,971 hm^3 of water utilized, 62.6% comes from surface sources and is stored in dams or diverted from rivers, whereas the remaining 37.4% comes from groundwater that is extracted from aquifers through deep wells. It is very important to mention that there are 653 aquifers in Mexico; 374 of them are located in the study area (Table 11.14), and of these, 36.3% are overexploited (CONAGUA 2018).

Table 11.13 Average water volume utilized for energy generation excluding hydropower and self-supplying industry in northern Mexico (period 2005–2015)

State	Energy generation (hm^3)			Industry (hm^3)		
	Total	Surface	Groundwater	Total	Surface	Groundwater
Aguascalientes	0.0	0.0	0.0	12.3	2.0	10.3
Baja California	194.5	0.0	194.5	80.9	68.7	12.2
Baja California Sur	4.0	0.0	4.0	11.6	1.9	9.7
Coahuila	74.2	47.1	27.1	73.9	1.1	72.8
Chihuahua	28.0	0.0	28.0	53.8	6.0	47.8
Durango	11.4	0.0	11.4	18.8	2.0	16.8
Nuevo Leon	1.5	0.0	1.5	81.8	0.0	81.8
San Luis Potosi	33.7	14.0	19.7	34.1	10.1	24.0
Sinaloa	0.0	0.0	0.0	43.4	35.4	8.0
Sonora	6.7	5.1	1.6	93.7	2.7	91.1
Tamaulipas	54.2	51.2	3.0	110.3	101.0	9.3
Zacatecas	0.0	0.0	0.0	42.5	1.0	41.5
Northern Mexico	408.2	117.4	290.8	657.0	231.8	425.3

Table 11.14 Total number of aquifers and overexploited aquifers located in Northern Mexico (2017)

State	Number of aquifers		
	Total	Overexploited	Percentage
Aguascalientes	5	5	100%
Baja California	48	18	38%
Baja California Sur	39	17	44%
Chihuahua	61	19	31%
Coahuila	28	9	32%
Durango	29	10	34%
Nuevo Leon	23	11	48%
San Luis Potosi	19	10	53%
Sinaloa	14	2	14%
Sonora	60	18	30%
Tamaulipas	14	3	21%
Zacatecas	34	14	41%
Northern Mexico	374	136	36.3%

11.6 Drought

Mexico has dealt with drought throughout its history; for example, early pre-Hispanic documents indicate crop losses and diminishing water levels in reservoirs associated with a deficit in precipitation, causing food scarcity and hunger. The collapse of the Mayan civilization may have been associated with a very long period of drought in the ninth century (Gill 2000). Society adopted some measures to cope with drought during the fifteenth century (hydraulic works, water storage systems, crop calendars, and irrigated terraces) and realized the relationship between meteorological phenomena and the crop cycle. From the sixteenth century to the early nineteenth century, droughts affected the colonial economy; the most severe drought recorded for the northern states occurred in 1808–1811 (Padilla et al. 1980). In the twentieth century, several episodes of drought occurred in Mexico. The most important occurred in 1948–1954, 1960–1964, 1970–1978 and 1993–1998 (CENAPRED 2014), mainly affecting the northern states. In particular, the 1996 drought, with four years of below-normal rainfall, produced severe conditions in northern Mexico: More than 4.6 million hectares of cropland was damaged, and six million hectares remained unplanted. In the livestock sector, 300,000 head of cattle died and 700,000 others were sold at very low prices. The farm losses were estimated at more than US$1 billion (CENAPRED 2014).

In recent years, droughts in Mexico have resulted in damage costs as high as US$1.5 billion for the 2010–2012 three-year drought event. Water scarcity caused losses of 450,000 head of cattle and low yields of several crops, especially maize and wheat (CENAPRED 2014). The immediate consequence of this damage was the rising price of agricultural products, affecting the poorest social sectors. In addition, droughts generated conflicts among producers and caused unemployment in rural areas, which increased the migration to urban areas. The greatest damage occurred in the northern states (Aguascalientes, Chihuahua, Coahuila, Durango, Nuevo Leon, Baja California, Sonora, Sinaloa, and Zacatecas).

The occurrence of droughts has been faced with various actions, while some farmers sow their fields, praying for rain or some kind of government relief, others decide to abandon their land and migrate to the big cities of Mexico or the USA.

The most recent drought episodes in northern Mexico have shown the vulnerability of Mexican agriculture to climate variability, and the Mexican Government has performed some tasks to improve water-resources management, including improvements in infrastructure and seasonal climate forecasting.

Escalante-Sandoval and Nuñez-Garcia (2017) obtained the long-term meteorological drought features for a simulated scenario called "Normal Conditions." The results reveal that the region suffers a drought on average every 4.1 years with a mean duration of 2.2 years. Based on their findings, it was possible to estimate the per capita renewable water resources for a drought episode absent climate change (Table 11.15), which would reduce available water resources from 1356 m^3/year (current conditions) to 732 m^3/year in 2099.

Table 11.15 Total and per capita renewable water resources for an average drought episode in northern Mexico (1950–2013)

State	(1)	(2)	(3)	(4)	(5)	(6)
Aguascalientes	330	83	247	256	204	171
Baja California	1788	1179	609	513	393	299
Baja California Sur	613	296	317	803	408	83
Coahuila	1486	152	1333	502	404	325
Chihuahua	4502	1851	2650	1213	1017	856
Durango	8026	7297	730	4548	3846	3408
Nuevo Leon	2116	1464	651	416	316	226
San Luis Potosi	6335	4757	1578	2301	1985	1807
Sinaloa	5563	3601	1962	1864	1609	1433
Sonora	3769	1258	2511	1285	996	756
Tamaulipas	4701	3927	773	1327	1081	889
Zacatecas	2080	1271	810	1320	1159	1063
Northern Mexico	41,308	27,136	14,172	1258	1004	732

(1) Total renewable water resources in the period 1950–2013 (in hm^3/year)
(2) Total mean natural surface runoff in the period 1950–2013 (in hm^3/year)
(3) Total mean aquifer recharge in the period 1950–2013 (in hm^3/year)
(4) Per capita renewable water resources estimated for the year 2015 (in m^3/year)
(5) Per capita renewable water resources estimated for the year 2039 (in m^3/year)
(6) Per capita renewable water resources estimated for the year 2099 (in m^3/year)

11.7   Climate Change

The Fifth Assessment Report (AR5) of the Intergovernmental Panel on Climate Change (IPCC) stated that climate change is real, and its effects will reach all levels of the population (IPCC 2013). Significant changes in climatic extremes increasingly affected a substantial portion of the global land area during the second half of the twentieth century (Frich et al. 2002). Global climate is determined by meteorological variables, mainly temperature and precipitation (Loukas et al. 2007).

In Mexico, the "Instituto Nacional de Ecologia y Cambio Climatico" performed a regional climatic analysis using projections from 15 Global Circulation Models of the Coupled Model Intercomparison Project Phase 5 under the prevailing conditions of the Representative Concentration Pathways (RCPs) to replace the previous scenarios of the Special Report on Emission Scenarios (INECC, 2014). RCPs define the trajectories of the modification in the net solar irradiance (W/m^2) of the tropopause due to an increase in the concentration of greenhouse gases and other forcing agents for the year 2099. The four RCPs include a mitigation scenario leading to a very-low forcing level (RCP 2.6), a pre-2099 emission stabilization scenario (RCP 4.5), a post-2099 emission stabilization scenario (RCP 6) and a very-high baseline emission scenario (RCP 8.5). According to these projections, precipitation in northern Mexico is expected to decline significantly. For the stabilization scenario RCP4.5, precipitation will decrease 13%, and temperature will increase 1.3 °C.

Table 11.16 Total and per capita renewable water resources in northern Mexico for the RCP 4.5 scenario (2015–2039)

State	(1)	(2)	(3)	(4)	(5)	(6)
Aguascalientes	395	155	239	288	262	244
Baja California	2210	1524	686	592	530	486
Baja California Sur	802	489	313	912	725	533
Coahuila	1719	670	1048	549	501	467
Chihuahua	7636	5259	2377	1967	1828	1725
Durango	10,137	9425	712	5487	5111	4857
Nuevo Leon	2952	2353	599	543	484	441
San Luis Potosi	8478	6851	1627	2955	2775	2657
Sinaloa	6960	4949	2011	2241	2107	2013
Sonora	4870	2594	2276	1558	1401	1287
Tamaulipas	6609	5862	747	1769	1624	1519
Zacatecas	2601	1911	690	1592	1506	1449
Northern Mexico	55,367	42,043	13,324	1686	1453	1346

(1) Total renewable water resources expected in the period 2015–2039 (in hm^3/year)
(2) Total mean natural surface runoff expected in the period 2015–2039 (in hm^3/year)
(3) Total mean aquifer recharge expected in the period 2015–2039 (in hm^3/year)
(4) Per capita renewable water resources estimated for the year 2020 (in m^3/year)
(5) Per capita renewable water resources estimated for the year 2030 (in m^3/year)
(6) Per capita renewable water resources estimated for the year 2040 (in m^3/year)

Furthermore, for the very-high baseline emission scenario RCP 8.5, the region will get 23.8% less rain, and the temperature will increase 4.5 °C.

Escalante-Sandoval and Nuñez-Garcia (2017) simulated the possible alterations in rainfall patterns by considering the projections of two climatic change scenarios "RCP 4.5 for the near future (2015–2039) and RCP 8.5 for the far future (2075–2099)". For each former scenario, the modifications to the average climatic conditions measured during the period 1950–2013 and the presence of a drought episode were considered. The expected volume of the renewable water for the various scenarios is shown in Tables 11.16, 11.17, 11.18 and 11.19.

11.8 Expected Water Stress

Due to the arid conditions in northern Mexico, the water resources in the region are already stressed; nevertheless, uncontrolled population growth would increase food demand and, consequently, desirable food production, forcing water resources to be augmented and aggravating the stress level. Magaña (2006) projected the stress level for the year 2030 considering the key factors that affect water demand, including, population, gross domestic product, water and food demand, among others. The results indicated the water stress would increase between 7% and 17%, for northern Mexico. However, performing a similar exercise, by considering water availability

Table 11.17 Total and per capita renewable water resources in northern Mexico for the RCP 8.5 scenario (2070–2099)

State	(1)	(2)	(3)	(4)	(5)	(6)
Aguascalientes	259	102	157	142	138	134
Baja California	1996	1376	619	368	354	333
Baja California Sur	563	343	220	169	130	76
Coahuila	682	266	416	161	156	149
Chihuahua	3971	2735	1236	802	782	755
Durango	7274	6763	511	3200	3152	3088
Nuevo Leon	1664	1327	338	203	193	178
San Luis Potosi	5952	4810	1142	1742	1722	1697
Sinaloa	5624	3999	1625	1502	1479	1449
Sonora	3030	1614	1416	673	647	608
Tamaulipas	4122	3656	466	834	812	779
Zacatecas	1219	896	323	638	631	623
Northern Mexico	36,356	27,887	8469	1107	711	644

(1) Total renewable water resources expected in the period 2070–2099 (in hm^3/year)
(2) Total mean natural surface runoff expected in the period 2070–2099 (in hm^3/year)
(3) Total mean aquifer recharge expected in the period 2070–2099 (in hm^3/year)
(4) Per capita renewable water resources estimated for the year 2070 (in m^3/year)
(5) Per capita renewable water resources estimated for the year 2080 (in m^3/year)
(6) Per capita renewable water resources estimated for the year 2099 (in m^3/year)

Table 11.18 Total and per capita renewable water resources in northern Mexico under drought conditions for the RCP 4.5 scenario (2015–2039)

State	(1)	(2)	(3)	(4)	(5)	(6)
Aguascalientes	292	88	204	213	193	180
Baja California	1547	1039	507	415	371	340
Baja California Sur	501	268	233	570	453	333
Coahuila	976	174	802	312	285	265
Chihuahua	3301	1470	1832	850	790	746
Durango	6943	6352	591	3758	3501	3327
Nuevo Leon	1710	1226	483	314	280	255
San Luis Potosi	5688	4352	1337	1983	1862	1783
Sinaloa	4973	3306	1667	1601	1506	1438
Sonora	3094	1210	1883	990	890	818
Tamaulipas	3840	3240	600	1028	944	883
Zacatecas	1720	1126	594	1053	997	959
Northern Mexico	34,586	23,852	10,734	1053	908	841

(1) Total renewable water resources expected in the period 2015–2039 (in hm^3/year)
(2) Total mean natural surface runoff expected in the period 2015–2039 (in hm^3/year)
(3) Total mean aquifer recharge expected in the period 2015–2039 (in hm^3/year)
(4) Per capita renewable water resources estimated for the year 2020 (in m^3/year)
(5) Per capita renewable water resources estimated for the year 2030 (in m^3/year)
(6) Per capita renewable water resources estimated for the year 2040 (in m^3/year)

Table 11.19 Total and per capita renewable water resources in northern Mexico under drought conditions for the RCP 8.5 scenario (2070–2099)

State	(1)	(2)	(3)	(4)	(5)	(6)
Aguascalientes	236	86	149	129	126	122
Baja California	1501	1017	485	277	267	251
Baja California Sur	444	255	188	133	102	60
Coahuila	467	104	364	110	107	102
Chihuahua	1919	886	1034	387	378	365
Durango	5784	5327	456	2545	2506	2456
Nuevo Leon	1197	897	300	146	139	128
San Luis Potosi	5088	4030	1057	1489	1472	1451
Sinaloa	4407	2992	1416	1177	1159	1136
Sonora	2234	979	1255	497	477	448
Tamaulipas	2930	2520	409	593	577	554
Zacatecas	1107	793	314	579	573	565
Northern Mexico	27,314	19,885	7428	832	534	484

(1) Total renewable water resources expected in the period 2070–2099 (in hm³/year)
(2) Total mean natural surface runoff expected in the period 2070–2099 (in hm³/year)
(3) Total mean aquifer recharge expected in the period 2070–2099 (in hm³/year)
(4) Per capita renewable water resources estimated for the year 2070 (in m³/year)
(5) Per capita renewable water resources estimated for the year 2080 (in m³/year)
(6) Per capita renewable water resources estimated for the year 2099 (in m³/year)

under the RCP 4.5 scenario for the period 2015–2039 and population projections in the states for the year 2030 (CONAPO, 2015) the water stress condition would become far worse, with the stress for the whole region increasing from the current conditions of 50.7% (Table 11.20) to 81.4% (Table 11.21). In the case of scenario RCP 8.5, this regional rate would increase to 177.4% (Table 11.22).

11.9   Conclusions

In northern Mexico, climate change could produce serious modifications in the meteorological variables. For the scenario RCP4.5, precipitation will decrease 13%, and temperature will increase 1.3 °C. Furthermore, for the scenario RCP 8.5, the region will get 23.8% less rain, and the temperature will increase 4.5 °C. For both scenarios, a decrease in water flow is expected.

The effects of climate change will reduce the total renewable water resources from 76,554 hm³/year (2015) to 55,367 hm³/year for the RCP 4.5 (2039) and to 36,356 hm³/year for the RCP 8.5 (2099). Moreover, the irrigation requirements will increase from 34,048 hm³/year (2015) to 39,730 hm³/year (RCP 4.5) or 57,226 hm³/year (RCP 8.5). These demands require augmenting the water supplied by dams and wells located in aquifers already overexploited.

Table 11.20 Water stress in northern Mexico for the year 2015

State	Uses of water (hm^3)					Renewable water (hm^3)	Water stress %
	Public	Irrigation	Industry	Energy	Total		
Aguascalientes	128	480	12	0	619	514	120.5
Baja California	188	2587	83	192	3050	2989	102.0
Baja California Sur	65	339	14	5	422	1264	33.4
Chihuahua	490	4589	54	28	5161	3151	43.4
Coahuila	240	1648	76	74	2038	11,888	64.7
Durango	170	1367	17	12	1566	13,370	11.7
Nuevo Leon	512	1473	83	1	2069	4285	48.3
San Luis Potosi	180	1338	34	31	1583	10,597	14.9
Sinaloa	409	8990	43	0	9442	8682	108.8
Sonora	770	6130	110	16	7026	7018	100.1
Tamaulipas	335	3710	115	55	4215	8928	47.2
Zacatecas	124	1397	72	0	1593	3868	41.2
Northern Mexico	3611	34,048	713	413	38,784	76,554	50.7

Table 11.21 Water stress in northern Mexico for year 2039 considering RCP 4.5 scenario (2015–2039)

State	Uses of water (hm^3)					Renewable water (hm^3)	Water stress %
	Public	Irrigation	Industry	Energy	Total		
Aguascalientes	150	559	12	0	720	395	182.6
Baja California	225	3058	83	192	3559	2210	161.1
Baja California Sur	94	423	14	5	536	802	66.8
Chihuahua	553	5137	54	28	5772	1719	75.6
Coahuila	278	1892	76	74	2320	7636	135.0
Durango	191	1545	17	12	1766	10,137	17.4
Nuevo Leon	615	1783	83	1	2481	2952	84.0
San Luis Potosi	200	1494	34	31	1759	8478	20.8
Sinaloa	453	10,566	43	0	11,062	6960	159.0
Sonora	914	7199	110	16	8240	4870	169.2
Tamaulipas	385	4510	115	55	5065	6609	76.6
Zacatecas	136	1565	72	0	1773	2601	68.2
Northern Mexico	4195	39,730	713	413	45,052	55,367	81.4

Table 11.22 Water stress in northern Mexico for year 2099 considering RCP 8.5 scenario (2070–2099)

	Uses of water (hm^3)					Renewable	Water
State	Public	Irrigation	Industry	Energy	Total	water (hm^3)	stress %
Aguascalientes	193	798	12	0	1003	259	387.6
Baja California	326	4602	83	192	5203	1996	260.7
Baja California Sur	633	747	14	5	1398	563	248.2
Chihuahua	701	6310	54	28	7093	682	178.6
Coahuila	374	2567	76	74	3091	3971	453.5
Durango	229	1978	17	12	2236	7274	30.7
Nuevo Leon	949	2887	83	1	3920	1664	235.5
San Luis Potosi	231	1816	34	31	2113	5952	35.5
Sinaloa	537	15,596	43	0	16,176	5624	287.6
Sonora	1320	10,601	110	16	12,046	3030	397.6
Tamaulipas	504	7398	115	55	8072	4122	195.8
Zacatecas	155	1927	72	0	2154	1219	176.7
Northern Mexico	6153	57,226	713	413	64,505	36,356	177.4

The current condition of water stress is 50.7%, and it will increase to 81.4% for the RCP 4.5 (2039) and to 177.4% for the RCP 8.5 (2099).

The lack of water will have negative impacts not only on agriculture but also on domestic uses. So, knowledge of the volumes utilized by specific users, along with the availability of rainwater, surface water, and groundwater, should constitute the basis upon which to establish adequate conditions for the sustainable management of the water resources in northern Mexico.

References

Allen GR, Pereira L, Raes D, Smith M (2006) Evapotranspiracion del cultivo: Determinacion de los requerimientos de agua en los cultivos. Estudios FAO Riego y Drenaje 56. ISSN 0254-5293. (In Spanish)

Al-Bakri J, Suleiman A, Abdulla F, Ayad J (2010) Potential impact of climate change on ran fed agricultura of a semi-arid basin in Jordan. Phys Chem Earth 35:125–134

Arredondo T, Huber E (2011) Chapter 51: Impacts of Drought on Agriculture in Northern Mexico. In: Coping with global environmental change. Disasters and security. Springer, Berlin, pp 875–891

Asha K, Munisamy G, Bhat A (2012) Impact of climate on rainfed agriculture in India: a case study of Dharwad. Int J Sci Dev 3(4):368–371

Calzadilla A, Zhu T, Rehdanz K, Tol R, Ringler C (2014) Climate change and agriculture: impacts and adaptation in South Africa. Water Resour Econ 5:24–48

CENAPRED (2014) Serie "Impacto socioeconomico de los desastres naturales". Centro Nacional de Prevencion de Desastres. SEGOB, Mexico. (In Spanish)

CONAGUA (2016) Estadisticas del Agua en Mexico 2016. www.conagua.gob.mx

CONAGUA (2018) Disponibilidad por acuiferos. https://www.gob.mx/conagua/acciones-y-programas/disponibilidad-por-acuiferos-66095. Accessed 20 Nov 2017

CONAPO (2015) Estimaciones y proyecciones de la poblacion en Mexico 2010–2030. Consejo Nacional de Poblacion. www.conapo.gob.mx. Accessed 15 Nov 2017

Escalante-Sandoval C, Nuñez-Garcia P (2017) Meteorological drought features in northern and northwestern parts of Mexico under different climate change scenarios. J Arid Land 9(1):65–75

Frich P, Alexander LV, Della-Marta P, Gleason B, Haylock M, Klein-Tank AMG, Peterson T (2002) Observed coherent changes in climatic extremes during the second half of the twentieth century. Clim Res 19:193–212

Gill RB (2000) The great Maya droughts: water, life, and death. University of New Mexico Press, Albuquerque. ISBN 0-8263-2194-1. OCLC43567384

Gouju X, Fengju Z, Zhengji Q, Runyuan W, Juying H (2013) Response to climate change for potato water use efficiency in semi-arid areas of China. Agric Water Manag 127:119–123

Hellin J, Bellon M, Hearne S (2014) Maize landraces and adaptation to climate change in Mexico. J Crop Improv 28(4):484–501

INECC (2014) Escenarios de Cambio Climatico. Available at: http://escenarios.inecc.gob.mx/

INEGI (2016) Instituto Nacional de Geografia y Estadistica. Mexico. (In Spanish) http://www3.inegi.org.mx/sistemas/productos/inegi.org.mx/geo/contenidos/geodesia/default.aspx. Accessed 15 Nov 2017

IPCC (2013) Summary for policymakers. In: Stocker TF, Qin D, Plattner G-K, Tignor M, Allen SK, Boschung J, Nauels A, Xia Y, Bex V, Midgley PM (eds) Climate change 2013: the physical science basis. Contribution of working group I to the fifth assessment report of the intergovernmental panel on climate change. Cambridge University Press, Cambridge, UK/New York

Loukas A, Vasiliades L, Tzabiras J (2007) Evaluation of climate change on drought impulses in Thessaly, Greece. Eur Water 17(18):17–28

Magaña-Rueda VO (2006) Informe sobre escenarios futuros del sector agua en Mexico. Instituto Nacional de Ecologia. Secretaria del Medio Ambiente y Recursos Naturales. (In Spanish)

Mongui H, Majule A, Lyimo J (2010) Vulnerability and adaptation of rain fed agriculture to climate change and variability in semi-arid Tanzania. Afr J Environ Sci Technol 4(6):371–381

Mundo-Molina M (2015) Climate change effects on evapotranspiration in Mexico. Am J Clim Chang 4:163–172

Padilla G, Rodriguez L, Castorena G, Florescano E (1980) Analisis historico de las sequias en Mexico. SARH, Mexico. (In Spanish)

Pedroza-Gonzalez E, Hinojosa-Cuellar G (2014) Manejo y distribucion de agua en los distritos de riego. IMTA, Jiutepec

Pope K et al (2001) Origin and environmental setting of ancient agriculture in the lowlands of Mesoamerica. Science 292:1370–1373

SIAP (2017) Servicio de Informacion Agroalimentaria y Pesquera. Secretaria de Agricultura, Ganaderia, Desarrollo Rural, Pesca y Alimentacion. Mexico. (In Spanish) http://wwwsiapgobmx/. Accessed 20 Nov 2017

Valverde P, Serralheiro R, Carvalho M, Maia R, Oliveira B, Ramos V (2015a) Climate change impacts on rain fed agriculture in the Guadiana river basin (Portugal). Agric Water Manag 150:35–45

Valverde P, Serralheiro R, Carvalho M, Maia R, Oliveira B, Ramos V (2015b) Climate change impacts on irrigated agriculture in the Guadiana river basin (Portugal). Agric Water Manag 152:17–30

Chapter 12
Dam-Operation Policy During Hurricane Season Using Regional Flows with Canonical Correlation Analysis

Juan Pablo Molina-Aguilar, Alfonso Gutierrez-López, and Ivonne Cruz Paz

Abstract The period of hurricanes extends along the Mexican coast from May to October and year after year causes considerable economic and human damage. An important part of risk management during a disaster of this type is the operation of dams. Many of these structures are used for water supply and electric-power generation; however, their maximum flow-control capacity is of great importance for flood-risk management.

The hydrometric data of 15 stations located within the basin of the El Caracol dam are used. Maximum precipitation data are also used in 27 stations within the basin. Note that the series used contain data for years with hurricane records. With these data, a regional frequency analysis is performed using the Gradex approach, which highlights the impact of precipitation from hurricane events.

Using a matrix of physiographic, climatological, and environmental basin characteristics, in addition to the characteristics of the reservoir and the operation of the dam, a principal component analysis EOF is performed to prioritize the variables that must be included in the operation policy of the dam. With the satellite images recorded every 15 min, the precipitation intensities of the hurricanes are calculated using the hydro-estimator technique and correlated with the flow rates measured by performing a canonical correlation analysis. The results show that downscaling between the precipitation data coming from the satellite images and the measured flow rates on the surface, it is possible to propose a policy for the supply and control operation of dams. The conclusion is that the hydro-estimator techniques together with the canonical correlation analysis are adequate procedures to propose the operation policies in dams during the hurricane season.

Keywords Hurricanes · Dam operation · Principal component analysis · Canonical correlation analysis

J. P. Molina-Aguilar · A. Gutierrez-López (✉) · I. C. Paz
Universidad Autonoma de Queretaro, Water Research Center, Centro de Investigaciones del Agua-Queretaro (CIAQ), International Flood Initiative, Latin-American and the Caribbean Region (IFI-LAC), International Hydrological Programme (IHP-UNESCO), Queretaro, Mexico
e-mail: alfonso.gutierrez@uaq.mx

© Springer Nature Switzerland AG 2020
J. A. Raynal-Villasenor (ed.), *Water Resources of Mexico*, World Water Resources 6,
https://doi.org/10.1007/978-3-030-40686-8_12

12.1   Introduction

In Mexico, there are approximately 5000 dams that are administered by the National Water Commission (Comision Nacional del Agua); the Federal Electricity Commission (CFE); the International Boundary and Water Commission (CILA), Section Mexico; Associations of Users and Private Owners; representatives of state, municipal and Mexico City governments, as well as formal and informal users.

Therefore, it is necessary to have an Emergency Action Plan (EAP) where all the institutions, government sectors and users mentioned above, must collaborate in implementing, in a coordinated way, the necessary actions to perform an evacuation procedures and/or make a warning alert in case of an emergency. This would help to avoid and/or to reduce the loss of human lives and goods and/or environmental damage in the geographical environment of the dams that are already rated as high risk with formal classification with high damage potential. For dams classified as high risk or with a high damage potential in the event of failure, the responsible entity must perform a periodic or intermediate safety inspection every year (see NMX-AA-175/2-SCFI-2016 Safe operation of dams. 2. Security inspections) and inform the the National Water Commission (Comision Nacional del Agua).

Meanwhile, for the diversion dams, storage dams or flood control dams with dam bodies with heights above the river channel less than 15 m and with a capacity to the Ordinary High Water Level (NAMO) less than 250,000 cubic meters, and that are classified preliminarily at low risk or whose formal classification of consequences for their potential for damage in case of failure it is low (see NMX-AA-175-SCFI-2015 Safe operation of dams, Part 1.- Risk analysis and classification of dams), they do not require the preparation of an EAP. Instead, they should only have a directory of the local, state, or federal authorities to inform them that the dam is in a state of emergency.

In all these cases, the dams need to have an adequate operating policy, especially for flood control in the hurricane season.

The period of hurricanes extends along the Mexican coasts from May to October, and year after year hurricane causes significant economic and human losses. An important part of risk management during a disaster of this type is the adequate operation of dams. Many of these structures are used for water supply and electric-power generation; however, their maximum flow-control capacity is of great importance for flood-risk management.

To collect rainfall and other hydrometric data, Mexico has about little more than 5000 weather stations and 1000 flow measurement sites. So, it is very useful to have real-time predictions of precipitation during the hurricane season. Thus, the hydro-estimator technique is used to correlate the operation policies of the dams with the trajectory and intensity of hurricane precipitation.

Note that the series used contain data from years with hurricane records. With these data, a regional frequency analysis is performed using the Gradex approach, which highlights the impact of precipitation from hurricane events. Using a matrix of physiographic, climatological, and environmental basin characteristics, in addition to the characteristics of the reservoir and the operation of the dam, a canonical correlation analysis is performed to prioritize the variables that must be included in the operation policy of the dam.

With satellite images recorded every 15 min, the precipitation intensities of the hurricanes are calculated using the hydro-estimator technique and correlated with the flow rates measured through a process named canonical correlation analysis.

The results show that the downscaling between the precipitation coming from satellite images and measured flow rates on the surface is possible to propose a policy of operation for the supply and control dams. The conclusions are intended to establish the technique of hydro-estimator in combination with canonical correlation analysis as an appropriate procedure to propose dam operation policies during the hurricane season.

12.2 Theoretical Framework

12.2.1 Canonical Correlation Analysis

The canonical correlation analysis can be considered as the logical generalization of the multiple regression analysis whose objective is the simultaneous correlation of p-dependent variables with respect to q-independent variables, establishing the linear combination of both sets of variables, through random vectors that maximize the value of the correlation coefficient (Hair et al. 2009). Considering the two sets of variables involved in the development of hydro-meteorological phenomenon, the first of which groups the independent variables $X = (x_1, x_2, \ldots, x_p)$ and in the second the dependent variables are grouped $Y = (y_1, y_2, \ldots, y_q)$ through V and W that are two compound lineal variables, which are named canonical variables (Crespin 2016).

These canonical variables are the product of the linear combination by multiplying the transpose of the weight vectors a and b by the set of dependent variables s and independent variables s, respectively.

$$V = a^T X = (a_1, a_2, \ldots, a_p) \begin{pmatrix} x_1 \\ x_2 \\ \cdot \\ x_p \end{pmatrix} = a_1 x_1 + a_2 x_2 + \ldots + a_p x_p \qquad (12.1)$$

$$W = b^T Y = (b_1, b_2, \ldots, b_q) \begin{pmatrix} y_1 \\ y_2 \\ \cdot \\ y_q \end{pmatrix} = b_1 y_1 + b_2 y_2 + \ldots + b_q y_q \qquad (12.2)$$

To do this, the sample covariance matrix of size p x q for the first and second set of variables is defined as

$$C = \begin{pmatrix} C_{XX} & C_{XY} \\ C_{YX} & C_{YY} \end{pmatrix} \qquad (12.3)$$

where

$$C_{YX} = C_{XY}{}^{T} \qquad (12.4)$$

Similarly, the canonical correlation between both composite variables is defined as

$$corr(V, W) = \frac{cov(V, W)}{\sqrt{var(V)}\sqrt{var(W)}} = \frac{a^{T} C_{XY} b}{\sqrt{a^{T} C_{XX} a}\sqrt{b^{T} C_{YY} b}} \qquad (12.5)$$

Defining as restrictions of the problem that

$$Var(V) = a^{T} C_{XX} a = 1 \qquad (12.6)$$

$$Var(W) = b^{T} C_{YY} b = 1 \qquad (12.7)$$

So, the problem of canonical correlation is reduced to maximize

$$Corr(V, W) = a^{T} C_{XY} b \qquad (12.8)$$

To realize the maximization of the canonical correlation, a Lagrange function of two parameters, τ_1 and τ_2, is defined as

$$L(a, b) = a^{T} C_{XY} b - \tau_1 \left(a^{T} C_{XX} a - 1\right) - \tau_2 \left(b^{T} C_{YY} b - 1\right) \qquad (12.9)$$

From which, we obtain the partial derivatives of the Lagrange function in terms of the weight vectors solution a and b.

$$\frac{\partial L(a, b)}{\partial a} = C_{XY} b - 2\tau_1 C_{XX} a = 0 \qquad (12.10)$$

$$\frac{\partial L(a, b)}{\partial b} = C_{XY}^{T} a - 2\tau_2 C_{YY} b = 0 \qquad (12.11)$$

Both partial derivatives make a possible construction of a system of equations that enable us to determine the unknown vectors.

$$C_{XY}b = 2\tau_1 C_{XX}a \qquad (12.12)$$

$$C_{XY}^T a = 2\tau_2 C_{YY}b \qquad (12.13)$$

According to Eqs. (12.6) and (12.7), proceed to multiply Eqs. (12.12) and (12.13) by the transposed vectors to respect the initial approach.

$$(C_{XY}b = 2\tau_1 C_{XX}a)a^T$$
$$a^T C_{XY}b = 2\tau_1 a^T C_{XX}a$$
$$a^T C_{XY}b = 2\tau_1 \qquad (12.14)$$
$$(C_{XY}^T a = 2\tau_2 C_{YY}b)b^T$$
$$b^T C_{XY}^T a = 2\tau_2 b^T C_{YY}b$$

$$b^T C_{XY}^T a = 2\tau_2 \qquad (12.15)$$

Developing the left side of Eqs. (12.14) and (12.15) yields a scalar pattern and because

$$\left(a^T C_{XY}b\right)^T = b^T C_{XY}^T a = 1, \qquad (12.16)$$

both equations are equivalent as defined by $\lambda = 2\tau_1 = 2\tau_2$. Therefore, we can rewrite Eqs. (12.12) and (12.13) as

$$C_{XY}b = \lambda C_{XX}a \qquad (12.17)$$

$$C_{XY}^T a = \lambda C_{YY}b \qquad (12.18)$$

This allows reduction of the system using the method of algebraic substitution clearing the vector a of Eq. (12.17) to substitute in Eq. (12.18)

$$C_{XY}^T \left[\frac{1}{\lambda} C_{XX}^{-1} C_{XY}b\right] = \lambda C_{YY}b$$
$$C_{XY}^T C_{XX}^{-1} C_{XY}b = \lambda^2 C_{YY}b$$
$$C_{YY}^{-1} C_{XY}^T C_{XX}^{-1} C_{XY}b = \lambda^2 b$$
$$C_{YY}^{-1} C_{XY}^T C_{XX}^{-1} C_{XY}b - \lambda^2 b = 0 \qquad (12.19)$$

In an analogous way, by clearing vector b from Eq. (12.18) to substitute in Eq. (12.17), we obtain

$$C_{XX}^{-1}C_{XY}C_{YY}^{-1}C_{XY}^{T}a - \lambda^2 a = 0 \tag{12.20}$$

The solution of Eqs. (12.19) and (12.20) makes it possible to obtain the vectors a and b that represent the eigenvectors.

Once the two eigenvectors are defined, the value of their structure coefficient (r) is defined for C_{XY}, setting the option to define triple solutions according to the number of eigenvectors, that is, solutions from each eigenvector studied and their respective weight vectors (λ_i, a_i, b_i). Based on this, the canonical solutions are proposed by using the eigenvector in turn and the corresponding canonical variables (λ_i, V_i, W_i). Consequently, $\lambda_i = \mathrm{corr}(V_i, W_i)$.

12.2.2 Satellite Images

The characteristics of the digital satellite images used correspond to the Geostationary Operational Environmental Satellite GOES-13 with coverage of the Mexican territory. The satellite GOES-13 is able to capture images of the channels of water vapor, infrared and visible available every 15 min at a temporal frequency (Meza et al. 2014) in a raw and colorless format.

The images have a .pcx extension with a BZ2 compression and tagged with a nomenclature including the date and time when they were taken, e.g., the image 201409150015.pcx corresponds to the image captured on 15 September 2014 at 00 h15 m (Meza et al. 2014).

The images as far as year 2003 have such assigned reference, in their upper left-hand corner the coordinate latitude 36.4768° and −122.2590° longitude, while their lower right-hand corner the coordinate latitude 14.1118° and longitude −79.0817°.

Since 2010, the images have a resolution of 2.36 km, with 817 pixels of latitudinal resolution and 1280 pixels of longitudinal resolution – every single one with eight bits resolution. Regarding their infrared spectrum, the images have brightness values ranging in value from 0 to 255, also known as their digital level, while the water vapor images range from digital level from 113 to 255.

12.2.3 Reading of Brightness

The temporal and spatial evolution of weather phenomena are recorded in the magnitude of the brightness value in each of the pixels that are part of the captured satellite images. The GOES-13 images used have 1,045,760 pixels, their interpretation and segmentation by threshold was made using the Sat Viewer® tool developed at the Water Research Center (CIAQ) of the Autonomous University of Querétaro (UAQ). The tool mentioned consists of a display screen for bmp format using a selection tool that generates a table of referenced brightness with nomenclature Bi, j for its identification; the subscript i corresponds to the column and j corresponds to the row of location of the pixel.

The measurements made on the interested pixels are extracted in a text file with extension .txt, which allows to calculate the temperature at the top of the clouds and make forecasts of the intensity of rain.

12.2.4 Disaggregation of Satellite Images

The characteristic features of recording and measuring instruments often present limitations for the appropriate use and comparison of information associated with weather phenomena. As previously mentioned in Sect. 12.2.2, digital images from GOES-13 have a 15-min time resolution, while the records from automatic weather stations in Mexico are 10 min long. This difference in temporal resolution required the disaggregation of the information obtained by the satellite.

This methodology is based on standardizing and normalizing the data series that come from the interpretation of the satellite images. For this purpose, first the statistical data of the series are defined, then the standardization is performed by subtracting the mean value from each of the data of the series and finally the standardization is obtained by dividing each of the results by the standard deviation. With the described operations, all the data of the series present value in the range of -1 to 1. The resulting series is characterized by a mean value equal to zero and a standard deviation with a value equal to 1.

The new series are used to determine the values that correspond to the 10-min time resolution, to be able to match both sources of information. However, this treatment is only applied to data greater than zero. Once the required values have been determined, the first step is to make an inverse process to eliminate the standardization of data based on the statistics of the series defined at the beginning of the process.

The hydro-estimator technique represents a law of regression of the estimated rain intensity with respect to the maximum temperature of the clouds. This technique requires corrections for humidity, growth and temperature gradient at the top of the clouds, to improve the estimations made (Vicente et al. 1998); this technique considers a vertical path of precipitation from the cloud to the surface where it is measured.

$$R = 1.1183x10^{11} \exp\left(-3.6382x10^{-2}T^{1.2}\right) \tag{12.21}$$

$$\text{IF B} > 176 \text{ then T} = 418 - B$$

$$\text{If B} \leq 176 \text{ then T} = 330 - (B/2)$$

R is the intensity predicted in mm per hour, and T is the temperature at the top of the clouds measured in degrees Kelvin (195° K to 260° K) as a function of the brightness of the pixel (B).

12.3 Evolution of the Capacities of the Dams

The National Water Commission publishes the daily volumes of each of the dams in Mexico. Although it does not constitute the official policy of the dam, it is only applied by the authorities based on experience, and the historical records represent the temporal evolution of the filling and handling of the dam. Tables 12.1 and 12.2 show an example of the evolution of the capacities of the main dams in the states of Guerrero and Michoacán.

Table 12.1 Evolution of the capacities of the dams in Michoacán state in hm^3

Dam	20/10/ 2015	21/10/ 2015	22/10/ 2015	23/10/ 2015	24/10/ 2015	25/10/ 2015
Agostitlan	16.26	16.24	16.24	16.24	16.24	16.24
Aristeo Mercado	11.99	11.99	11.99	11.92	12.22	12.38
Coíntzio	77.62	77.62	77.62	77.62	77.16	78.68
De Gonzalo	5.6	5.68	5.77	5.95	6.26	6.4
El Bosque	199.99	199.87	200	200.22	200.68	200.68
Infiernillo	5557.07	5548.79	5562.59	5609.49	5755.72	6015.06
Jaripo	9.78	9.78	9.78	9.82	10.84	10.49
Los Olivos	5.93	5.88	5.88	5.84	11.95	11.03
Melchor Ocampo	209.57	209.57	209.36	209.36	209.79	210
Pucuato	9.75	9.69	9.69	9.69	9.69	9.69
San Juanico	57.72	57.72	57.6	57.97	61.82	61.82
Tepuxtepec	420.46	420.06	419.65	419.24	418.43	418.02
Zicuiran	35.65	35.65	35.61	35.85	37.07	36.9

Table 12.2 Evolution of the capacities of the dams in Guerrero state in hm^3

Dam	20/10/ 2015	21/10/ 2015	22/10/ 2015	23/10/ 2015	24/10/ 2015	25/10/ 2015
Andres Figueroa	102.81	102.76	102.72	102.76	102.94	103.95
El Gallo	389.23	388.25	388.25	388.25	390.21	396.02
Ing. Carlos Ramirez Ulloa	1421.35	1425.89	1414.10	1398.49	1388.00	1393.46
La Calera	9.8	9.84	9.91	10.36	10.66	17.91
Revolucion Mexicana	107.38	107.38	107.86	108.07	108.35	108.28
Valerio Trujano	25.73	25.67	25.56	25.44	25.33	25.28
Vicente Guerrero	249.5	249.3	249.2	249.2	249.2	249.2

12.4   Methodology

To carry out the temporal analysis of the meteorological variables, we proceeded to identify the satellite images that coincide with the study period on 1–16 June 2012. They were decompressed, converted to bmp format and, through the application Sat-Viewer®, the brightness value of the coordinate pixel georeferenced was extracted in each of them for the principal dams along the Pacific Coast (Table 12.3, and Figs. 12.1 and 12.2). Later, the cold cloud-top temperature was determined to estimate the precipitation based on the hydro-estimator.

Once the dams under study have been located, the hurricane to be studied is selected—in this case Hurricane Carlotta, which started on 13 June 2012 and ended on 16 June 2012. The trajectory of this hurricane is shown in Fig. 12.3. Next, the hydro-estimator is applied to each of the satellite images, from which the precipitation intensity R is obtained, in mm/h. Table 12.4 shows the proposed intervals for the

Table 12.3 Location of the study dams in a satellite image whose size is 1280 x 817 pixels

| | | | | 1280 x 817 | |
State	Dam	Longitude	Latitude	x	y
Guerrero	Ing. Carlos Ramírez Ulloa	−99.995	17.948	418	154
Guerrero	Valerio Trujano	−99.466	18.296	440	169
Michoacán	Infiernillo	−101.893	18.272	349	166
Michoacán	Tepuxtepec	−100.229	20.001	423	239

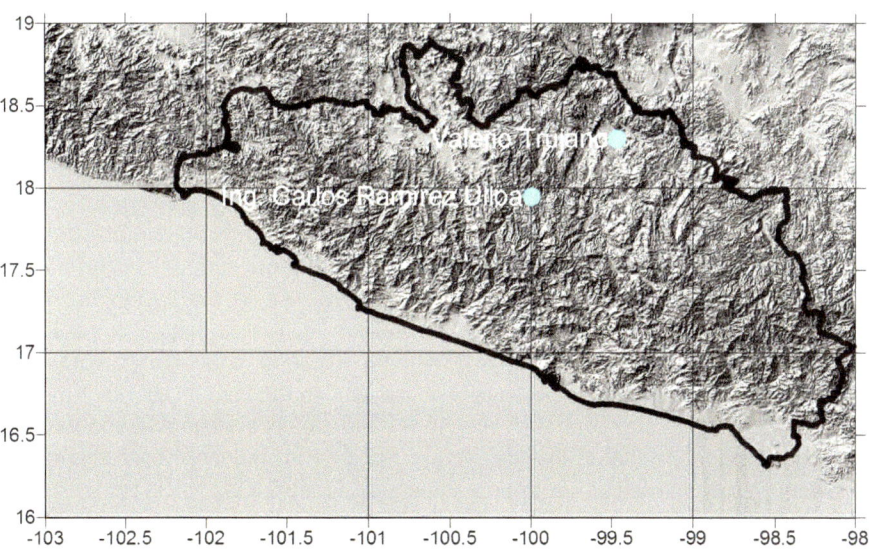

Fig. 12.1 Location map of the Carlos Ramirez Ulloa and Valerio Trujano dams in Guerrero State

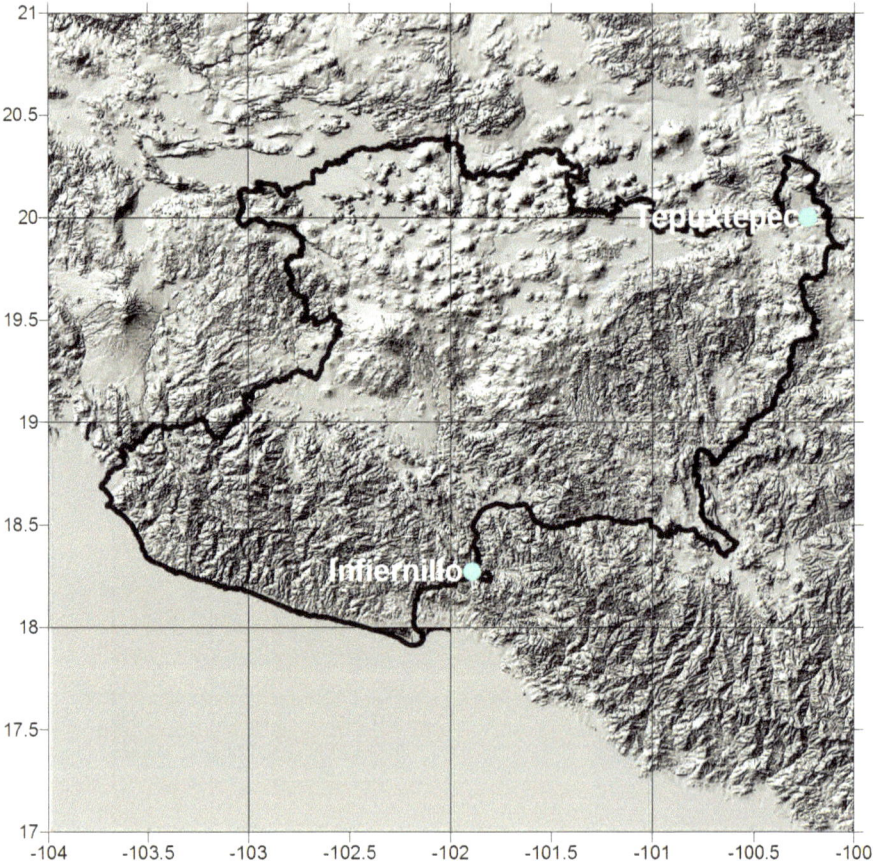

Fig. 12.2 Location map of the Infiernillo and Tepuxtepec dams in Michoacán State

intensity values obtained from the hydro-estimator. This activity makes it possible to obtain images of the hurricane's precipitation potential by adding colors.

   Relevant to the policy of operation of the dams, there is access to information on the evolution of the storage of the reservoir during the entire period of occurrence of the hurricane, as shown in Table 12.3. With the data of the evolution of the precipitation obtained with Eq. (12.21), we made a data matrix, where the evolution variables of the hurricane are correlated with the evolution of the storage volume of the dam. To characterize the hurricane, we use data on the latitude and longitude of displacement of the hurricane and the velocity and atmospheric pressure in the eye. To characterize the satellite images, we use the areas of influence by counting the number of pixels in each of the proposed intervals. Finally, a canonical analysis of all

Fig. 12.3 Track of Hurricane Carlotta

Table 12.4 Scale of rainfall intensities proposed for Eq. (12.21)

Scale	Geometric area	Brightness digital level	rainfall intensities mm/h
Light blue	Z6	100–125	0.002–0.013
Blue	Z5	126–150	0.014–0.070
Green	Z4	151–175	0.075–0.362
Yellow	Z3	176–200	0.386–8.649
Orange	Z2	201–225	9.832–205.123
Red	Z1	226–235	greater than 200

Vicente et al. (1998)

the variables is carried out to forecast the evolution of the volumes of the dam and to have a tool to infer the future operation policy. In this case, the data of the first 48 h of the hurricane's evolution are used, and with the canonical analysis, the final 12 h were predicted in the hurricane's passage.

12.5   Results

With the proposed methodology, a canonical analysis of two dams in the state of Guerrero and two dams in the state of Michoacán was carried out, with the results as shown in Table 12.5. Figures 12.4, 12.5, 12.6, 12.7 show the development of Hurricane Carlotta and its respective precipitation-intensity scales. It is important to mention that estimates of the precipitation intensities were made, beginning with the onset of the hurricane's impact on the Mexican coast and up to the cessation of destruction. During this time, the storage volume variation is related canonically to the parameters of movement and the severity of the hurricane. Through the proposed methodology, we sought to forecast the dams' storage during the last moments of the hurricane. This technique enables formulating a prognostic equation correlating the evolution of the storage of the dam and the severity of the rains caused by the hurricane.

12.6   Conclusions

The results of the forecast for the operation policy of the two dams show that the canonical analysis is a reliable tool for prognosis with multivariate and multiscale variables. The results show that the correlation between the variables is greater than 90%. Figure 12.8 shows that the last 12 h of forecast are very similar to the actual capacities recorded in the dam during the passage of Hurricane Carlotta. The proposed methodology enables formulating forecast equations based on historical data that can describe the behavior of the dam capacities during the passage of a hurricane.

Table 12.5 Evolution of the capacities of the dams in Guerrero state in hm^3

		Capacities (hm^3)			
State	Dam	June 13th	June 14th	June 15th	June 16th
Guerrero	Ramirez Ulloa	1012.6	989.6	987.2	969.3
Guerrero	Valerio Trujano	11.91	11.84	11.8	11.77
Michoacán	Tepuxtepec	283.28	283.28	282.84	282.94
Michoacán	Infiernillo	2817.94	2817.94	2827.65	2872.65

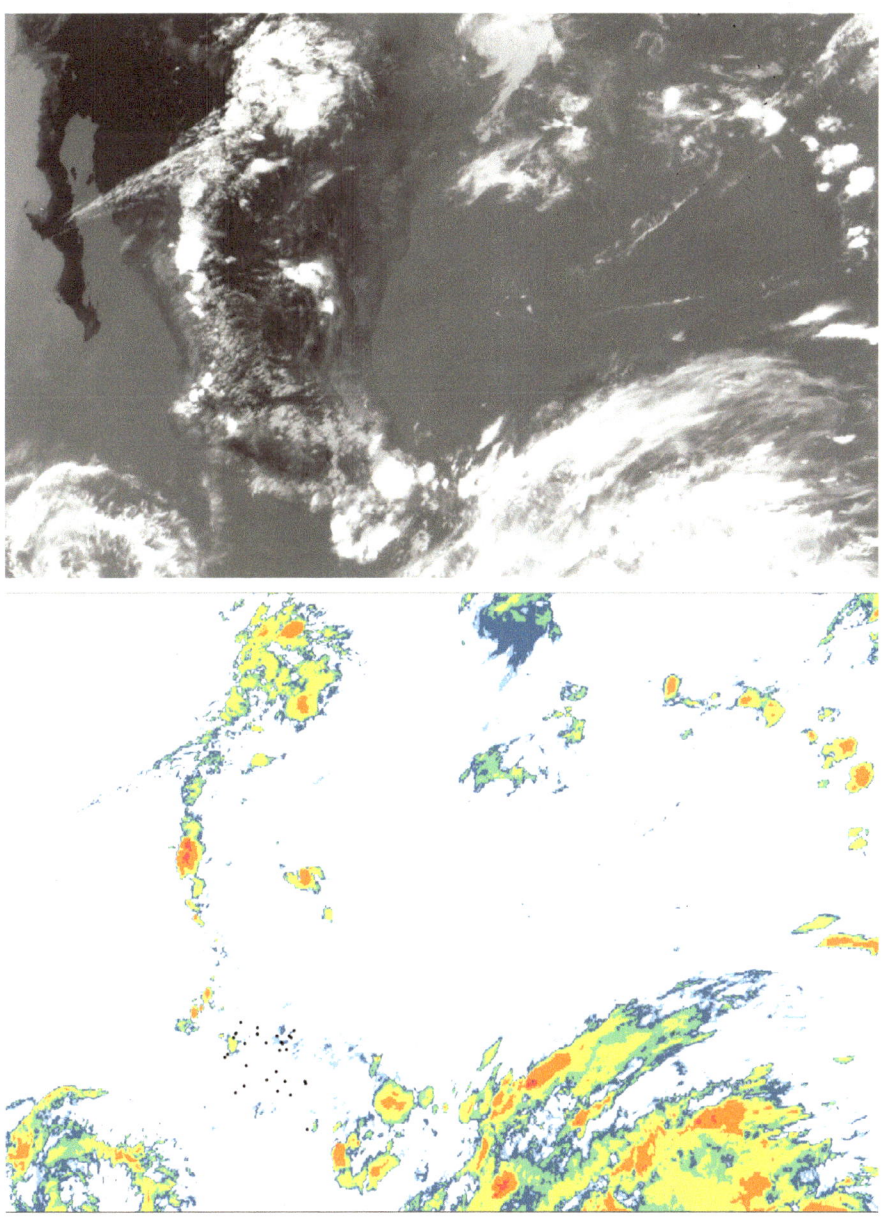

Fig. 12.4 Image and track of Hurricane Carlotta 201206132201

Fig. 12.5 Image and track of Hurricane Carlotta 201206151601

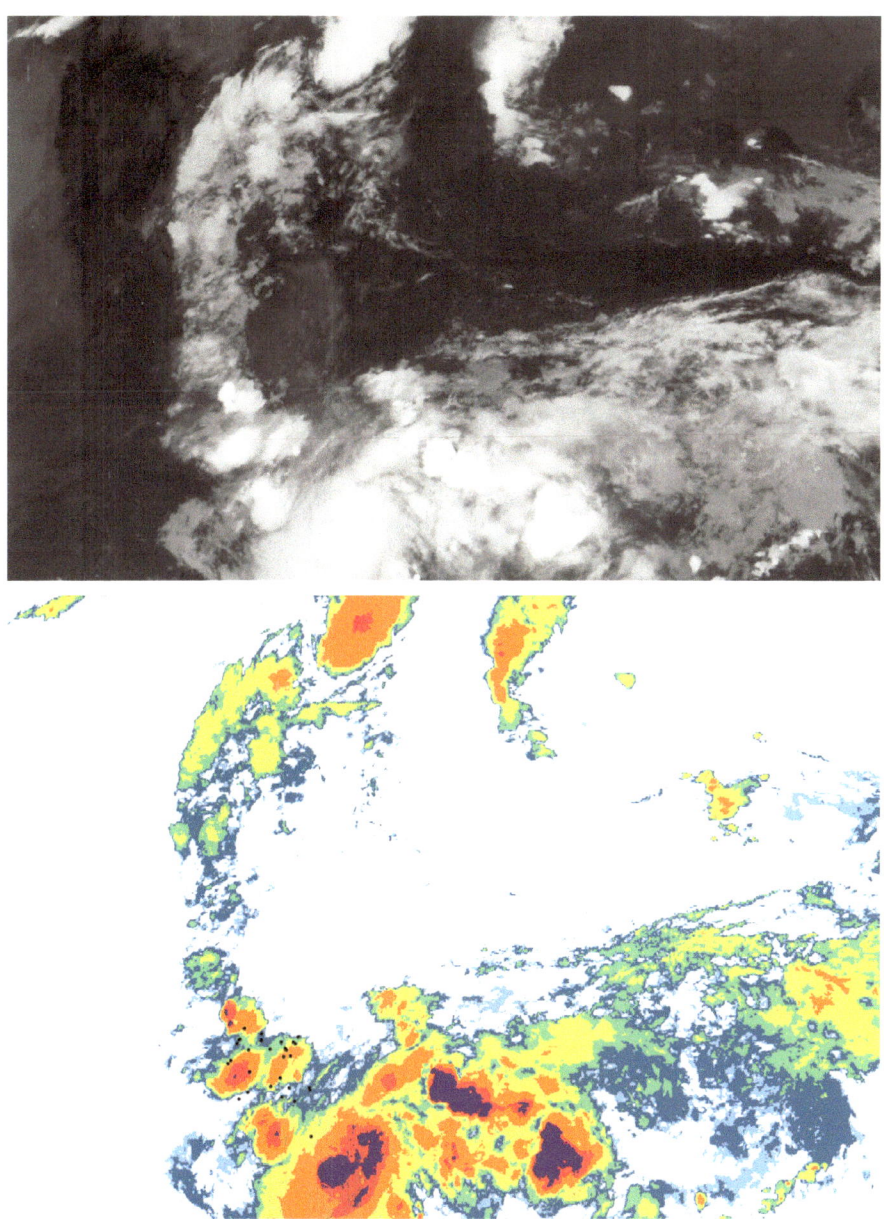

Fig. 12.6 Image and track of Hurricane Carlotta 201206160401

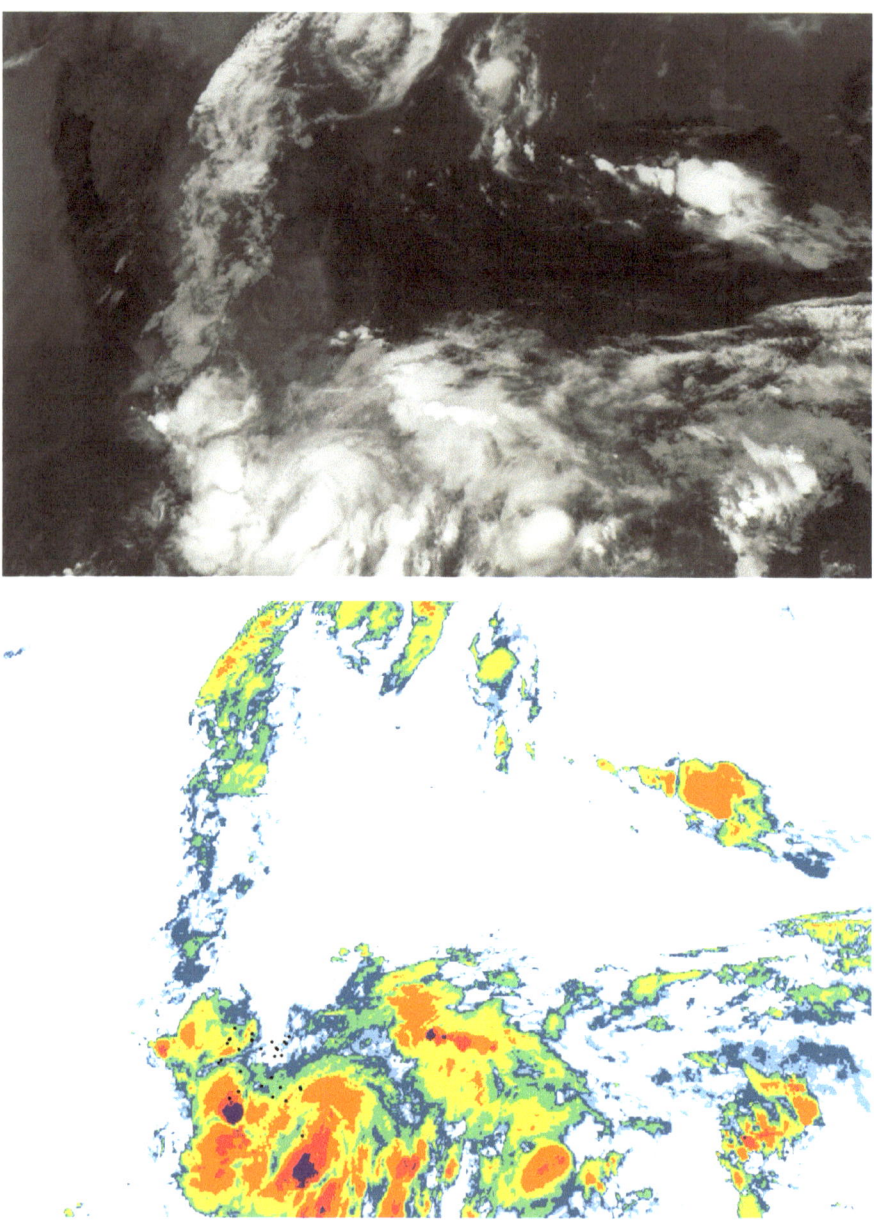

Fig. 12.7 Image and track of Hurricane Carlotta 201206161001

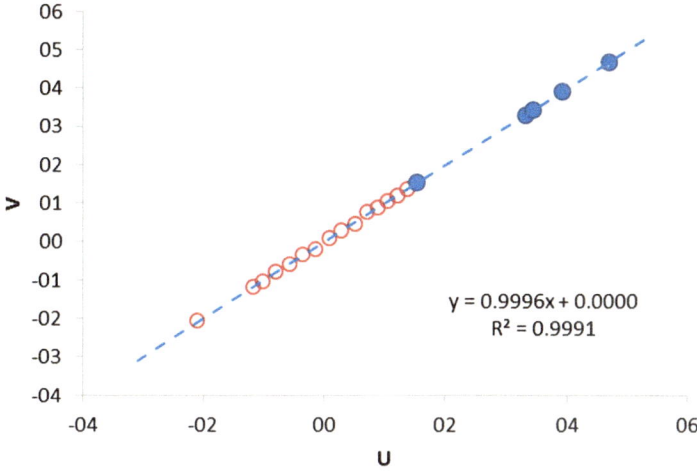

Fig. 12.8 Dams operation policy of Valerio Trujano dam

References

Albanil EA, Pascual RR, López QM, Quiroz AC, Chablé PLA (2015) Annual report 2015. Summit report in Mexico. General Coordination of the National Meteorological Service. Meteorology and Climatology Management. Sub management of Forecast to Medium and Long Term

Andenberg MR (1973) Cluster analysis for applications. Academic Press, New York

Beran MA, Brilly M, Becker A, Bonacci O (1990) Regionalization in hydrology, IAHS Publication no. 191. International Association of Hydrological Sciences, Wallingford, p 260

Brown P, Russell B (2001) Siting and maintenance of weather stations. Turf irrigation management series: III. College of Agriculture and Life Sciences. The University of Arizona

Closas AH, Arriola EA, Kuc CI, Yellow MR, Jovanovich EC (2013) Multivariate analysis, concepts and applications in Educational Psychology and Psychometrics. Approaches vol. 25 no.1 Libertador San Martín jun

Crespin EEO (2016) Análisismultivariante: aplicaciones con SPSS, 1st edn. UFG Editores, San Salvador, 304 p

De la Fuente FS (2011) Conglomerate analysis. Faculty of Economics and Business Sciences. Autonomous University of Madrid

Garcia-Osorio C, Fyfe C (2005) Visualization of high-dimensional data via orthogonal curves. J Univ Comput Sci 11(11):1806–1819

Hair JF, Anderson RE, Tatham RL, Black W (2009) Multivariate data analysis, 7th edn. Prentice Hall, Englewood Cliffs

Jolliffe IT (2002) Main component analysis, 2nd edn. Springer, New York, p 487

Medina GG, Grangeda GJ, Ruiz CJA, Báez G (2008) Use of meteorological stations in agriculture. Information brochure 50. National Institute of Forestry, Agriculture and Livestock Research. http://www.zacatecas.inifap.gob.mx/publicaciones/Uso_de_estaciones_meteorologicas_en_la_agricultura.pdf. Accessed 27 February 2017

Meza RM, Curiel HL, González TDME, Álvarez MC (2014) Satellite images and spatial distribution of rainfall in the state of Querétaro. Case: Storm of August 18, 2014. University network for characterization of hydrometeorological, fluvial and coastal risks. Nthe. Electrical magazine of scientific, technological and information dissemination of the state of Querétaro. Periodicity: quarterly—2016 No.14.

National Water Commission (CNA) (2012) National Meteorological Service: 135 years of history in Mexico. http://www.conagua.gob.mx/CONAGUA07/Publicaciones/Publicaciones/CGSMN-1-12.pdf. Accessed 3 March 2017

Prieto GRE (2006) Classification statistical techniques. An example of Cluster analysis. Monograph to obtain the title of Industrial Engineer. Institute of Basic Sciences and Engineering. Autonomous University of the State of Hidalgo

Vicente GA, Scofield RA, Menzel WP (1998) The operational GOES infrared rainfall estimation technique. Bull Am Meteorol Soc 79:1883–1898

Wilks DS (2006) Statistical methods in the atmospheric sciences, International geophysics series, 2nd edn. Academic Press, Amsterdam

Chapter 13
Hydrologic and Hydraulic Works of the Aztec Civilization

Jose A. Raynal-Villasenor

Abstract In less than 200 years, the Aztec civilization progressed from a nomadic tribe to a highly developed society, producing wonderful architectonic developments and creating an amazing capital city: *Mēxihco-Tenochtitlan*. This city amazed the Spanish conquerors when they first saw it in 1519 AD. The Aztec produced many art products, like sculptures, paintings, and poetry. They made their capital city very comfortable by means of streets, aqueducts, and water-supply and flood-control works. Water was very important in all aspects of their lives. In their religion, the water god, *Tlaloc*, had the same level of importance as their most important god called *Huitzilopochtli*, the Sun god and the so-called left-handed hummingbird. This chapter describes the main hydrologic and hydraulic works of the Aztecs, and it explains the relationship between their water gods and their culture.

Keywords Aztec civilization · Hydrologic and hydraulic works · Aqueducts · Flood-control systems · *Chinampas*

13.1 Introduction

Biswas (1970) coined the term *hydraulic civilizations* to refer to those that were founded near a source of water, like the Egyptians by the Nile River, the Babylonians by the Tigris and Euphrates rivers, the Chinese by the Yellow River, and the Romans by the Po River. The Aztec civilization has enough characteristics to be considered a *hydraulic* civilization, given the fact that it was born, grew, and collapsed surrounded by water.

The greatness of the city of *Mexico-Tenochtitlan* is well described by the words contained in the *Chimalpahin-Quauhtlehuanitzin*: "As long as the world exists, the

J. A. Raynal-Villasenor (✉)
Department of Civil and Environmental Engineering, Universidad de las Americas Puebla, Cholula, Puebla, Mexico
e-mail: josea.raynal@udlap.mx

© Springer Nature Switzerland AG 2020
J. A. Raynal-Villasenor (ed.), *Water Resources of Mexico*, World Water Resources 6,
https://doi.org/10.1007/978-3-030-40686-8_13

Fig. 13.1 *Mexico-Tenochtitlan.* (Thelmadatter 2008a)

glory and honor of Mēxihco-Tenochtitlan will never be forgotten." This description is best understood by a picture like the one presented in Fig. 13.1.

No wonder the Spanish conquerors were so amazed when they looked, for the first time, upon Aztec's capital city that they believed they were seeing a mirage. By the time the Spanish conquerors arrive at *Mēxihco-Tenochtitlan*, in the year 1519 AD, such a city was bigger, population wise, than London, Madrid, Rome, or any other main European settlement.

In less than 200 years, from 1325 AD to 1519 AD, the Aztecs progressed from a tribe of nomads to a highly advanced civilization that constructed an empire with a capital city with a multitude of wonderful constructions, like pyramids, palaces, roads, canals, aqueducts, etc. Moreover, they developed a format to record their history by artistic drawings now called codex (see Fig. 13.2).

13.2   Valley of Mexico Geographic Setting

The Valley of Mexico is part of a vast area designated as Mesoamerica (see Fig. 13.3) (Sémhur 2011).

The Valley of Mexico is 8000 km^2 in area, its limits being latitude 20°09′12″ in the south and 19°01′18″ in the north and, in longitude, the meridian 98°31′58″ in the east and 99°30′52″ in the west (Bribiesca-Castrejon 1960). The closed basin is well defined by several topographic barriers. It is interesting to note that the natural outlet

Fig. 13.2 The Aztec exodus. (Boturini Codex)

Fig. 13.3 Mesoamerica. (Sémhur 2011)

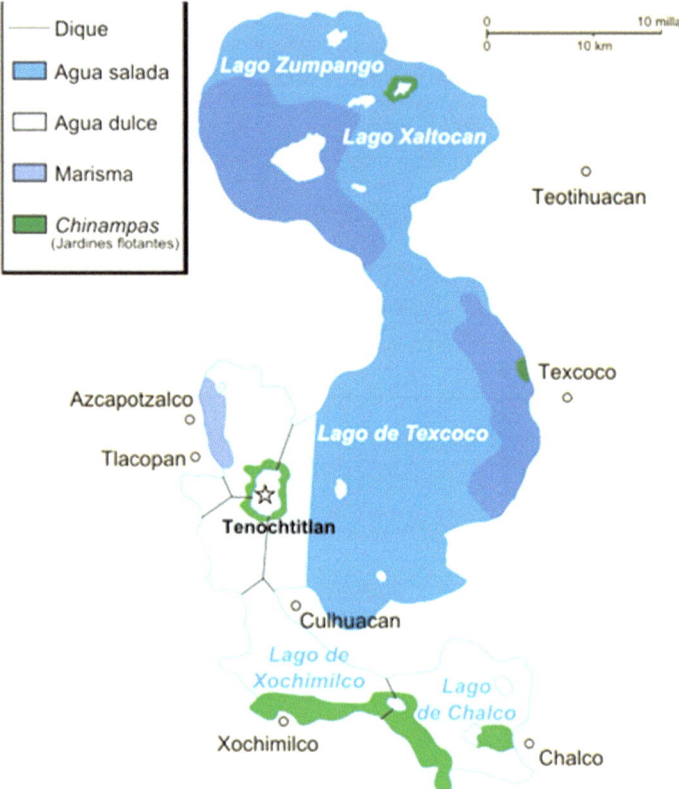

Fig. 13.4 Lakes of the Valley of Mexico circa 1519. (Historicair 2007)

of the basin of the Valley of Mexico was originally located in the south, in the direction to the city of *Cuautla*, but an ancient tectonic disturbance, caused by the uplift of the *Ajusco* mountain range, closed the water flow through the natural outlet and transformed the Valley of Mexico basin into a closed basin resulting in the growth of several lakes that have been extensively modified throughout the last 500 years.

The set of lakes that formed the water bodies in the Valley of Mexico included (see Fig. 13.4):

1. The *Lake of Mexico* in the center of the Valley of Mexico *Citlaltepetl*, which was the one with the second lowest water level in the Valley of Mexico.
2. *Lakes Chalco* and *Xochimilco* in the south, which in ancient times formed just one lake. These lakes once had high water levels in the Valley of Mexico.
3. *Lake Texcoco* in the east, which had the lowest water level and was the largest lake in the Valley of Mexico. This lake still exists today, but its size is much smaller than it was 500 years ago.

Table 13.1 Relative elevations of the lakes of the Valley of Mexico with respect to the elevation of the *Lake of Texcoco*

Name of the lake	Elevation (*varas*)
Zumpango	6. 620
Xaltocan	3.474
Saint Christopher	3.597
Mexico	1.907
Xochimilco	3.119
Chalco	3.820
Texcoco	0

Bribiesca-Castrejon (1960)
1 *vara* = 0.838 m

4. In the north were the lakes of *Zumpango*, the one with the highest water levels in the Valley of Mexico, and *Saint Christopher* and *Xaltocan* with water levels very similar to the lakes of *Chalco* and *Xochimilco*.

Table 13.1 shows the relative elevations of the lakes with respect to the *Lake of Texcoco*. During ancient intense rainy seasons, all the lakes joined to form one lake with an area of 2000 km^2 (Bribiesca-Castrejon 1960).

13.3   The Foundation of Mēxihco-Tenochtitlan

The Aztec had been slaves at a mythological place called *Aztlan*. Up to now, nobody knows where this place was located, although some historians have located *Aztlan* in *Mexcaltitan*, the so-called Mexican Venice, a city in the state of Nayarit in Western Mexico. Others locate *Aztlan* as far as Arizona in the USA.

It took 150 years for the Aztecs to reach the place where they founded *Mēxihco-Tenochtitlan,* as their capital city. The Aztecs arrived in the Valley of Mexico in the year 1215 AD, and this journey is illustrated in the Boturini Codex (see Fig. 13.2).

Tradition was that the exact place for such a foundation had to be identified by a magical sign foretold to a priest in a dream induced by the main god *Huitzilopochtli*, when they still were living at *Aztlan*. This then convinced the Aztecs to start looking for the place containing the following imagery: an eagle devouring a snake standing over a *nopal*, which is a species of cactus, the *nopal* above a rock and everything surrounded by water (see Fig. 13.5).

Tibon (1986) gave the following interpretation of the composition of such a magical sign: the exact spot where the sign will be revealed was the place where the heart of *Copil*, the son of goddess *Malinalxochitl* who was the head of the Lunar Branch of the Aztec religion, was buried by the priests of the Solar Branch of such a religion (UNAM 2012). They had conducted such a burial, after *Copil* was killed in combat with priests of *Huitzilopochtli*, the Sun god. Then, the nopal emerged from *Copil*'s heart and was known among the Aztecs as the human-hearts tree. The rock beneath the nopal has the meaning of a navel, the center of everything: a human body or an animal or Planet Earth; the eagle and Sun are synonyms in the Aztec language,

Fig. 13.5 The sign for the foundation of *Tenochtitlan.* (Raynal-Villasenor 2018a)

and the snake is commonly associated with darkness. The eagle devouring the snake represents the triumph of light over darkness. But, it also represents the victory of the Solar Branch over the Lunar Branch of the Aztec religion.

According to Tibon (1986), the location of the magical sign was the terrestrial reflection of the navel of the rabbit that is observed in the Moon, formed by the asteroids that have struck the Moon since ancient times. The Moon's rabbit was represented on the Earth's surface by the group of lakes that existed in the Valley of Mexico up to the sixteenth century (see Fig. 13.4). The resemblance between the Moon's rabbit and the form of the lakes in the Valley of Mexico is quite remarkable.

Tibon (1986) constructed, based on a previous interpretation, the translation into Spanish of the first word of the name of the capital city of the Aztec Empire, *Mēxihco-Tenochtitlan*: the place of the reflection on Earth of the navel of the rabbit that is on the Moon. The second word, *Tenochtitlan*, is simply the place of *Tenoch*, the first ruler of the Aztecs, who abolished the priest council that guided the Aztecs along their journey from *Aztlan* to *Mēxihco-Tenochtitlan*. *Tenoch* was honored by having his name as part of the name of the capital city of the Aztec Empire. He also started the tradition of having a single ruler, as the maximum authority in the Aztec Empire is called *tlatoani*. The *tlatoani* was neither a king nor an emperor because it was a title that rarely was inherited by a relative.

13.4 Aztec Mythology

Huitzilopochtli, the Sun god and the so-called left-handed hummingbird, was the supreme deity among the Aztec gods. His temple was at the top of the Great Temple (*Templo Mayor*), which had red decoration at the top (see Fig. 13.1).

The conception of *Huitzilopochtli*, the Sun god, was produced when a feather of a dove fell on her mother's lap. Given that *Coatlique*, her mother, has been a widow for several years, *Coyolxauhqui*, one of her daughters, conspired to kill her mother along with the 400 *Surianos*, her brothers. But *Huitzilopochtli*, who became an adult at the very moment of his birth, started shooting arrows with his bow, and some of the *Surianos* were killed. The rest that survived from *Huitzilopochtli*'s attack were so scared of him that they flew into the sky and became the stars. *Coyolxauhqui* was not so lucky: she was killed by *Huitzilopochtli* who beheaded her and knocked her down from the top of the pyramid where the fight had taken place, and she was dismembered by the impact with the steps of the pyramid as she fell from the top of the pyramid (Tibon 1986).

Back (1978, 1981) coined and defined hydro-mythology as "hydro-mythology is a study of hydrologic inspired folklore, myths, or legends that can be used to deduce beliefs of early people concerning water."

There are many gods related to water in Aztec mythology. Only five of them will be described here: *Tlaloc, Chalchihuitlicue, Metzli, Atlacoya,* and *Amimitl*.

The Aztec gods were classified according to the 13 Lords of the Day and with their associated "bird" (Cartwright 2017). The god and associated bird are shown in Table 13.2.

Table 13.2 Classification of Aztec gods according to the 13 Lords of the Day Cartwright (2017)	God's name	Associated bird
	Xiuhtecuhtli/Huehueteotl	Blue hummingbird
	Tlaltecuhtli	Green hummingbird
	Chalchiutlicue	Hawk
	Tonatiuh	Quail
	Tlazolteotl	Eagle
	Teoyaomiqui/Mictlantecuhtli	Screech owl
	Centeotl-Xochipilli	Butterfly
	Tlaloc	Eagle
	Quetzalcoatl	Turkey
	Tezcatlipoca	Horned owl
	Mictlantecuhtli/Chalmecatecuhtli	Macaw
	Tlahuizcalpantecuhtli	Quetzal
	Ilamatecuhtli	Parrot

Table 13.3 Classification of Aztec gods according to the nine Lords of the Night

God's name	Augury
Xiuhtecuhtli/Huehueteotl	Unfavorable
Itztli	Unfavorable
Piltzintecuhtli-Tonatiuh	Excellent
Centeotl	Excellent
Mictlantecuhtli	Favorable
Chalchiutlicue	Favorable
Tlazolteotl	Unfavorable
Tepeyolohtli-Tezcatlipoca	Favorable
Tlaloc	Favorable

Cartwright (2017)

The Aztec gods were also classified according to the nine Lords of the Night and with their associated augury (Cartwright 2017). The god and associated augury are shown in Table 13.3.

Among the Aztec water gods, only *Chalchiutlicue* and *Tlaloc* appeared in both lists, which gives another clue about their importance within the Aztec pantheon.

Tlaloc (see Fig. 13.6) is the god who has been worshipped since the time of the city of *Teotihuacan* (200 BC–600 AD), when the *Teotihuacans* were the first inhabitants of the Valley. *Tlaloc* was the god of rain and lightning as well, and he was the god of floods and droughts, too. He was the ruler of the third Sun.

Tlaloc used to live in a place, like paradise, called the *Tlalocan*. This place was located on the top of the mountains, and such a place was abundant in all foods. *Tlaloc* shared this magnificent place with a maize goddess, child servants, and dwarfs, called *tlaloques* (Back 1981). To produce rain, *Tlaloc* had four different water reservoirs from which the rain fell down to Earth. The best water comes from the first recipient which contained good water that produced maize and fruits. The second recipient produced cobwebs and blight. The third recipient contained water that produced frost that killed plants. The fourth recipient contained water that produced droughts and absences of crops. The waters of the four recipients were aligned with the cardinal points: East, West, North, and South; only the east recipient water bestowed good results on people and crops. Sometimes, the servants of *Tlaloc*, called *tlaloques*, smashed their jars very hard with a stick so that thunder was produced, and the pieces from the broken jars cast below became thunderbolts.

At the top of the Main Temple (*Templo Mayor*), the biggest construction in the central part of Fig. 13.1, there were two constructions, like temples, for the performance of religious rituals: the one on the right with red decorations at the top was devoted to *Huichilopochtli*, the Sun god and the maximum deity of the Aztec religion; on the left with blue decorations at the top, there was the one for *Tlaloc*, the main water god. Given that the Aztecs constructed *Tlaloc*'s temple of the same size and the same height as *Huichilopochtli*'s temple, the Sun and main god, it was a clear statement on how valuable water was to them.

The Aztecs had ceremonies devoted to *Tlaloc*, which included sacrifices of figurines, jewelry, or selected human victims. *Tlaloc*, as the main water god, was

Fig. 13.6 *Tlaloc*, the main water god in Mesoamerica (Laud Codex)

of paramount importance to the Aztecs, so *Tlaloc* shared the top of the pyramid of the Major Temple (*Templo Mayor*); indeed, he had his temple next door to (see the upper-left part of Fig. 13.1) the temple of *Huitzilopochtli*, the Sun god, the most important divinity in Aztec mythology.

Chalchihuitlicue was the water goddess of the living and flowing waters, and she usually was represented by a jade skirt and was the sister and/or consort of *Tlaloc*. She was the water goddess of the sea, rivers, lakes, and springs (see Fig. 13.7). Among the Aztecs, the Gulf of Mexico was known as the waters of *Chalchihuitlicue*. Such a water goddess used to be a very popular and very feared deity among the Aztecs. Many homes in *Mēxihco-Tenochtitlan* devoted a special place for *Chalchihuitlicue*, where they offered gifts to appease her temperamental mood. She has a close association with maize and with snakes, and she was also a cleanser (Back 1981).

Metzli was the goddess who had the power to produce storms and floods (see Fig. 13.8). She was the Moon's goddess. She has the power to control water, too.

Fig. 13.7 *Chalchihuitlicue*, goddess of the living and flowing waters (Borbonic Codex)

Fig. 13.8 *Meztli*, goddess of the Moon (Borgia Codex)

Atlacoya was the goddess who produced droughts; she was known as the sad waters goddess, too (see Fig. 13.9).

Amimitl was the god of lakes with the power to control storms (see Fig. 13.10). He also was the fishermen's protector.

Fig. 13.9 *Atlacoya*,
goddess of droughts.
(Gwendal Uguen 2018a)

Fig. 13.10 *Amimitl*, god of
the lakes. (Gwendal Uguen
2018b)

13.5   Aztec Major Hydrologic and Hydraulic Works

Only four classes of hydrologic and hydraulic works will be considered here: aqueducts to water supply, flood-control structures, agricultural system based on the construction of *chinampas*, and leisure baths.

13.5.1 Aqueducts

The Aztecs realized from the very beginning that securing water for their city was of paramount importance, and to do that they constructed these several aqueducts:

1. The aqueduct of *Chapultepec* was the first aqueduct, also named "the great aqueduct." This aqueduct used the flowing waters coming from the spring located at the *Chapultepec* hill. It was built during the reign of *Chimalpopoca*, one of the rulers called "tlatoanis" of the Aztec Empire. During the reign of *Moctezuma-Ilhuicamina*, this aqueduct was reconstructed and improved. The Spanish colonial remains of this aqueduct can be seen in modern Mexico City on *Chapultepec* Avenue (see Fig. 13.11).

   The aqueduct of *Chapultepec* was a marvelous hydraulic structure for its time, consisting of a twin set of pipes: one for the daily operation and the other as a reserve to be used when the one that was in operation needed cleaning and/or repair. It was made of compacted soil and wood in the section needed to allow the passing of canoes that were the means of transportation in the lakes of the Valley of Mexico (Palerm 1973).

   Chapultepec's aqueduct started its operation in 1466 AD, providing the required freshwater to mitigate the thirst of the city of *Mēxihco-Tenochtitlan*.

   Netzahualcoyotl, the King of *Texcoco*, a neighboring kingdom to the Aztec Empire, was the mind behind the construction of the aqueduct of *Chapultepec*. He was the most well-known engineer of Mesoamerica, and he was a celebrated poet, too.

   The Aztecs needed to obtain, from the mainland, the required construction materials (wood, stone, and lime) to build the aqueduct of *Chapultepec*. These materials had to come from the *Tepanecs*, a tribe settled down along the shore of the *Lake of Mexico*. As usual, the bargaining between the Aztecs and the neighboring tribes involved war, and with the *Tepanecs* there was no exception

Fig. 13.11 Remains of *Chapultepec*'s aqueduct in Mexico City. (Raynal Villaseñor 2018b)

that such bargaining ended with the murder of *Chimalpopoca*, the Aztec ruler. After that, the Aztecs declared war on the *Tepanecs*, conquered them, and finally got access to the materials they needed to build the aqueduct of *Chapultepec*.

2. The aqueduct of *Acuecuexcatl* was the second aqueduct built by the Aztecs. Since the beginning, it was marked by bad premonitions, but more water was needed to fulfill the increased needs of the flourishing city of *Mēxihco-Tenochtitlan*.

   The Aztecs needed the waters coming from the *Acuecuexcatl* spring, but this spring was in the neighboring kingdom of *Coyoacan* in the Southern Valley of Mexico. Its king *Tzotzoma* warned the Aztecs about the irregular flows of water coming from the *Acuecuexcatl* spring, when the Aztecs approached *Tzotzoma* to ask his permission to use the waters of the spring to build an aqueduct. *Ahuizotl*, the Aztec's ruler at that time, was very angered by the warning of *Tzotzoma* because he thought that he was refusing to give permission to use the waters of the spring for the aqueduct, and he declared war over the kingdom of *Coyoacan*, this war ending with the hanging of *Tzotzoma*.

   Once *Ahuizotl* eliminated *Tzotzoma*, the aqueduct was constructed, and its operation started in the year 1500 AD. But since the beginning of the operation of the aqueduct, the amount of water coming from the *Acuecuexcatl* spring produced uncontrollable volumes of water, as was warned by *Tzotzoma* and soon it produced a devastating flood in the city of *Mēxihco-Tenochtitlan*.

   Ahuizotl, then Aztec ruler, called Netzahualpilli, the son of Netzahualcoyotl, both kings of Texcoco, for help to control the flooding of Mēxihco-Tenochtitlan.

   Netzahualpilli used divers to block the flowing waters coming from the spring of *Acuecuexcatl*. *Ahuizotl* paid for his error with his life, when he fell during the flood, due to a slippery floor, and he suffered fatal injuries to his head; he died few days after his fall.

13.5.2 Flood-Control Works

Another important aspect of the daily life in the city of *Mēxihco-Tenochtitlan* was that of flood control because the city was surrounded by water. So, flood-control works became very important from the very beginning of the city of *Mēxihco-Tenochtitlan*, and they had to protect the lives and goods of the Aztec society from the excess of water produced mainly in the rainy season of summer and fall.

The Aztecs started their regional flood-control works by constructing several roads for the city of *Mēxihco-Tenochtitlan* to the mainland. Such roads had openings where the canoes could circulate and transport goods and people, and these roads played an important role as structural measures for flood control (Raynal-Villasenor 1987).

Fig. 13.12 *Netzahualcoyotl*'s levee. (Museo de la Ciudad de Mexico 2018)

The first road that was constructed was the *Tacuba* road during the reign of *Chimalpopoca,* starting operation in about 1418 AD.

The second road was that of Tlaltelolco-Atzcapotzalco, which was built during the reign of Itzcoatl, the successor to Chimalpopoca.

When the Aztecs conquered the lands to the south of the Valley of Mexico, the *Coyoacan–Xochimilco* road was built.

These three roads confined the freshwaters of the west and southwest regions of the Valley of Mexico, so fishing improved significantly (Raynal-Villasenor 1986).

Everything was fine in the city of *Mēxihco-Tenochtitlan* until year 1449 AD. In that year, an extremely rainy season occurred producing a large flood with a water depth of 2 m, so no one could walk in the city, and the activities of the Aztec society were greatly affected (Raynal-Villasenor 1986).

Moctezuma-Ilhuicamina, the ruler at that time, called for help to *Netzahualcoyotl*, the King of *Texcoco* and his cousin.

Netzahualcoyotl started regional hydrological works that would protect the city of *Mēxihco-Tenochtitlan* from the floods coming from the surrounding lakes. He convinced *Moctezuma-Ilhuicamina* to construct the *Netzahualcoyotl* dike or dike of the Indians (see the thin brown line in the middle of Fig. 13.12). This hydrologic flood-control work was the most important flood-control work of the Aztec culture. The dike was 16 km long and 20 m wide. *Netzahualcoyotl*'s dike is a structural flood-control measure, as was pointed out by Raynal-Villasenor (1986), in its purpose and in conception. Some historians argue that this dike was a water-quality structural measure.

Netzahualcoyotl's dike was complemented by the *Cuitlahuac* road, which marked a physical division of the waters of the lakes of *Chalco* and *Xochimilco*, and by the *Mexicaltzingo* road, which marked a physical division between the waters of the lakes of *Mexico* and *Xochimilco* (Raynal-Villasenor 1986).

Because all these roads had to allow canoe traffic, they had gates through which the canoes were able to pass and also to control the flow of water between the lakes. *Moctezuma-Ilhuicamina*, the acting ruler of the Aztecs at that time, ordered the building of the *Tepeyac* road. This road had a twofold purpose: the first one was to contain the waters coming from lakes in the north and the second was to open communication with the lands in the north of the Valley of Mexico. All these flood-control works comprised the first comprehensive flood-control project in Mesoamerica (Raynal-Villasenor 1986).

13.6 Agricultural System Based on the Construction of "Chinampas"

Even though the Aztecs were not the inventors of *chinampas*, they applied and developed such a hydroponic system for agriculture production of food and fiber on a massive scale (see Fig. 13.13). The required steps for the construction of *chinampas* were described (Aghajanian 2007).

The *chinampas* served a twofold purpose: as a means of construction and for land-reclamation measures from the lake. The system worked so well that *Mexico-Tenochtitlan*, the capital city of the Aztec Empire, had more than 100,000 inhabitants by the time the Spanish conquerors reached it in the year 1519 AD.

Fig. 13.13 *Chinampas.* (Thelmadatter 2008b)

An interesting fact is that the "chinampas" still are a good agriculture system in marshy areas. The last remains of this agriculture system can be observed nowadays at *Xochimilco*, a suburb located in Southern Mexico City.

13.7 Leisure Baths

Netzahualcoyotl, the King of *Texcoco*, a neighboring kingdom and an ally of the Aztec Empire, built two leisure baths in the fifteenth century, one for himself and another for his wife, at the top of a hill.

To convey the waters required for both baths, he constructed an aqueduct (see Figs. 13.14 and 13.15). The baths' location can still be visited at *Tetzcotzingo*, a small town in the State of Mexico near Mexico City.

An interesting fact is that the bath of *Netzahualcoyotl*'s wife (see Fig. 13.15) was carved into the stone, describing almost a perfect circle in its transversal shape. The inhabitants of the Valley of Mexico had no knowledge of metal tools, and they did not have the animals to carry all the materials that they used for the construction of their buildings.

Fig. 13.14 Bath of King *Netzahualcoyotl*. (Misaelos 2008)

Fig. 13.15 Bath of King *Netzahualcoyotl*'s wife. (Toltecayotl.org 2014)

13.8   Conclusions

The Aztecs evolved from a group of nomads to become a highly sophisticated society in less than 200 years. They were capable of developing beautiful architectonic buildings, well-designed urban settings consisting of roads, canals and aqueducts, and some other excellent hydraulic and hydrologic works, more advanced by far than those in any other part of the world.

The Aztec culture had a wonderful integration of society and the environment: they developed their human activities in harmony with the subtle existing natural conditions in the Valley of Mexico.

The Aztec culture had been branded so intrinsically with the sign of water that its civilization was born, lived, flourished, and perished surrounded by water.

References

Aghajanian A (2007) Chinampas: their role in Aztec empire-building and expansion, Lulu.com, 112 p. http://www.amazon.com/gp/product/1430306815/ref=ox_sc_act_title_1?ie=UTF8& m=ATVPDKIKX0DER. Accessed 19 Mar 2018

Back W (1978) Archaeology, hydrogeology and hydromythology in the new world. Paper presented at the international symposium on implications of hydrogeology to other earth sciences, University of Montpellier, Montpellier, France

Back W (1981) Hydromythology and ethnohydrology in the new world. Wat Resour Res 17 (2):257–287

Biswas AK (1970) History of hydrology. North Holland Publ. Co., Amsterdam

Bribiesca-Castrejon JL (1960) Hydrologic history of the Valley of Mexico. Revista Ingenieria Hidraulica 1960:43–61. (In Spanish)

Cartwright M (2017) Aztec Pantheon, https://www.ancient.eu/article/1034/aztec-pantheon/. Accessed 19 Mar 2018

Gwendal Uguen (2018a) https://www.flickr.com/photos/gwendalcentrifugue/8351782692/. Accessed 19 Mar 2018 Free for reproduction

Gwendal Uguen (2018b) https://www.flickr.com/photos/gwendalcentrifugue/7868347290. Accessed 19 Mar 2018. Free for reproduction

Historicair (2007) Valley of Mexico circa 1519. https://upload.wikimedia.org/wikipedia/commons/ 9/9a/Valley_of_Mexico_c.1519-fr.svg. Accessed 19 Mar 2018. Free for reproduction

Misaelos (2008) Vista panoramica desde uno de los baños de Nezahualcoyotl en Texcoco, Estado de México. https://commons.wikimedia.org/wiki/File:Ba%C3%B1os_de_Nezahualcoyotl. JPG#filehistory. Accessed 19 Mar 2018. Free for reproduction

Museo de la Ciudad de México (2018) Plano de Mexico Tenochtitlan, el valle y los lagos en el siglo XV Luis Covarrubias 1963. Photograph used with permission

Palerm A (1973) Pre-hispanic Hydraulic Works in the Lake System of the Valley of Mexico, SEP-INAH, Mexico City. (In Spanish)

Raynal-Villasenor JA (1986) Flood control works in the Aztec Empire. Proceedings of the Ninth National Congress on Hydraulics, Queretaro, Mexico, pp 639–650, (In Spanish)

Raynal-Villasenor JA (1987) The remarkable hydrologic works of the Aztec civilization. In: Water for the future: hydrology in perspective, IAHS pub. No. 164, pp 3–9

Raynal-Villasenor JA (2018a) Photograph taken across the North-East corner of Mexico City's Main Square (Zocalo)

Raynal-Villasenor JA (2018b) Photograph taken at Mexico City's Chapultepec Avenue

Sémhur (2011) Map of Mesoamerican civilizations cultural area. https://upload.wikimedia.org/ wikipedia/commons/9/9a/Mesoamerica_topographic_map-blank.svg. Accessed 19 Mar 2018. Free for reproduction

Thelmadatter (2008a) Model of the Aztec City of Tenochtitlan at the National Museum of Anthropology in Mexico City. https://upload.wikimedia.org/wikipedia/commons/9/9d/ TenochtitlanModel.JPG. Accessed 19 Mar 2018. Public domain

Thelmadatter (2008b) Scale models of chinampas used by the Aztecs in the lakes surrounding Tenochititlan on display at the museum of the Templo Mayor. https://upload.wikimedia.org/ wikipedia/commons/4/45/ChinampaScaleModel.JPG. Accessed 19 Mar 2018. Public domain

Tibon G (1986) The mystery of the foundation of the City of Mexico, lecture given at the Auditorium "Javier Barros Sierra" of the Universidad Nacional Autonoma de Mexico. Mexico City, Mexico. (In Spanish)

Toltecayotl.org (2014) Baños de Netzahualcoyotl en Texcoco. http://www.toltecayotl.org/tolteca/ index.php/2014-03-30-23-22-05/zonas-arqueologicas/4560-banos-de-nezahualcoyotl-en-tex coco. Accessed 19 Mar 2018. Free for reproduction

UNAM (2012) Great Nahuatl Dictionary [on-line]. Universidad Nacional Autonoma de Mexico, Mexico City. In Spanish. http://wwwgdnunammx. Accessed 19 Mar 2018

Chapter 14
Analysis of the Spatial Dependence of Rainfall Fields in the Southeast of Mexico, Using Directional Variograms

Alfonso Gutierrez-López, Marilú Meza-Ruiz, and Jose Vargas-Baecheler

Abstract A basic review regarding the origin of precipitation and its spatial and temporal distribution in the southern region of Mexico is presented. The origin of extreme rainfall events that caused severe damage in the Mexican states of Chiapas and Tabasco in October 2007 is analyzed in detail. The cartography of hydrologic variables is considered in its historical perspective, and the basis for the study of the spatial dependence of these variables in a geographical and physical environment is also introduced. An example of the interpretation of directional variograms is presented, and how this mathematical function can represent spatial variability and form rainfall patterns within a region through a kriging procedure is explained. Directional variograms for the rainy months in Chiapas and Tabasco are estimated. The analyses of satellite images confirm that theoretical variograms are the right tool to faithfully represent the spatial variability of a hydrologic phenomenon. The results presented in this paper show that there exist, in this region of Mexico, rain patterns that can be analyzed through geostatistical methods and that the selection of the variogram to carry out a kriging interpolation implies much more than a simple visual adjustment.

Keywords Kriging · Rainfall fields · Spatial interpolation · Directional variograms · Tabasco · Chiapas

A. Gutierrez-López (✉) · M. Meza-Ruiz
Universidad Autonoma de Queretaro, Water Research Center, Centro de Investigaciones del Agua-Queretaro (CIAQ), International Flood Initiative, Latin-American and the Caribbean Region (IFI-LAC), International Hydrological Programme (IHP-UNESCO), Queretaro, Mexico
e-mail: alfonso.gutierrez@uaq.mx

J. Vargas-Baecheler
Facultad de Ingeniería, Universidad de Concepción, Concepción, Chile

© Springer Nature Switzerland AG 2020
J. A. Raynal-Villasenor (ed.), *Water Resources of Mexico*, World Water Resources 6,
https://doi.org/10.1007/978-3-030-40686-8_14

14.1 Introduction

Rain is not a continuous phenomenon; indeed, rainfall occurs for short periods compared with the duration of the interval between these events. To characterize this intermittent rainfall, some authors (Garbutt et al. 1981; Goze 1990) analyzed separately the steady changing of the probability value that one day is rainy, and, on the other hand, the distribution of rainfall accumulated during a particular time interval. While this finding may respond to many operational requirements, the approach has two disadvantages. The first is that it is not possible to characterize the meteorological phenomena causing the rain of these events. The second is that a phenomenon known as hydrological abundance is presented, where the monthly cumulative may result from the sum of daily events or one major event followed by zero events. Hence, for a complete description of rainfall patterns, understood as determining the magnitude and frequency of an event, one must specify how events and the distribution of the intensities over time occur. At all times, these two concepts' magnitude and frequency vary depending on the time of year and according to some unquantifiable factors, such as the type of event or the position of the measuring point in the study area. However, they may be defined by a probability distribution of random variables; it is for this reason that the abundance of an event is usually characterized by a single parameter: the full depth of rain (Creutin and Obled 1982). The characterization of the meteorological phenomena causing the rain in these events, meanwhile, is studied somewhat for its complexity and the number of parameters that are involved. Then, it is interesting to incorporate this approach for the spatial analysis of precipitation. The space–time analysis of precipitation is necessary for hydrologic modeling of runoff, erosion, and the most extreme weather phenomena. When measurements are not available in a spatial scale appropriate breakdown, it is necessary to have a system of exact interpolation, which enables an accurate description of the phenomenon. Many interpolation procedures for rain have been proposed: Dubois et al. (1998) using geostatistical approaches, like ordinary kriging and co-kriging (Goovaerts 2000; Lloyd 2005) and other techniques based on splines (Hutchinson 1998) or genetic algorithms (Demyanov et al. 1998; Huang et al. 1998), for example. In this paper, geostatistical modeling is used in a multivariate way to incorporate additional information into the interpolation procedure. For example, the elevation is frequently used as supplemental information (Goovaerts 2000; Hevesi et al. 1992), where generally the correlation between elevation and precipitation depends on the mechanism by which rain is generated. In particular, the additional information that can enrich a hydrological model should include the type of precipitation observed during the interpolation as well as its spatial distribution. The southern region of Mexico is especially affected by convective systems and hurricanes during summers and autumns, as well as by heavy rainfall caused by cold fronts in the wintertime. Recalling the heavy rains that occurred in October 2007 in this region, the city of Villahermosa in Tabasco state was heavily damaged by floods caused by Cold Front No. 4 (FF-4), with over 300 mm rainfall within 24 h. Undoubtedly, the occurrence of this extreme precipitation had a strong correlation not only with lifting and relief but also with effects of geographical orientation of the cold front. Thus, in this study, the hypothesis that the sheet of precipitation presents a spatial correlation with the direction, orientation, and shape of the FF-4 arises. Using a geostatistical tool, specifically the directional variogram, the spatial dependence of this phenomenon is quantified. In addition, satellite imagery of rainfall events from that area is used to verify the results. If the fact that the spatial structure of the studied variable is defined by the directional variograms is accepted, then we can analyze the behavior of the variable in the space environment. This means that the concept of anisotropy is introduced into precipitation analysis to identify various patterns, trends, and functions of spatial dependence in precipitation.

14.2   Mapping Fields of Rain

Early methods used to map the fields of rain made use of the skill of the technician assuming the subjectivity of the influence of certain factors, such as relief (orographic effects), for example, manually drawing isohyets. The informal criterion estimate was linked to the experience of the technician for the study of rainfall in a region. Thus, they gave a "smoother" isohyet. The quality of the drawings thus obtained, plotted with the cumulative rainfall over a given (month, year) period, was then only a function of the density measurement network. These procedures were analyzed quantitatively and a map was generated whose accuracy was not necessarily critically needed. In a more formal way, Thiessen in 1991 proposed a more formal method to calculate the spatial averages over a watershed. This procedure breaks the study area into polygons, whose boundaries are obtained by drawing bisectors between stations taken in pairs. In this way, the measures by the gauge value are affected by the area of the surfaces obtained by the polygons, and the criteria estimation follows the law called nearest neighbor. The rain function is represented as $Z(x, y)$ where $Z(x_i, y_i)$ are the values known at the experimental points. The average spatial variation in the region, with an area S equal to $t(x_i, y_i)$, is given by the weight:

$$\int \int_S Z(x, y) dx \, dy = \sum_i \alpha_i Z(x_i, y_i) \qquad (14.1)$$

Where

$$\alpha_i = S_i / S$$

S_i is the i-th surface of the basin S, contained within the limits of the area where $Z(x_i, y_i)$ is located.

This method is mentioned as an example of analysis of spatial features of rain, from pseudo-deterministic criteria, which are still used for calculating the spatial averages, but are not applied as a mapping technique, due to the discontinuity (failure) that occurs at the borders of the polygons. The evolution of information technology has made possible automating the plot of the isohyets, which ensures a correct interpretation of the climate. It is primarily an adjustment of the various types of interpolation functions on the measured values, making possible the problem of the point estimate. It is a purely numerical method where interpolation functions are chosen for their ability to be implemented in a computer. Among other functions, the polygons and trigonometric functions have been used as interpolators to calculate the value of rain, in the nodes of an irregular grid on a finite domain mesh from which the isohyets are drawn. However, the irregularity in the distribution of measuring stations over large areas usually leads to results that are often far from the desired values. For example, the strong tendency of polynomial functions to oscillate over large amplitudes produces aberrations in the point estimate, which must be corrected manually on the isohyet cards. To avoid this manual intervention, various methods have been used for numerical smoothing, for example, the method of least squares. Hydrologists recently researched functions that give a smooth appearance to the isohyets and have a relationship with the physiology of the area (Becerra and Gutierrez López 2006). This is analogous to the theory used by Matheron (1970) and Paihua (1978), which used spline interpolation functions. These functions are obtained by minimizing the curvature, which is considered a flexible function with the hypothesis that it is working in an elastic medium.

$$\iint_{\Omega} \left[\frac{\partial^2 S(x, y)}{\partial x^2} + \frac{\partial^2 S(x, y)}{\partial x \partial y} + \frac{\partial^2 S(x, y)}{\partial y^2} \right] dx\, dy \qquad (14.2)$$

It is an exact interpolator of $S(x, y)$ where the characteristics only depend on the position of the experimental points, which in our case, for rain for a certain period will give us a point estimate.

$$S(x, y) = Z^*(x, y) \text{ at each point of the domain}$$

$$S(x_i, y_i) = Z(x_i, y_i) \text{ in the experimental points}$$

Pseudo-deterministic methods for each event are studied independently of the others; implicitly, this means no coincidence between two successive events, characteristic of geostatistical methods, is considered. Moreover, the event space is considered homogeneous over time, that is, it is assumed that events are extracted from the same population. This is important as well as the hypothesis that statistically each event can, by itself, be representative of an area of events taking place. To proceed with the evaluation of specific values, it is necessary to use a linear operator, explaining the estimate $Z^*(x, y)$ as a form of weighting:

$$Z^*(x, y) = \sum_i \lambda_i(x, y)\, Z(x_i, y_i) \qquad (14.3)$$

which is an unbiased estimate that satisfies the condition:

$$E[Z^*(x, y) - Z(x_i, y_i)] = 0 \tag{14.4}$$

This equation represents exactly the average of all measurements of space events or accomplishments. The estimation criterion or optimal approach is to minimize the mean square error of estimation, at every point of the domain where the variables are counted rain.

$$E\left[\left\{Z^*(x, y) - Z(x_i, y_i)\right\}^2\right] \tag{14.5}$$

This is the principle of geostatistical methods: the calculation of the weighting coefficients $\lambda_i(x, y)$ of the Eq. (14.2) satisfies the conditions (14.4) and (14.5).

14.3   Description of the Study Area

The study area corresponds to the southeast of the Mexican Republic. It belongs to the hydrologic region numbers 23 and 30 and includes the states of Chiapas and Tabasco. The region is in the Isthmus of Tehuantepec and bordered on the south by the Pacific Ocean, on the north by the Gulf of Mexico, on the east by the Yucatan Peninsula, and on the west by the states of Oaxaca and Veracruz. Because of its location, this region is highly vulnerable to the effects of weather phenomena of cyclonic origin on both sides and cold fronts. This region is the rainiest part of the country, with an average rainfall of 2124 mm per year, with a maximum annual record of 4269 mm (Talisman station). Rainfall in the area is concentrated in the months from June to October (Table 14.1). In these 5 months, 71% of the annual rain falls. The rain is recorded at 92 conventional weather stations and two meteorological observatories. Information from these stations was provided by the South Border Basin Organization of the National Water Commission, and contains historical data up to December 2015.

Table 14.1 Monthly Average Rainfall representative of the Mexican Southeast

Month	Average Rainfall (mm)	Month	Average Rainfall (mm)
January	70.89	July	257.63
February	58.07	August	303.56
March	43.84	September	380.16
April	67.55	October	270.20
May	155.23	November	131.27
Jun	301.75	December	83.41
		Annual	2124.0

14.4 Paths of Precipitation Systems in Southeastern Mexico

Since the general objective of the research refers to the analysis of the fields of rain in the Southeast, in addition to the information available on the surface, as shown in Fig.14.5, the distribution of precipitation systems is used as viewed by meteorological satellites. By the close relationship between clouds and rain generation, the movement of these is directly related to the behavior of rain on surfaces. Valdes M. et al. (2005) conducted an exploratory study of mesoscale convective systems (SCMs) in Mexico that occurred in the period from mid-1996 to mid-1997. It was found that the paths of SCMs in a full rainy season (July) are predominantly parallel to the Pacific coast, a movement that is kept in the same direction as the rainy season. After September, the convective activity increases over the Gulf, and SCMs follow various directions in this region, unlike the Pacific coast where the SCMs decreased significantly, a trend that is maintained for the month of October when rainfall begins to withdraw mainly on these coasts. However, most of the SCMs that occurred in October took place in the Yucatan Peninsula. Changes in SCMs tracks happen during the fall. The presence of cold fronts phenomen is responsible for the increased probability of the generation of SCMs during the autumn and winter season. To illustrate the above simply, the IR satellite imagery, Channel 4, was provided by the National Weather Service (NWS) for the years 2004 and 2005. Two pictures (Figs.14.1 and 14.2), which are clearly identified, show the prevailing trajectories of SCMs related to the rainy season and the withdrawal from the same territory, parallel to the Pacific coast and preferably to the east, respectively. Cloudiness was also observed for the analyzed behavior of the SCMs (Valdes, et al. 2005).

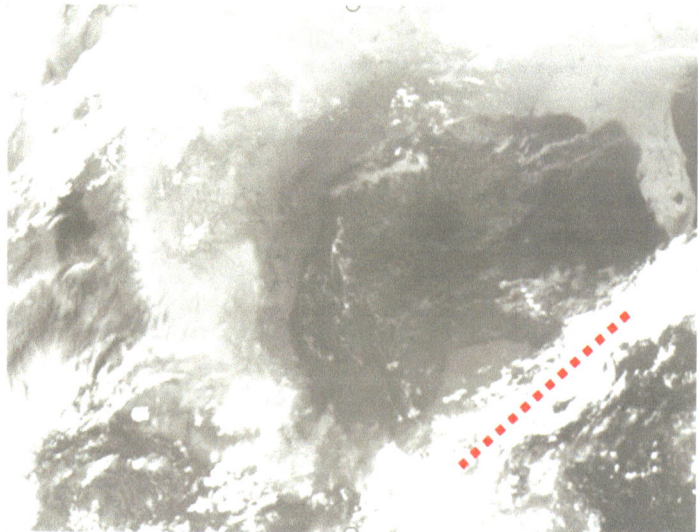

Fig. 14.1 System rainy July 16, 2004 10 h57 predominant direction NW–SE

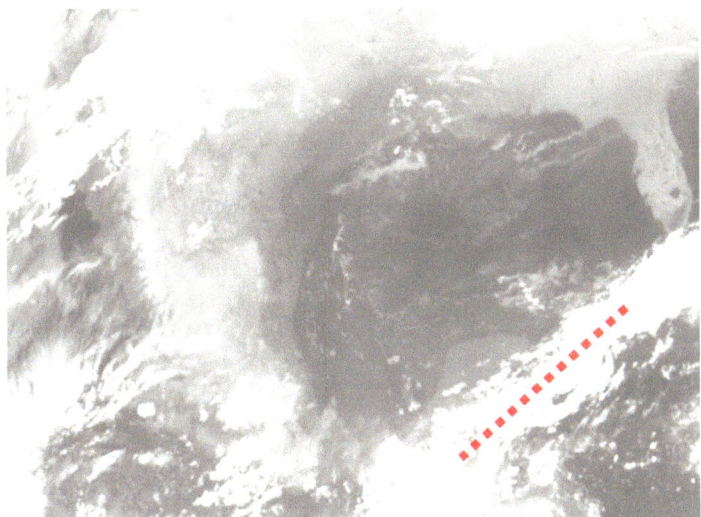

Fig. 14.2 System rainy October 17, 2004 09 h50 predominant NE–SW

14.5 Interpretation of Directional Variograms

A graphical analysis of the data variograms of rain in the region was carried out. Given the track record of those discussed in the previous section and the spatial dependence with concepts of the studied systems, one can say a priori that the monthly directional variograms reflected the behavior of rain systems in the region of study. This means that observing a directional variogram for the month of June will be significantly different in shape and dimensions from the directional variogram for the month of November. Thus, using the geographical convention shown in Fig. 14.1 and following the development of the example shown, four directional variograms for tolerances of 90°, 45°, and 15° were calculated—for each of the months of the year, plus the annual variogram. Regarding tolerance, Gutierrez-López et al. (2005) showed that to explain and describe in detail the rainfall pattern in the southern part of Mexico, a spatial function having a direction perpendicular to the coast is required and that, in turn, forms an angle of 45° with the axes of latitude and longitude. Therefore, the study of the spatial dependence of rain fields using directional variograms is conducted every month, using variograms in the directions (0°/180°, 45°/225°, 90°/270° 135°/315°) with a tolerance of 45°.

To find a mathematical function of spatial dependence, related to the type or classification of the rainy episodes, one should also consider aspects of change reflected in the variogram and one should consider the global aspects of the dynamics of the precipitation. Finding a spatial dependence is also based on the following factors:

(i) The existence of areas with heavy rainfall intensities
(ii) The spatial organization of the fields of daily and monthly rain

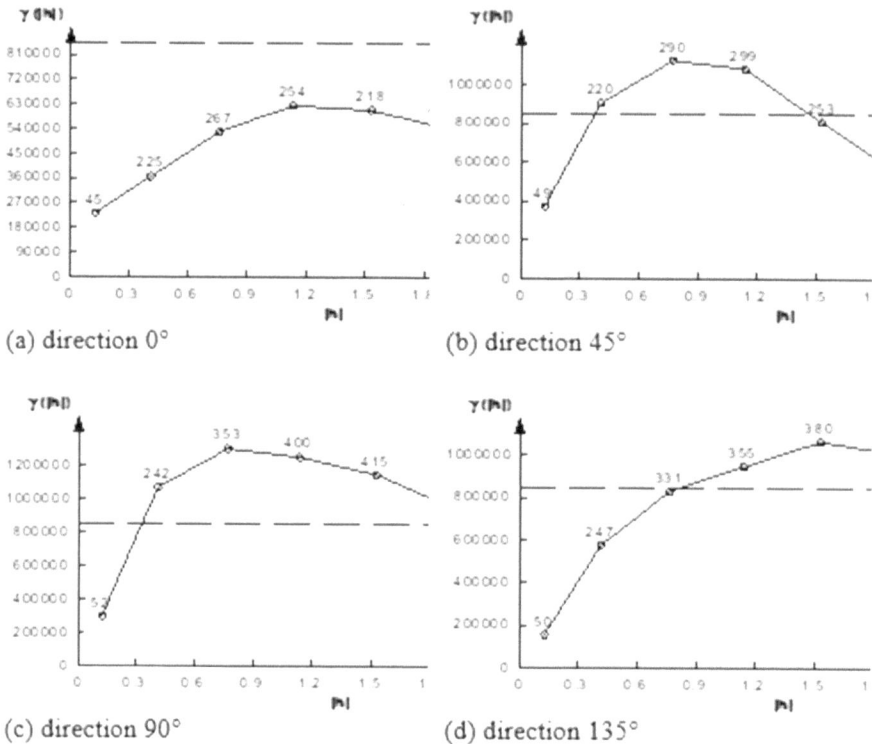

Fig. 14.3 Directional experimental variogram of annual precipitation, tolerance 45°

(iii) The existence of privileged directions of travel

(iv) The speed of evolution of phenomena

These factors are complemented by the calculation of directional variogram fields of rain, which quantify, to a certain extent, the spatial characteristics of rainfall fields identified qualitatively. Figures 14.3, 14.4, 14.5, 14.6, 14.7, 14.8 and 14.9 show the directional variograms for the months of June–November for this four-way analysis. Table 14.2 shows the type of directional variogram (theoretical model) with their best fit for each months. From a quick scan of these illustrations, it can be seen, first, that the variograms for the months of October and November, as the first group, have a pattern like the annual variogram behavior in the region. On the other hand, the variograms for the months of June, July, and August form a second group exhibiting similar behavior, while the variograms of September appear like a transition between the two groups. A principal component analysis (PCA) technique allows verifying the cluster analysis.

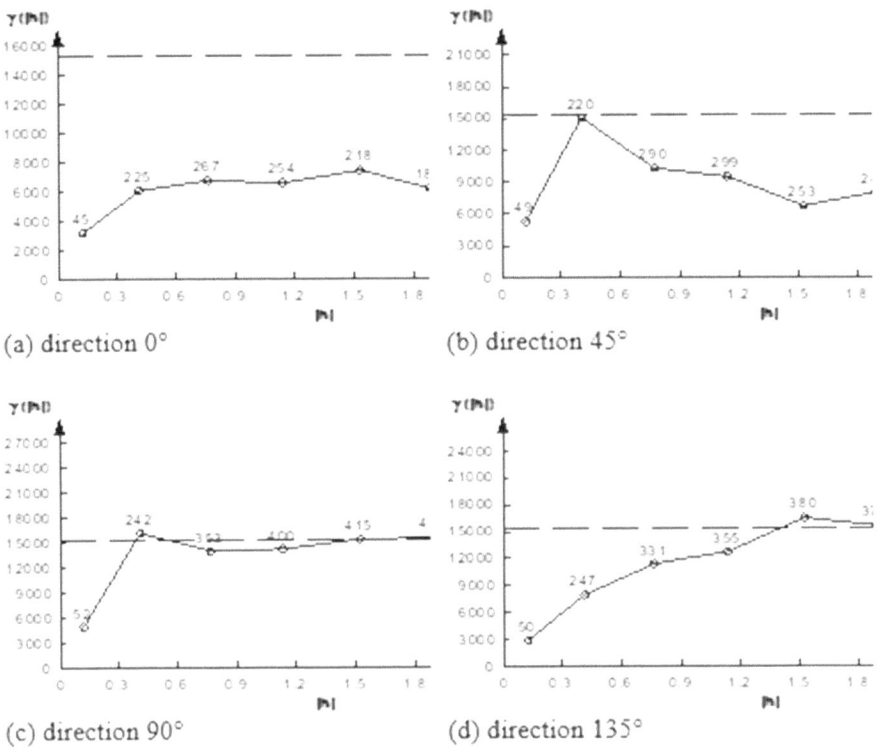

(a) direction 0°   (b) direction 45°

(c) direction 90°   (d) direction 135°

Fig. 14.4 Directional experimental variograms of precipitation (June), tolerance 45°

14.6   Discussion

The mountainous regions are in most cases water production areas and water storages. In an area of plains, it is easy to reconstruct the spatial structure of precipitation; however, in mountainous areas, the relief plays an important role because precipitation changes from one basin to another, depending on the topographic relief. Generally, the highest rainfall is present on the slopes exposed to rain systems. The topographic relief is a key parameter in defining the predominant type of precipitation in a region and in the spatial pattern of rainfall distribution (Becerra and Gutierrez-Lopez 2006). There are statistical links between precipitation and certain features of the topographic relief of a region, so that the definition of certain parameters is linked to topography, combined with the use of a technique of optimal interpolation for estimating rainfall anywhere in space, from a network of irregular measurement (Lebel and Laborde 1988). Many interpolation methods are used in mapping rainfall fields, however, Cokriging using topographic relief is the most precise. The first requires a study of the spatial structure of the variables studied. The second method seeks a relationship between rainfall parameters and their relevance

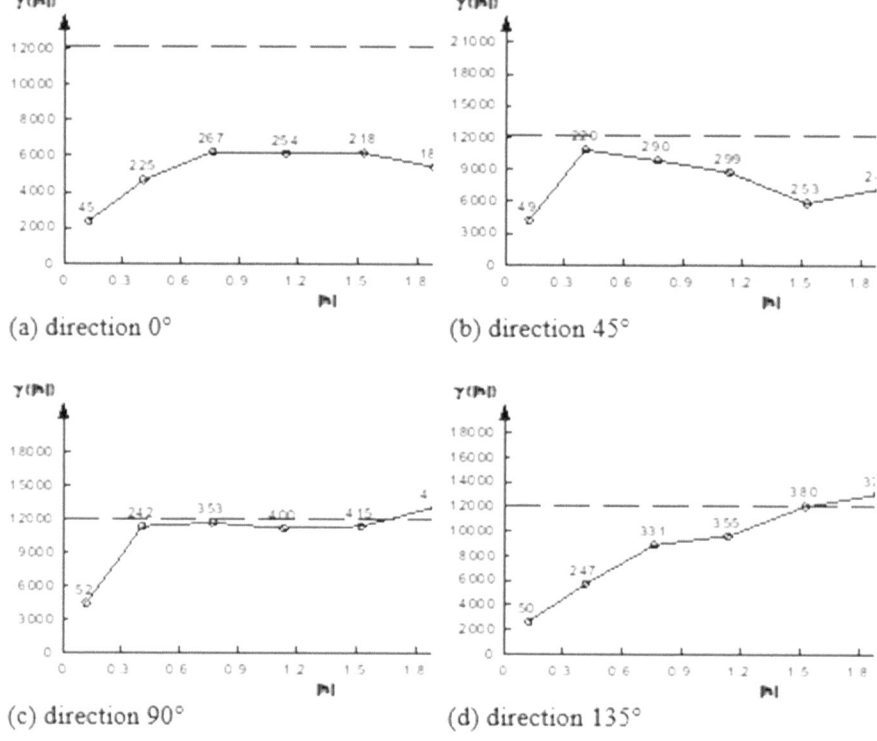

Fig. 14.5 Directional experimental variograms of precipitation (July), tolerance 45°

to explain the topographic rainfall parameters and the spatial distribution of rainfall. Given these considerations and as shown by directional variograms' adjustment, it is found that both the relief and the time of year (mechanism and dynamics of the atmosphere) are strongly correlated. A confirmation of this can be found in Table 14.2, where the months of June, July, and August show, for each of the locations, the same variogram-adjusted theoretical models. It is found that for the direction parallel to the coast (135°), for example, the theoretical variograms for those months approximated the potential model. The kernel parameter for these months ranges from 2625 to 2885; the range in recent months is 0.95 with a drift between 6232 and 9439. This reinforces the point made earlier in the sense that the trajectories of SCMs during the full rainy season are predominantly parallel to the Pacific Coast (Fig. 14.1). It is generally observed that for the summer months the adjusted theoretical models are predominantly of a Gaussian or power nature. For the direction (45°), which explains the higher spatial variability in this region, according to Gutierrez-Lopez et al. (2005), it has predominantly an adjustment to a Gaussian variogram. In summer, the value of the nugget effect parameter varies between 4290 and 5495; the range in recent months is 1.98–2.0 with a sill between 11,083 and 13,717. The similarity of these parameters and the theoretical variogram model,

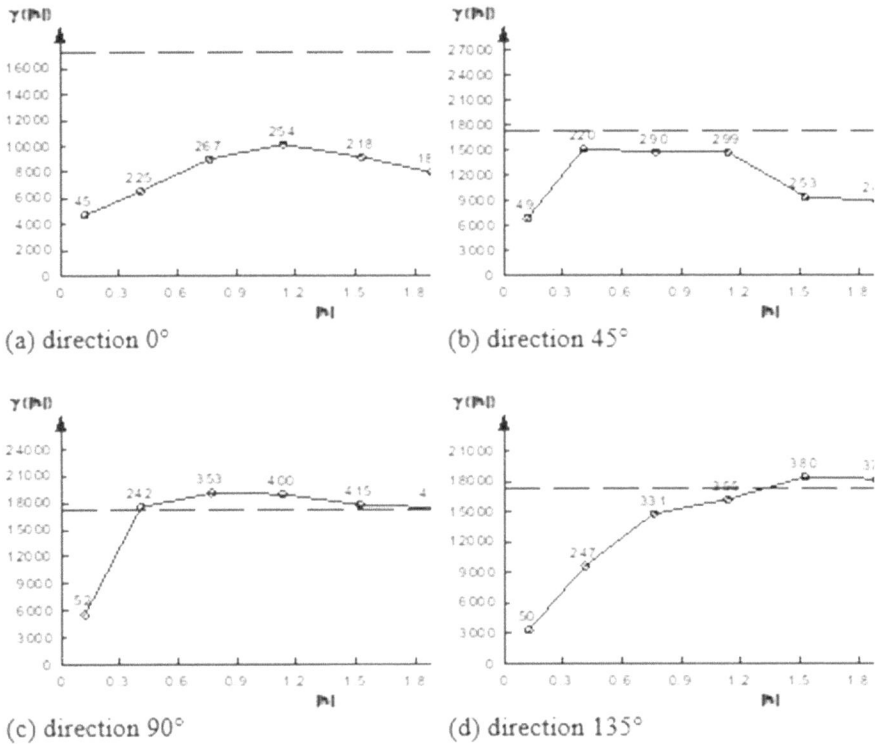

Fig. 14.6 Directional experimental variograms of precipitation (August), tolerance 45°

indicate a correlation between rainfall patterns and the topographic configuration of region. It is particularly interesting to note that, in cases when the location is related to latitude (0°), in the months of June and August alone, it showed a nugget effect (a geostatistical term used to describe the variability seen between samples that are closely spaced). This means that latitude is not a good parameter to represent the spatial variability in this region (Gutierrez-López et al. 2005). If we define a parameter ϕ, such as the relationship between a characteristic of the theoretical variogram with the monthly rainfall.

($\phi = hp/sill$), then the parameter will have to compare not only the type of variogram, but also its relationship with rainfall in the region. Thus, the value of this parameter's geostatistical association in the case of this first group of months varies between 0.022 and 0.027.

As it regards the second set of months (October and November) (see Table 14.3), the variogram that best fits were a spherical variogram and an exponential transition in September. Likewise, the results show that the mathematical structure of the variogram adequately represents the monthly change in precipitation. For the

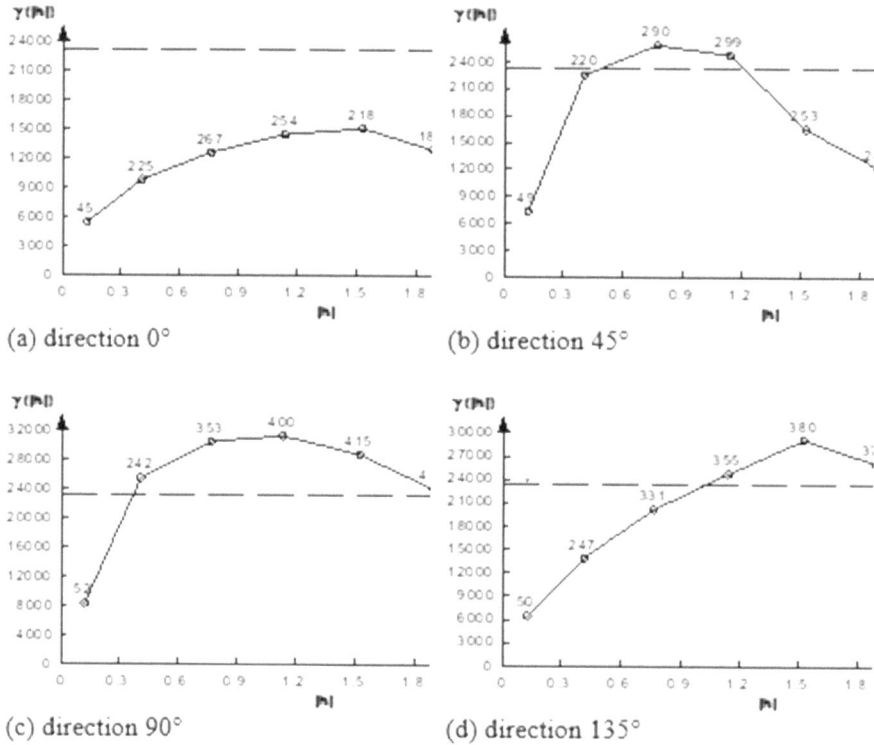

Fig. 14.7 Directional experimental variograms of precipitation (September), tolerance 45°

direction (45°), which explains the higher spatial variability in this region, we have predominantly an adjustment to a spherical variogram, where in these months, the value of the seed parameter varies between 2635 and 9016 with an intermediate value of 7188 in September. The range in these months is 1.11 to 1.53 with a drift between 9081 and 15,868. This reinforces the point made earlier in the sense of the trajectories of Cold Fronts in the autumn months (Fig. 14.7). For these months, the value of this parameter varies the geo-statistical association between 0.015 and 0.017, with a value equal to 0.0248 in September considered as the transition.

14.7 Conclusions

First, a review of methods was conducted to map areas of rain, followed by a treatment of the mathematical approach to quantify the spatial dependence of hydrological variables. The variogram as the tool was selected to carry this out. The genesis of precipitation systems in the southern region of the Mexican Republic was reviewed and spatial patterns of behavior were described. Regarding the rains

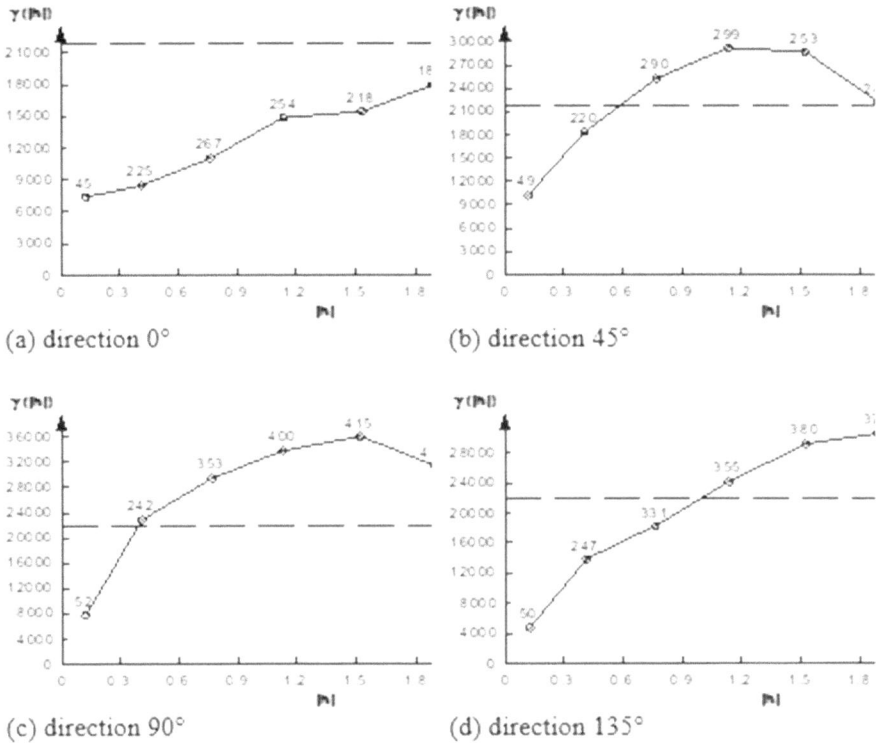

(a) direction 0°   (b) direction 45°

(c) direction 90°   (d) direction 135°

Fig. 14.8 Directional experimental variograms of precipitation (October), tolerance $45°$

that hit this area in October 2007, it was concluded that the hydrometeorological events associated with heavy rains in Villahermosa were due to a natural condition. The cool-front #4 traveled through the Gulf of Mexico and arrived in Yucatan, stuck, then returned, and remained stable in Tabasco. This phenomenon carries large amounts of humidity associated with the unsteady zone in the Caribbean which favored deep convection in the sea and into the continent. Where the condition of the Cold Front phenomenon became stationary, the high intensity of precipitation due to the topography of Chiapas and the configuration of the Isthmus of Tehuantepec. Similarly, the directional variograms were obtained for the rainy season, and it was concluded that the form and its mathematical formulation have a definite relationship with the spatial configuration of the systems. The variograms of October and November are the same and have a similar configuration as the annual variability. In all cases, the perpendicular directional variograms have the same spatial configuration. It is noteworthy that the results of this study make possible a correct mapping of the fields of rain and should be used to represent spatial variables associated with this phenomenon. That is, if you want to interpolate or map a spatial variable in the month of June, a Gaussian or a power variogram should be used. To represent spatially hydro-climatologically the months of October or November, the

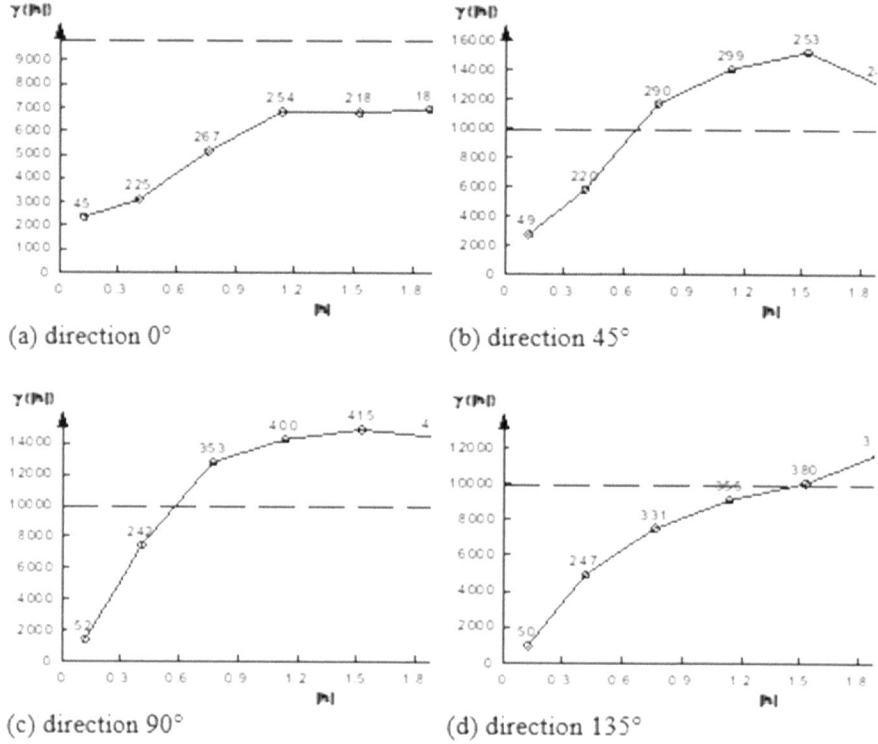

Fig. 14.9 Directional experimental variograms of precipitation (November), tolerance 45°

Table 14.2 Setting parameters of the theoretical directional variograms for the wet months

	Direction	Nugget	Range	Sill
Jun				
Nugget	0°	6819.73	–	–
v. Gaussian	45°	5494.93	2.036	13,717
v. Gaussian	90°	5410.67	1.948	12,705
v. Power	135°	2885.33	0.950	9049
July				
v. Power	0°	2864	0.980	2286
v. Gaussian	45°	4290	1.976	11,489
v. Gaussian	90°	5042	2.009	11,269
v. Power	135°	2625	0.95	6232
August				
Nugget	0°	7477	–	–
v. Gaussian	45°	5166	2.036	11,083
v. Gaussian	90°	5937	1.712	15,639
v. Power	135°	2655	0.940	9430

Table 14.3 Settings parameters of the theoretical directional variograms for the dry months

	Direction	Nugget	Range	Sill
Annual				
v. Spherical	0°	238,567	1.5491	404,090
v. Spherical	45°	278,687	0.6639	613,105
v. Exponential	90°	264,747	1.0800	682,777
v. Spherical	135°	174,307	1.0242	688,217
September				
v. Spherical	0°	3454	1.8147	14,557
v. Exponential	45°	7188	1.2424	15,316
v Exponential	90°	8608	0.8852	17,704
v. Exponential	135°	7081	1.2459	15,191
October				
v. Spherical	0°	3606.53	1.5220	11,902
v. Spherical	45°	9016.33	1.1065	15,868
v. Spherical	90°	7213.07	0.7082	18,032
v. Spherical	135°	4644.93	1.1185	17,311
November				
v. Spherical	0°	2109.69	1.5124	4057.35
v. Spherical	45°	2635.71	1.5341	9081.93
v. Spherical	90°	2272.05	1.5934	8763.81
v. Spherical	135°	2596.77	1.8103	8439.42

variogram is used in the kriging; for example, it would be the spherical variables. Finally, the association of a geostatistical parameter proposed in this paper can be used as a weight or weight in Eq. (14.2), which would give the interpolation equation the additional property to represent and consider the variability space of the phenomenon under study.

Acknowledgments The authors thank Jose Luis Carrasco Martinez for satellite imagery, Ms. Ivonne Cruz Paz for valuable assistance in the technical review of the manuscript, and Ms. Martha Roman Juarez for the capture and processing of the database of precipitation for this region.

References

Becerra R, Gutieerrez-López A (2006) Modelación hidrológica empleando isoyetas de relieve, una aproximación Geoestadística. In: Demuth S, Gustard A, Planos E, Scatena F, Servat E (eds) Climate variability and change—hydrological impacts, IAHS publication no. 308, pp 62–67
Creutin J, Obled C (1982) Objective analyses and mapping techniques for rainfall fields: an objective comparison. Water Resour Res 18(2):413–431
Demyanov V, Kanevski M, Chernov S, Savaliera E, Timonin V (1998) Neural network residual kriging application for climatic data. J Geogr Inf Decis Anal 2(2):215–232

Dubois G, Malczewski J, Cort MD (eds) (1998) Spatial interpolation comparison 97. J Geogr Inf Decis Anal 2 (1–2).. Special Issue

Garbutt DJ, Stern RD, Dennet MD, Elston J (1981) A comparison of the rainfall climate of eleven places in West Africa using a two-part model for daily rainfall. Arch Meteorol Geophys Bioklimatol B 29:137–155

Goovaerts P (2000) Geostatistical approaches for incorporating elevation into the spatial interpolation of rainfall. J Hydrol 228(1–2):113–129

Goze E (1990) Modèl estochastique de la pluviométrie au Sahel. Aplication à l'agronomie, Thèse de l'Université de Montpellier II, pp 119

Gutierrez-López A, Lebel T, Mejía R (2005) Estudio espacio-temporal del regimen pluviometrico en la zona meridional de la república mexicana. Revista Ingeniería Hidraulica en Mexico, IMTA XX(1):57–65

Hevesi JA, Flint AL, Istok JD (1992) Precipitation estimation in mountainous terrain using multivariate geostatistics. Part 1: structural analysis. J Appl Meteorol 31:661–676

Huang Y, Wong P, Gedeon T (1998) Spatial interpolation using fuzzy reasoning and genetic algorithms. J Geogr Inf Decis Anal 2(2):204–214

Hutchinson MF (1998) Interpolation of rainfall data with thin plate smoothing splines: II analysis of topographic dependence. J Geogr Inf Decis Anal 2(2):152–167

Lebel T, Laborde J (1988) A geostatistical approach for areal rainfall statistics assessment. Stoch Hydrol Hydraul 2:245–261

Lloyd CD (2005) Assessing the effect of integrating elevation data into the estimation of monthly precipitation in Great Britain. J Hydrol 308(1–4):128

Matheron G (1970) La théorie des variables régionalisées et ses applications. ESMP, Cahiers du Centre de Morphologie Mathématique Fasc 5

Paihua L (1978) Un ensemble de programmes pour l'interpolation des fonctions par fonction spline de type plaque-mince. Rapport de recherche No. 140, USMG-INPG, France

Valdes M, Cortez M, Pastrana J (2005) Un estudio exploratorio de los sistemas convectivos de mesoescala de Mexico. Inv Geogr 56:26–42

Chapter 15
Possible Scenarios of Global-Warming Impacts on Evaporation in Mexico

Jose A. Raynal-Villasenor, Maria E. Raynal-Gutierrez, and Bryan Zegarra-Ybarra

Abstract We constructed five possible scenarios to define the impacts of global climate change on evaporation and soil-water moisture-content deficiency in the Balsas River, Bravo River, Grijalva River, Lerma River, Panuco River, and Papaloapan River basins in Mexico. They are based on forecasts of the Hadley Centre that predicts for Mexico an increase in air temperature between 3 °C and 5 °C by the end of the twenty-first century. These basins are the most important watersheds in Mexico. We used increases in air temperature 1–5 °C above the current conditions to construct five possible scenarios for two surfaces: water and short grass. Then, we analyzed the impact of each of these five possible scenarios in the potential evaporation and soil-water moisture-content deficiency for both surfaces. The results showed that, with the increases in air temperature, just mentioned, they produced an increase in potential evaporation in open water surfaces of 2.5–12.1% for 1 °C and 5 °C of increase in air temperature, respectively. The evaporation on short-grass surfaces will increase 2.6–12.7% for 1 °C and 5 °C of increase in air temperature, respectively. Soil-water moisture-content deficiency in short-grass surfaces will increase 3.3–16.0% for 1 °C and 5 °C of increase in air temperature, respectively, all of these compared with current values. In the most adverse scenario, reductions in food production in the Bravo River basin will be in the order of 28%, composed of the 16% of additional water demanded by crops plus 12.1% of additional evaporation on surface-water reservoirs located in the watershed.

Keywords Global warming · Global climate change · Potential evaporation · Soil water moisture content deficiency · Water · Short grass · Possible reduction of food production

J. A. Raynal-Villasenor (✉) · M. E. Raynal-Gutierrez
Department of Civil and Environmental Engineering, Universidad de las Americas Puebla, Cholula, Puebla, Mexico
e-mail: josea.raynal@udlap.mx; maria.raynal@udlap.mx

B. Zegarra-Ybarra
Professional School of Environmental Engineering, Universidad Nacional de San Agustin, Arequipa, Peru

© Springer Nature Switzerland AG 2020   271
J. A. Raynal-Villasenor (ed.), *Water Resources of Mexico*, World Water Resources 6,
https://doi.org/10.1007/978-3-030-40686-8_15

15.1   Introduction

There is no doubt in the scientific community that we are in the middle of global climate change and that it has, as its main driving cause, the global warming produced by the increase of the contents of so-called greenhouse gases (GHG). The GHG include water vapor, carbon dioxide (CO_2), methane (CH_4), and nitrous oxide (NO). Since the start of the Industrial Revolution, CO_2, CH_4, and NO have been increasing in their concentrations in the upper atmosphere in comparison with the levels observed in mid-nineteenth century. On the other hand, water vapor, which by the way is the most abundant of the GHGs, has the most beneficial effects on Planet Earth. Due to its presence in the atmosphere, Earth has an average global temperature of 14 °C, and not the −18 °C that would be the case in the absence of the water vapor in the upper atmosphere. The GHGs produced by anthropogenic activities are the main sources causing global warming and the resulting global climate change (IPCC 2014; Showstack 2009).

Gosling et al. (2017) produced scenarios for the comparison of changes in river runoff from multiple global and catchment-scale hydrological models under global-warming scenarios of 1, 2, and 3 °C.

In recent years, the problem of global climate change has been losing influence in the Mexican government's public policies. This problem was well addressed in the National Development Plan 2013–2018 (Presidencia 2013), but in the actual administration, such influence has vanished, and the importance of its considerable impact on the food economy, tourism, and environmental conservation, among others, has been neglected. It is necessary to implement a new set of strategies to enhance the actual and future environmental conditions and to promote a good quality of life for Mexican society.

During the Calderon and Peña-Nieto presidential administrations, there had not been much progress in proposing adaptation and mitigation actions to counteract the effects of global climate change. But in the Calderon administration, the problem of global climate change was thoroughly considered and a Climate Change Federal Law was promulgated in 2012. Since then, several modifications and updates to such law have been proposed and accepted, CD (2018), but very little has been done in implementing adaptation and mitigation actions in Mexico.

The Balsas River basin has important hydropower plants, namely the Infierrnillo, Caracol, and La Villita dams. At the basin's very end is located the most important complex of steel production, Las Truchas, in Puerto Lazaro Cardenas in the state of Michoacan. Within the Balsas River basin is located the Cutzamala River basin, which provides 30% of the water supply of Mexico City, the capital city of Mexico (GDF 2018).

The Bravo River is the Mexican part of the Bravo/Grande River general basin. Located in Northeastern Mexico, it is the most important international basin that Mexico has. It is a key food production region for domestic consumption and for the export of vegetable produce. This watershed is extremely important for Mexico due to the demographic, economic, political, and social conditions existing in such areas. The Bravo/Grande River is a natural border between Mexico and the USA with a length of 1455 kilometers. As such, it is most important part of the border.

The water of the Bravo/Grande River in this border zone is distributed between both countries, in accordance with the 1944 Treaty of International Waters signed by these two countries. It is important to mention that the water in this watershed is totally allocated by the treaty; therefore, there is no additional water in the Bravo River/Grande basin to be considered.

The Grijalva River basin and the Usumacinta River are containing 30% of the total surface water of Mexico. Moreover, the Grijalva River basin has the most important set of hydropower plants of Mexico, namely the Nezahualcoyotl, Peñitas, Chicoasen, and La Angostura dams. The latter is the biggest dam in Mexico with a total capacity of 20,000 million m^3.

The Lerma River basin, in Central Mexico, is an area of paramount importance in food production for domestic consumption and for the export of vegetable produce. The Lerma River watershed contains important areas of several states of Mexico, such as the state of Mexico, Guanajuato, Michoacan, Queretaro, and Jalisco. Also, the Lerma River basin has an important role in the demographic, economic, political, and social aspects of Mexico as a country.

The Panuco River basin is one of the largest watersheds in Central Mexico, and it is where the wastewaters coming from Mexico City, Mexico's capitol city, are deposited without much treatment. The city of Tula was named as the most polluted city of the world because of these wastewaters and by the pollution generated by the petroleum refinery located there.

The Papaloapan River basin, once one of the most important fluvial waterways of Mexico, is now a very important area for agricultural products, like pineapples. It has in its basin two very important hydropower plants, the Temascal and Cerro de Oro dams.

This paper attempts to explain the possible consequences that global climate change will have on the potential evaporation in some of the regions of Mexico considered in this study. Two types of evaporative surfaces were chosen: water and short grass.

The estimation of potential evaporation in short-grass cover is used as the crop of reference, and its results may be used to quantify the potential evaporation for other crops that are used for human and animal consumption.

15.2   The Penman Equation

The main mathematical equation that was used in this study is Penman's formula, and the form used here is:

$$E = \frac{\Delta}{\Delta + \gamma}\left\{ R_a(1 - alb)\left(0.25 + \frac{0.5n}{N}\right) - \sigma T^4\left(0.1 + \frac{0.9n}{N}\right)(0.34 - 0.044)\sqrt{e}\right\}$$
$$+ \left(1 - \frac{\Delta}{\Delta + \gamma}\right)\left\{0.26\left(1 + \frac{U}{160}\right)(e_s(T) - e)\right\}$$

$$(15.1)$$

The definitions of each variable contained in the Eq. (15.1) are as follows (Steward and Roberts 1984):

E = potential evaporation on an open surface water, mm/day
alb = albedo of the surface considered, mm/day: alb = 0.05 for an open water surface, alb = 0.25 for a surface covered with short grass
Δ = slope of the saturated water-vapor curve, mb/°C
Ra = solar radiation measured at the top of the atmosphere, mm/day
n = actual number of hours of sunshine, hours
N = maximum number of hours of sunshine, hours
σT^4 = outgoing longwave radiation emitted by a surface at temperature T, mm/day
T = average temperature of the daily maximum and minimum temperatures, °C
   U = wind run per day, km/day
e = vapor pressure at mid-morning, mb
   $e_s(T)$ = saturated vapor pressure at temperature T, mb
γ = psychrometric constant (with a value of 0.67 mb/°C), mb/°C

This form of the Penman formula becomes very useful where there is no information on the net radiation and maximum number of hours of sunshine, and this information may be obtained from existing tables based on the latitude of the site (Steward and Roberts 1984). By using the average daily temperature, the values of outgoing longwave radiation, the slope of the saturated vapor-pressure curve and the saturation pressure of water vapor may be obtained from existing tables (Steward and Roberts 1984). For the proposed possibility of air temperatures increase, we used the formula for two cases: an open water surface and a short-grass-covered surface. The values of potential evaporation produced by the Penman formula are the upper limits of evaporation, but their evaluation may give a good approximation of the possible behavior of the actual physical process of evaporation.

15.3   Discussion of the Results

With the forecasts for the increase in air temperature reported by the Hadley Centre (2005) (see Fig. 15.1), which predicts an increase for Mexico of 3–5 °C by the end of the twenty-first century, such forecasted conditions were used to construct five possible scenarios of air-temperature increase for the Balsas River, Bravo River, Grijalva River, Lerma River, Panuco River, and Papaloapan River basins in Mexico. The locations of these watersheds are shown in Figs. 15.2, 15.3, 15.4, 15.5, 15.6 and 15.7.

Five possible scenarios were constructed, with increases in air temperature of 1–5 °C above current conditions, for the potential evaporation, in several strategically distributed sites in the Balsas River, Bravo River, Grijalva River, Lerma River, Panuco River, and Papaloapan River basins. The scenarios mentioned before contain the increases in air temperature reported in the RCP2.6, RCP4.5, RCP6.0, and RCP8.5 scenarios proposed by IPCC (2014).

The selected sites in the Balsas River basin of Mexico were the climatological stations at Tlaxcala, Puebla, Balcon del Diablo, Chilpancingo, and Lazaro Cardenas. We observed that the Balcon del Diablo climatological station was the site with more

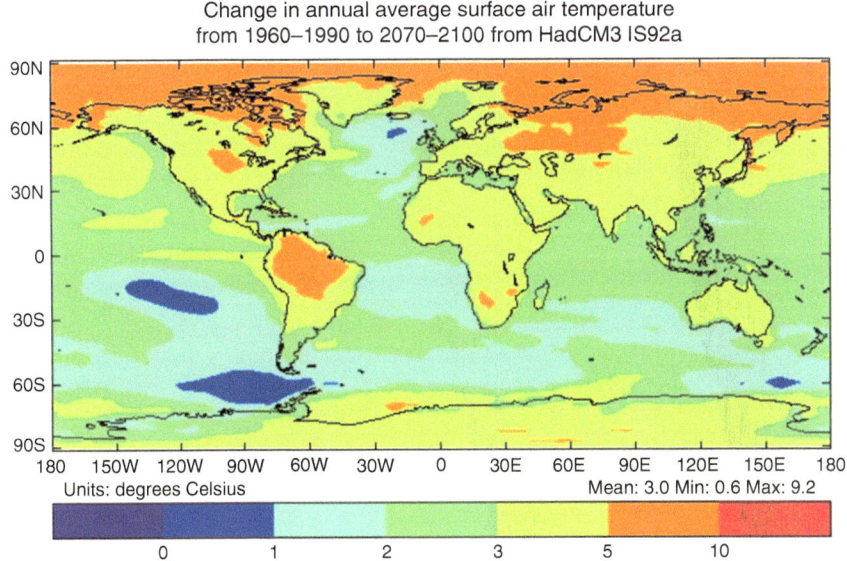

Hadley Centre for Climate Prediction and Research, The Met. Office

Fig. 15.1 Forecast of air temperature increase due to global warming at the end of the twenty-first century. (Hadley Centre 2005)

276   J. A. Raynal-Villasenor et al.

Fig. 15.2 Location of the Balsas River basin, Mexico. (INE 2016)

Fig. 15.3 Location of the Bravo River basin, Mexico. (INE 2016)

critical results for an increase of potential evaporation of all the climatological stations considered, so only results connected with it will be reported here.

The selected sites in the Bravo River basin were the climatological stations in the Conchos River sub-basin, namely Parral, La Boquilla, Camargo, Delicias, Chihuahua, and Ojinaga. The climatological stations of Venustiano Carranza Dam, Sabinas, and Monclova are in the Salado and Sabinas River sub-basins. Monterrey, Saltillo,

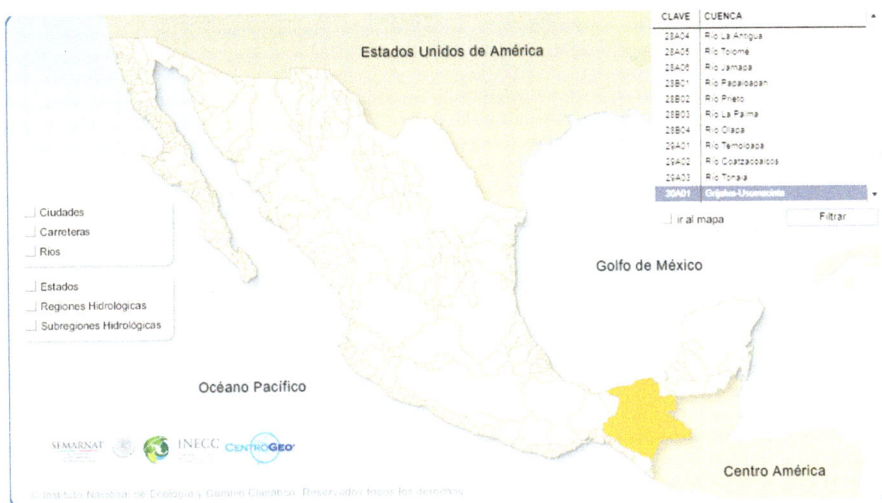

Fig. 15.4 Location of the Grijalva River basin, Mexico. (INE 2016)

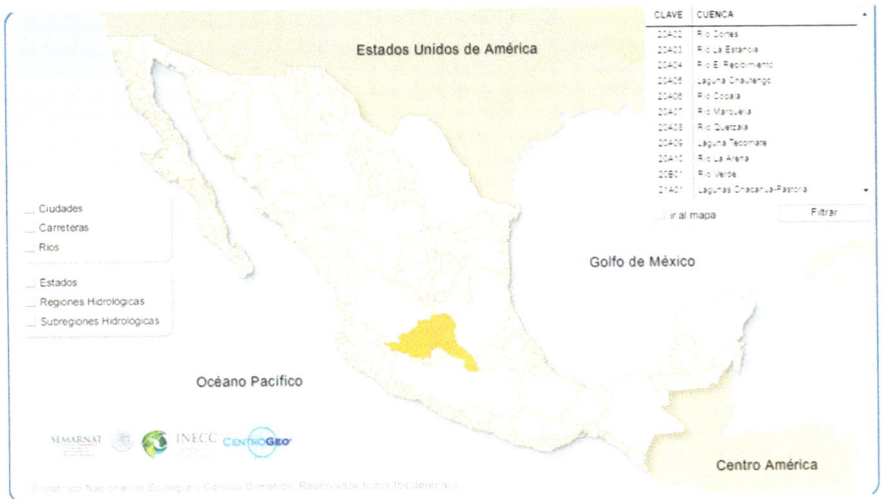

Fig. 15.5 Location of the Lerma River basin, Mexico. (INE 2016)

and El Cuchillo Dam climatological stations are in the sub-basin of the San Juan River. La Amistad Dam, Falcon Dam, Piedras Negras, Matamoros, and Nuevo Laredo are in the main Bravo River Basin. We observed that the climatological station of Parral was the site with the most critical results for an increase of potential evaporation among all the climatological stations considered in this study. So, only the results of Parral climatological station will be reported here.

Fig. 15.6 Location of the Panuco River basin, Mexico. (INE 2016)

Fig. 15.7 Location of the Papaloapan River basin, Mexico. (INE 2016)

The selected sites in the Grijalva River basin were Chicoasen, Malpaso, La Angostura, and Villahermosa climatological stations. We observed that the Villa-hermosa climatological station was the site with the most critical results for an increase of potential evaporation of all the climatological stations considered in this study, so only its results will be reported here.

The selected sites in the Lerma River basin were the climatological stations at Toluca, Querétaro, Celaya, and Guadalajara. We observed that the Toluca climatological station was the site with most critical results for an increase of potential evaporation of all the climatological stations considered, so only its results will be reported here (Raynal-Villasenor et al. (2016).

The selected sites in the Panuco River basin, Mexico, were the climatological stations at San Luis Potosi, Tampico, Matehuala, and Tula. We observed that the Tampico climatological station was the site with most critical results for an increase of potential evaporation of all the climatological stations considered in this study, so only its results will be reported here.

The selected sites in the Papaloapan River basin were the climatological stations at Veracruz, Tlacotalpan, Coatzacoalcos, and Tuxtepec. We observed that the Veracruz climatological station was the site with most critical results for an increase of potential evaporation of all the climatological stations considered in this study, so only its results will be reported here.

The results obtained by using Penman´s equation for the six selected sites in the study are shown in Table 15.1 for water-covered surfaces and in Table 15.2 for short-grass-covered surfaces.

The results obtained for the Balcon del Diablo climatological station in the Balsas River basin are graphically shown in Fig. 15.8.

The results obtained for the Parral climatological station in the Bravo River basin are graphically depicted in Fig. 15.9.

The results obtained for the Villahermosa climatological station in the Grijalva River basin are graphically shown in Fig. 15.10.

The results obtained for the Toluca climatological station in the Lerma River basin are graphically depicted in Fig. 15.11.

Table 15.1 Results for the percentage increments of evaporation due to the five possible scenarios of increased air temperatures of water-covered surface

Watershed	Climatological station	Air-temperature increments				
		1 °C	2 °C	3 °C	4 °C	5 °C
Balsas River	Balcon del Diablo	2.6	5.2	7.7	10.6	12.6
Bravo River	Parral	2.8	5	7.5	9.9	12.2
Grijalva River	Villahermosa	2.1	4.3	6.4	9.1	10.8
Lerma River	Toluca	2.8	5	7.5	9.9	12.2
Panuco River	Tampico	2.2	4.4	6.6	8.8	11
Papaloapan River	Veracruz	2.5	4.7	7	9.8	11.5

Table 15.2 Results for the percentage increments of evaporation due to the five possible scenarios of increased air temperatures of short grass-covered surface

Watershed	Climatological station	Air-temperature increments				
		1 °C	2 °C	3 °C	4 °C	5 °C
Balsas River	Balcon del Diablo	2.8	5.4	8	12	13.1
Bravo River	Parral	2.9	5.1	7.8	10.3	12.7
Grijalva River	Villahermosa	2.3	4.5	6.8	8.6	11.5
Lerma River	Toluca	2.9	5-5.1	7.8	10.3	12.7
Panuco River	Tampico	2.3	4.7	7	9.4	11.8
Papaloapan River	Veracruz	2.6	5	7.4	10	12.2

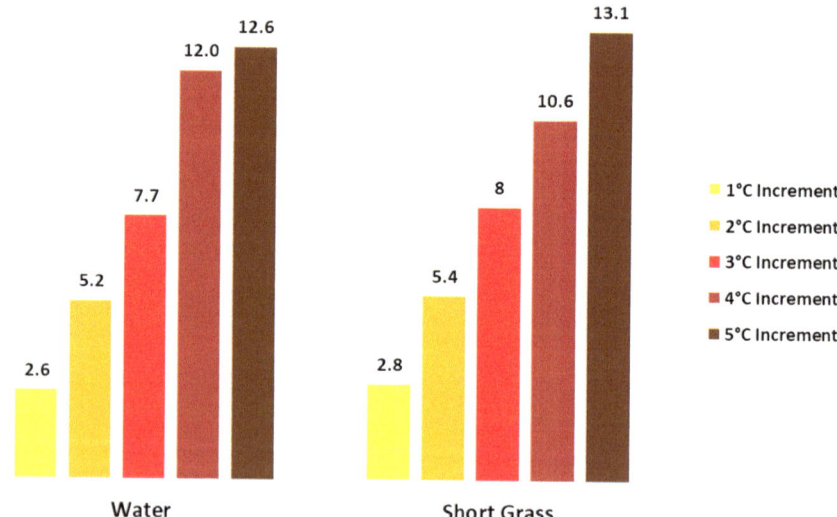

Fig. 15.8 Percentage of increment in potential evaporation in climatological station at Balcon del Diablo, Mexico (water- and short-grass-covered surfaces)

The results obtained for the Tampico climatological station in the Panuco River basin are graphically shown in Fig. 15.12.

The results obtained for the Veracruz climatological station in the Papaloapan River basin are graphically depicted in Fig. 15.13.

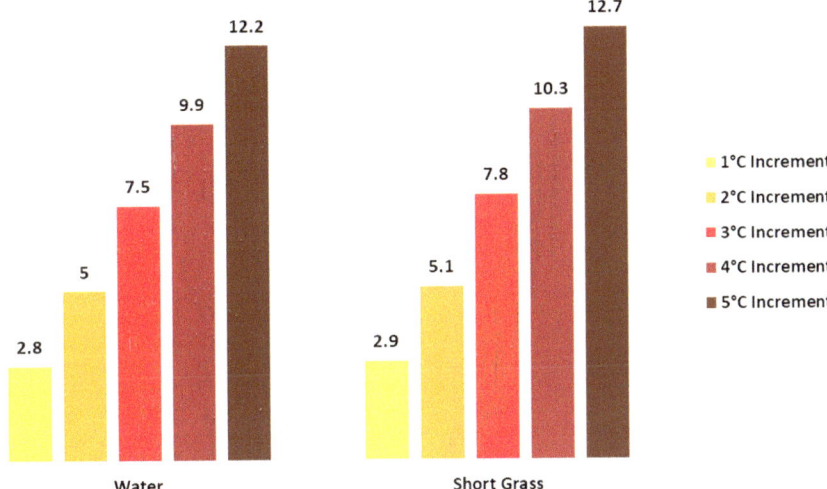

Fig. 15.9 Percentage of increment in potential evaporation in climatological station at Parral, Mexico (water- and short-grass-covered surfaces)

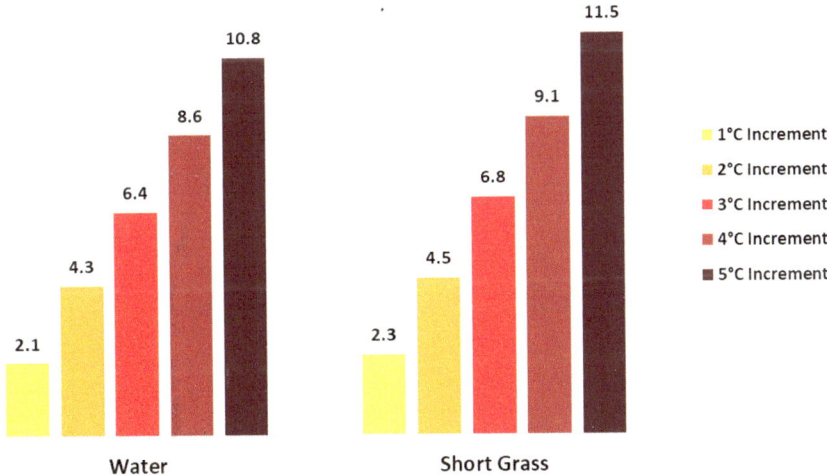

Fig. 15.10 Percentage of increment on potential evaporation in climatological station Villahermosa, Mexico (water- and short-grass-covered surfaces)

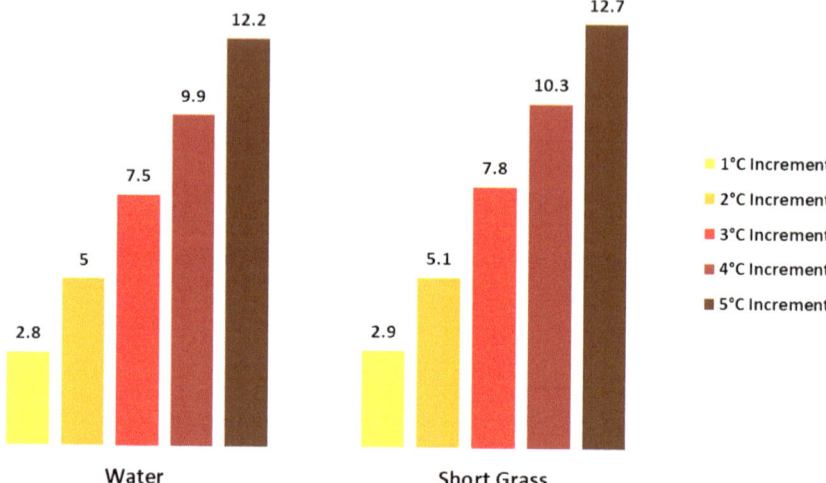

Fig. 15.11 Percentage of increment on potential evaporation in climatological station Toluca, Mexico (water- and short-grass-covered surfaces)

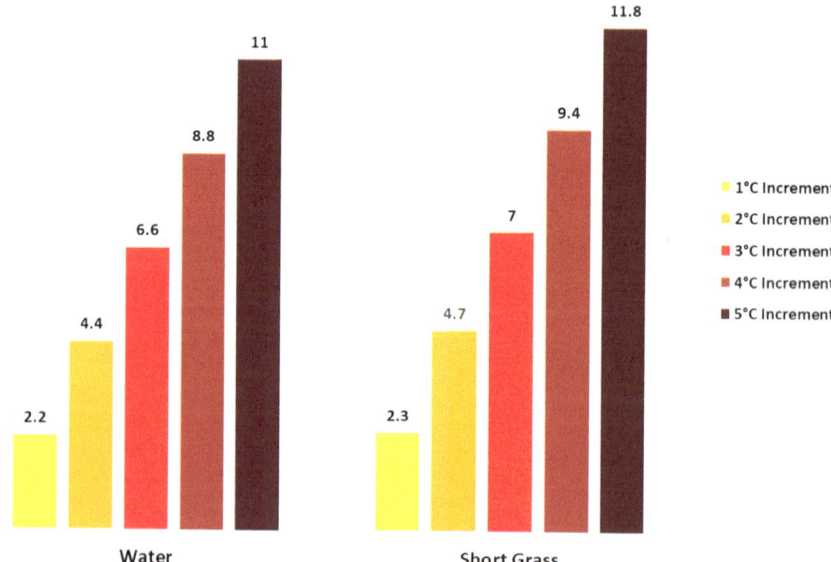

Fig. 15.12 Percentage of increment on potential evaporation in climatological station Tampico, Mexico (water- and short-grass-covered surfaces)

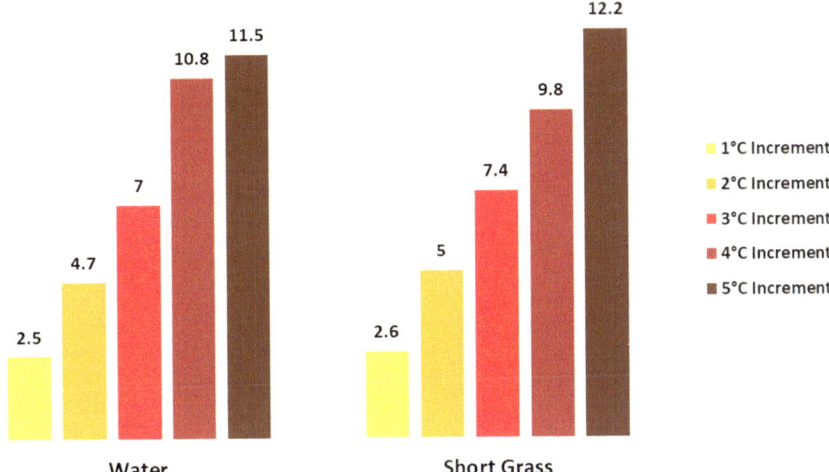

Fig. 15.13 Percentage of increment on potential evaporation in climatological Veracruz station, Mexico (water- and short-grass-covered surfaces)

15.4   Summary and Conclusions

With the construction of five possible scenarios in each of the river basins (Balsas River, Bravo River, Grijalva River, Lerma River, Panuco River, and Papaloapan River) with increases in air temperature above current conditions from 1 °C and up to 5 °C, respectively, the results obtained showed increased increments in the potential evaporation values that ranged 2.1–2.9% for 1 °C, 4.3–5.4% for 2 °C, 6.4–8% for 3 °C, 9.9–10.6% for 4 °C, and 11–12.7% for 5 °C air temperature increases compared to current values—for the two surfaces considered (water and short grass).

This would result in an expected reduction of food production of about 2–13% in such watersheds, for global warming scenarios with 1–5 °C increases in air temperature above current conditions; this will be caused by an increased water demand of crops in such basins.

The results obtained in this study showed that the Mexican government must pay attention to the possible consequences of the scenarios constructed here and must devise action plans to mitigate the adverse impacts that global warming may generate in the Balsas River, Bravo River, Grijalva River, Lerma River, Panuco River, and Papaloapan River watersheds.

References

Cámara de Diputados (CD) (2018) Ley General de Cambio Climatico www.diputados.gob.mx/LeyesBiblio/pdf/LGCC_130718.pdf. Accessed 15 Aug 2018

Gobierno del Distrito Federal (GDF) (2018) ¿Como se abastece la Ciudad de Mexico? http://www.transparenciamedioambiente.df.gob.mx/index.php?option=com_content&view=article&id=132%3Aabastecimiento&catid=57%3Aimpactos-en-la-vida-cotidiana&Itemid=415. Accessed 20 Jul 2018

Gosling SN, Zaherpour J, Mount NJ, Hattermann FF, Dankers R, Arheimer B, Breuer L, Ding J, Haddeland I, Kumar R, Kundu D, Liu J, van Griensven A, Veldkamp TIE, Vetter T, Wang X, Zhang X (2017) A comparison of changes in river runoff from multiple global and catchment-scale hydrological models under global warming scenarios of 1 °C, 2 °C and 3 °C. Clim Chang 141(3):577–595. https://doi.org/10.1007/s10584-016-1773-3

Hadley CentreHadley Centre (2005) Climate change and the greenhouse effect. [Report]. http://reefrelief.org/wp-content/uploads/climate_greenhouse2.pdf. Accessed 20 Jul 2018

Intergovernmental Panel on ClimateClimate Change (IPCC) (2014) 5th. Assessment, Report http://www.ipcc.ch/report/ar5/index.shtml. Accessed 20 Jul 2018

Instituto Nacional de Ecología y Cambio Climatico (INECC) (2017) Sistema Nacional de Consulta de las Cuencas Hidrograficas de Mexico. http://cuencas.inecc.gob.mx/cuenca/. Accessed 20 Jul 2018

Instituto Nacional de Ecología (INE) (2016) Sistema Nacional de Consulta de las Cuencas Hidrográficas de México. http://cuencas.ine.gob.mx/cuenca/. Accessed Jul 15 2018

Presidencia (2007) Plan Nacional de Desarrollo 2007–2010. In Spanish http://pnd.calderon.presidencia.gob.mx/index.php?page=documentos-pdf. Accessed 20 Jul 2018

Presidencia (2013) Plan Nacional de Desarrollo 2013–2018. In Spanish http://www.dof.gob.mx/nota_detalle.php?codigo=5299465&fecha=20/05/2013. 2016. Accessed 20 Jul 2018

Raynal-Villasenor JA, Raynal-Gutierrez ME, Sanchez-Sanchez C (2016) Possible scenarios of global climate change on the evaporation in the Lerma River Basin, Mexico. Int J Eng Sci Innov Technol 5(5):7–14

Showstack R (2009) Climate change now apparent and unequivocal, new report warns. EOS 90 (26):223

Steward JB, Roberts JM (1984) Evapotranspiration, Chapter 9. In: Lecture notes of the course on estimation of hydrological variables. Institute of Hydrology, Wallingford

Index